Enterobacter sakazakii

Emerging Issues in Food Safety
SERIES EDITOR, Michael P. Doyle

Enterobacter sakazakii

EDITED BY

Jeffrey M. Farber

Bureau of Microbial Hazards
Food Directorate
Health Canada
Ottawa, Ontario, Canada

AND

Stephen J. Forsythe

School of Biomedical and Natural Sciences
Nottingham Trent University
Clifton Lane
Nottingham, United Kingdom

ASM
PRESS
WASHINGTON, D.C.

Address editorial correspondence to ASM Press, 1752 N St., N.W., Washington, DC 20036-2904, USA

Send orders to ASM Press, P.O. Box 605, Herndon, VA 20172, USA
Phone: 800-546-2416; 703-661-1593
Fax: 703-661-1501
E-mail: books@asmusa.org
Online: http://estore.asm.org

Library of Congress Cataloging-in-Publication Data

Enterobacter sakazakii / edited by Jeffrey M. Farber and Stephen J. Forsythe.
     p. ; cm.—(Emerging issues in food safety)
   Includes bibliographical references and index.
   ISBN 978-1-55581-460-1 (hardcover : alk. paper)
   1. Enterobacter sakazakii.   2. Infant formulas—Contamination.
   I. Farber, Jeffrey M.   II. Forsythe, S. J. (Steve J.)   III. American
   Society for Microbiology.   IV. Series.
       [DNLM: 1. Enterobacter sakazakii.   2. Enterobacter
       sakazakii—pathogenicity.   3. Enterobacteriaceae Infections—
       prevention & control.   4. Food Contamination—
       prevention & control.   5. Infant Formula—standards.
       QW 138.5.E5 E605 2008]

   QR82.E6E54 2008
   579.3'4—dc22

                                                      2007025067

   10   9   8   7   6   5   4   3   2   1

Cover illustration: *Enterobacter sakazakii* attached to an enteral feeding tube. (Courtesy of Stephen J. Forsythe.)

Contents

v

Contributors

N. ABDULLAH SANI
Food Science Programme, School of Chemical Sciences and Food Technology, Faculty of Science and Technology, Universiti Kebangsaan Malaysia, 43600 Bangi, Selangor, Malaysia

ANNA B. BOWEN
Foodborne and Diarrheal Diseases Branch, Division of Bacterial and Mycotic Diseases, National Center for Infectious Diseases, Centers for Disease Control and Prevention, Atlanta, GA 30333

CHRISTOPHER R. BRADEN
Foodborne and Diarrheal Diseases Branch, Division of Bacterial and Mycotic Diseases, National Center for Infectious Diseases, Centers for Disease Control and Prevention, Atlanta, GA 30333

PIETER BREEUWER
Applied Science and Quality Assurance Department, Nestlé Product Technology Center Konolfingen, Nestlé Strasse 3, CH-3510, Konolfingen, Switzerland

JOAN BRENNAN
Department of Clinical Dietetics, The Hospital for Sick Children, Toronto, Ontario, Canada

JEAN-LOUIS CORDIER
Nestlé Nutrition, Avenue Reller 22, CH-1800 Vevey, Switzerland

CATHERINE DAUGA
Plateforme 4, Génopole, Institut Pasteur, 75724 Paris cedex 15, France

SUSAN DELLO
Department of Food and Nutritional Services, The Hospital for Sick Children,
Toronto, Ontario, Canada

S. ESTUNINGSIH
Department of Clinical, Reproduction, and Veterinary Pathology, Bogor Agricultural
University, Jalan Agatis, IPB Campus, Darmaga, 16681, Bogor, Indonesia

SÉAMUS FANNING
Centre for Food Safety, School of Agriculture, Food Science, and Veterinary
Medicine, University College Dublin, Belfield, Dublin 4, Ireland

JEFFREY M. FARBER
Bureau of Microbial Hazards, Health Products and Food Branch, Health Canada,
Ottawa, Ontario, Canada

JOHN J. FARMER III
Silver Hill Associates, 1781 Silver Hill Road, Stone Mountain, GA 30087

STEVE FORSYTHE
School of Biomedical and Natural Sciences, Nottingham Trent University,
Nottingham, United Kingdom

RAQUEL LENATI
Bureau of Microbial Hazards, Health Products and Food Branch, Health Canada,
Ottawa, Ontario, Canada

DEBORAH L. O'CONNOR
Department of Clinical Dietetics, The Hospital for Sick Children, and
Department of Nutritional Sciences, The University of Toronto, Toronto, Ontario,
Canada

FRANCO PAGOTTO
Bureau of Microbial Hazards, Health Products and Food Branch, Health Canada,
Ottawa, Ontario, Canada

LAURIE STREITENBERGER
Department of Infection Prevention and Control, The Hospital for Sick Children,
Toronto, Ontario, Canada

STACY TOWNSEND
School of Biomedical and Natural Sciences, Nottingham Trent University, Clifton
Lane, Nottingham, United Kingdom NG11 8NS

Series Editor's Foreword

Enterobacter sakazakii emerged as a major concern for manufacturers of infant formula because of its association with deaths of infants, largely premature or in intensive care units (ICUs), fed contaminated, reconstituted powdered infant formula that was typically temperature abused in ICUs. For these infants, the mortality rate was high, ranging from 40 to 80%. Since its recognition as a severe pathogen for specific high-risk populations, a wealth of information regarding the detection, ecology, survival characteristics, and pathogenicity of *E. sakazakii* has been generated.

Jeffrey Farber and Steve Forsythe, two leaders in studying the food-associated aspects of *E. sakazakii*, have combined forces with many internationally recognized *E. sakazakii* experts to provide an impressive compilation of the state-of-the-science surrounding this pathogen. Topics that are addressed include taxonomy, isolation and identification, epidemiology, pathogenicity, production and use practices of infant formula that influence *E. sakazakii* contamination and growth, and regulatory issues. I know of no other source that provides as much relevant, up to date information on the food safety aspects of *E. sakazakii* than this book. It is an invaluable resource for those of us interested in the microbiological safety of foods.

MICHAEL P. DOYLE, Series Editor
Emerging Issues in Food Safety

Preface

Enterobacter sakazakii is an emerging opportunistic pathogen that has caused much concern in the food industry, as well as in regulatory and academic communities. In fact, there has been more work done in the last 2 to 3 years than in the past 25 years. As an example, in the last 2 years alone there have been at least 85 publications on this organism.

Our experience indicated that there was no single source that had all the up-to-date information on this very unique pathogen. Thus, we have set out to compile the latest information on *E. sakazakii*, all under one cover.

This bacterium is unique in the sense that it is widespread in the environment and for the most part affects only a small subset of the population, i.e, infants under 1 year of age, although cases do occur in adults and the incidence of disease caused by this organism is underreported. It also survives the drying process very well and thus can be found in dry food products such as skimmed milk powder, lactose, starch, lecithin, orange powder, and banana powder, all ingredients that can be added to powdered infant formula. In addition, because the organism affects the youngest and most vulnerable segment of our population, the issue has been raised to the highest levels of attention and effort.

As mentioned above, the research community has made great strides in almost all areas associated with the biology of this organism. In addition, the powdered infant formula industry has done a tremendous job in reducing the incidence of contaminated formulae in the marketplace. Advances in methodology have been quite dramatic, and we are now able to more readily detect this organism than ever before. However, advancements will still need to be made to detect the increasingly lower levels of this organism that we will see in the future in powdered infant formula. There are various collaborative studies under way, and these will help to further advancements in this area. One can also already see rapid advances being made in understanding the pathogenesis of this organism, and in this area we think there will be tremendous progress.

At the international level, attention also has been drawn to this issue, and two major risk assessment meetings have taken place, which have resulted in two publications. In the Codex Alimentarius, the Committee on Food Hygiene (CCFH) has been involved in terms of trying to draft a revised document on hygienic practices for powdered formulae for infants and young children, which is currently at Step 3 in the Codex process. In addition, WHO has recently published a guidance document for the safe preparation, storage, and handling of powdered infant formula.

In the next 5 to 10 years, our understanding of this organism and how and why it causes human illness will increase dramatically, and regulatory agencies will be able to use the outputs of this information to devise better risk management strategies that are based on sound science. For example, bacterial taxonomy is being revolutionized by the use of DNA sequence analysis, which may not always agree with past divisions according to sugar fermentation profiles, etc. The family *Enterobacteriaceae* originally was defined according to such phenotyping, and therefore the application of DNA sequence analysis is revising our perspective of this family. The genus *Enterobacter* is comprised of a phenotypically diverse group of bacteria, and thus it is not surprising that it has been recently proposed that *E. sakazakii* actually should be taken out of the genus *Enterobacter* and placed into a new genus, "*Cronobacter*," consisting of four new species and one genomospecies. All the species would still be considered pathogenic for neonates. Thus, from a regulatory point of view, not much would change. However, in the future, if the proposed nomenclature changes are accepted, this may allow us to more specifically target the species of concern. For example, if we find out that there are differences in virulence in some of the new species, different microbiological criteria possibly could be set. Work in this area also could lead to the identification of other members of the *Enterobacteriaceae* that are able to cause disease in infants when present in sufficient numbers in powdered infant formula.

This book, which we trust you will enjoy, hopefully will be a useful reference tool for many years to come for regulators, industry members, and academics who are interested in this organism or who are involved in some way with powdered infant formula used for the feeding of infants.

JEFFREY M. FARBER
STEPHEN J. FORSYTHE

Enterobacter sakazakii
Edited by Jeffrey M. Farber and Stephen J. Forsythe
© 2008 ASM Press, Washington, D.C.

Taxonomy and Physiology of *Enterobacter sakazakii*

1

Catherine Dauga and Pieter Breeuwer

CLINICAL SIGNIFICANCE AND TAXONOMIC CONSIDERATION OF THE GENUS *ENTEROBACTER*

The *Enterobacter* species, representing a large and heterogeneous group within the *Enterobacteriaceae*, have increasingly been identified as pathogens over the past several decades. At this time, the genus *Enterobacter* includes 14 species recognized taxonomically (Kämpfer et al., 2005). Among them, nine species corresponding to the *E. cloacae* complex, *E. sakazakii* and *E. gergoviae*, are clinically significant in humans in causing generally nosocomial infections.

Enterobacter cloacae is the member of the genus most often encountered in clinical laboratories. It has become an important nosocomial pathogen (Hoffmann and Roggenkamp, 2003). This species is clearly associated with pulmonary, bloodstream, and other extraintestinal infections of humans. This species shows a great genetic heterogeneity, with 12 clusters defined from phylogenetic trees based on three different housekeeping genes and seven new species that are now individualized within the *E. cloacae* complex:

1. *Enterobacter asburiae* was deduced from the former enteric group 17 and was described and named in 1986 (Brenner et al., 1986). It plays a minor role as a clinical pathogen and has been described as a cause of primary lung infection in immunocompromised patients (Stewart and Quirk, 2001), and it has been isolated from blood in neonatal septicemia in an intensive care unit (Fernandez-Baca et al., 2001).

CATHERINE DAUGA, Plateforme 4, Génopole, Institut Pasteur, 75724 Paris cedex 15, France. PIETER BREEUWER, Applied Science and Quality Assurance Department, Nestlé Product Technology Center Konolfingen, Nestlé Strasse 3, CH-3510 Konolfingen, Switzerland.

2. *Enterobacter cancerogenus*, originally designated enteric group 19, was first ascribed to the genus *Erwinia*. After extensive taxonomic investigations, it has been transferred to the genus *Enterobacter* as a synonym of *Enterobacter taylorae* (Farmer et al., 1985). *E. cancerogenus* infections in humans reported in the literature refer to osteomyelitis, urinary tract infections, open wound infections, open fracture infections, and traumatic cut infections (Garazzino et al., 2005).

3. *Enterobacter dissolvens* is the relative closest to *E. cloacae*, because the DNA relatedness of the two species is up to 82%. In addition, *E. dissolvens* formed a neighboring lineage of *E. cloacae* on phylogenetic trees based on housekeeping genes (Hoffmann and Roggenkamp, 2003). It is now proposed as a subspecies and is named *E. cloacae* subsp. *dissolvens* (Hoffmann et al., 2005b).

4. *Enterobacter hormaechei*, initially described as enteric group 75, now represents a valid species including three subspecies, subsp. *oharae*, subsp. *hormaechei*, and subsp. *steigerwaltii*, corresponding to three small phylogenetic lineages (O'Hara et al., 1989; Hoffmann et al., 2005c). Most of the *E. hormaechei* strains were isolated from clinical samples, rarely in the context of a nosocomial outbreak (Davin-Regli et al., 1997).

5. *Enterobacter kobei* corresponds to the former enteric group 69. The type strain of *E. kobei* fell into a cluster, including the strain CDC1347-71, used as reference strain for *E. cloacae* by several authors. Further taxonomic investigations will have to define both species more clearly. This species is rarely reported in clinical settings, but its phenotypic similarity to *E. cloacae* might have led to its misidentification and could explain the underestimation of its clinical relevance (Hoffmann et al., 2005a).

6. *Enterobacter ludwigii* was recently individualized as a new species from the *E. cloacae* complex by using 16 strains isolated from clinical specimens (Hoffmann et al., 2004).

7. *Enterobacter nimipressuralis*, with a DNA relatedness to *E. cloacae* of up to 67%, was transferred from the genus *Erwinia* to the genus *Enterobacter* and now belongs to the *E. cloacae* complex (Hoffmann and Roggenkamp, 2003). This species, isolated from elm trees with a disease called wetwood, has never been described as a human pathogen.

Out of the *E. cloacae* complex, *Enterobacter gergoviae* is a species frequently encountered in the clinic. It causes sepsis as well as urinary tract and pulmonary infections (Brenner et al., 1980). This species has also recently been described as a cause of a nosocomial outbreak in a neonatal intensive care unit (Ganeswire et al., 2003).

Finally, *Enterobacter sakazakii*, previously known as "yellow-pigmented *Enterobacter cloacae*," was described as a new species in 1980 (Farmer et al., 1980). *E. sakazakii* had biochemical reactions very similar to those of *E. cloacae* but was distinguished from it based on differences in DNA relatedness, yellow pigment production, some biochemical traits, and antibiotic susceptibilities. The first cases of neonatal meningitis believed to have been caused by *E. sakazakii* were reported in 1958 (Urmenyi and Franklin, 1961). A clear epidemiologic correlation was proved between *E. sakazakii* isolated from patients and powdered (dry) infant formulas involved in two outbreaks that occurred in neonatal intensive care units (Clark et al., 1990). This species is now considered to be an opportunistic pathogen implicated in food-borne diseases, causing necrotizing enterocolitis and, most notably, neonatal sepsis and meningitis (Food and Agriculture Organization-World Health Organization, 2004).

PHYLOGENETIC COMMENTS AND TAXONOMIC IMPLICATIONS

Phylogenetic Position of *Enterobacter* Species and *Enterobacter sakazakii* among Environmental Strains by 16S rRNA Gene Sequence Analysis

16S rRNA gene-based phylogenetic trees offer a useful advantage to compare both easily recognizable species from different genera as well as difficult-to-identify or noncultivable isolates from the environment.

On the 16S rRNA gene phylogenetic tree (Fig. 1), the genus *Enterobacter* appeared to be polyphyletic, with several lineages closely related to *Citrobacter*, as previously described (Boy and Hansen, 2003; Iversen et al., 2004b). However, uncertainties about relationships between species were also observed. *E. hormaechei*, as well as some strains of *E. sakazakii* and *Citrobacter amalonaticus*, shared the same phylogenetic group, but this group was weakly validated by bootstrap values. It is the same problem for the association between *E. pyrinus* and the phylogenetic branch including *Citrobacter koseri* (formerly *C. diversus*). Only vague genetic clusters including sequences from species of both genera were generated on the 16S rRNA gene phylogenetic tree. This lack of phylogenetic information was probably due to the low variability of the gene.

The *E. sakazakii* species was quite heterogeneous and, as previously reported, had four clusters of strains found on the phylogenetic tree based on 16S rRNA genes (Iversen et al., 2004b). Each cluster showed a high within-group similarity (99.85%, 99.83%, 100%, and 99.43% for clusters 1 to 4, respectively). Cluster 1 and cluster 2 were neighboring and weakly limited. Cluster 3 and cluster 4, validated by high bootstrap values, were phylogenetically different from the rest of the sequences.

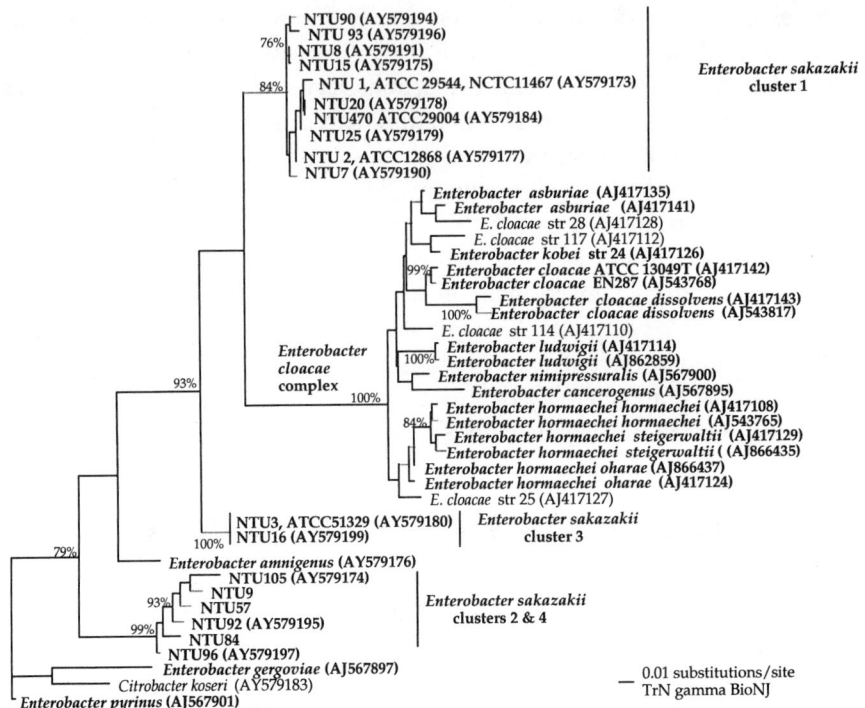

Figure 1 Neighbor-joining tree of *E. sakazakii* and related clinical and environmental isolates based on partial 16S rRNA gene (550 bp) sequences. Sequences were obtained from EMBL or were kindly provided by S. J. Forsythe. Alignment was performed by ClustalX 1.83. The phylogenetic tree was built using the K81 evolutionary model, a gamma parameter at 0.76, and the bioNJ algorithm, implemented in PAUP 40b10 software.

Close relationships were observed between one isolate from the microbiota of the ant lion, *Myrmeleon mobilis,* and sequences of *E. sakazakii* cluster 4. Similar relationships were suspected between *E. pyrinus* and symbionts or bacteria isolated from microbiota of insects, such as *Myrmeleon mobilis* or the plant mite (Acari) *Metaseiulus occidentalis.* Microbial communities of many insects include *Enterobacteriaceae,* and their role is interesting to elucidate. Some strains produce toxins which possibly contribute to prey paralysis or death and would have a very important ecological role (Dunn and Stabb, 2005).

The presence of *E. sakazakii* in the environment is poorly documented. However, a recent phylogenetic analysis of 16S rRNA genes showed that an environmental isolate from a coastal ecosystem near Barcelona (Spain) belonged to cluster 1 of *E. sakazakii.* This strain was found in a polluted coastal area, suggesting that many *Enterobacteriaceae,* including *E. sakazakii,* can survive in the marine environment (Agogué et al., 2005).

Finally, the phylogenetic tree included uncultured *Enterobacteriaceae* detected in the normal flora of infants. These *Enterobacteriaceae* shared the same phylogenetic groups as *Enterobacter* species, but none of them were in any of the *E. sakazakii* clusters (Songjinda et al., 2005).

Phylogenetic Position of *Enterobacter* Species and *E. sakazakii* among Other *Enterobacteriaceae*

For a closely related species or genus, the resolution of the 16S rRNA gene sequence analysis in taxonomic studies has been questioned. Therefore, *Enterobacter* species relationships were described by using phylogenetic trees based on two housekeeping protein-coding genes, *gyrB* and *hsp60*. These encode for DNA gyrase subunit B and heat shock protein 60, respectively.

gyrB *sequence analysis*

The phylogenetic tree based on *gyrB* sequences enabled the positioning of *E. sakazakii* among *Enterobacteriaceae* members (Fig. 2).

The *gyrB*-based phylogenies are largely consistent with the classification of most of the *Enterobacteriaceae*. Six prominent clusters were found, and groupings were in agreement with the genus assignment of species. Separate clusters were obtained for the genera *Salmonella, Escherichia, Shigella, Klebsiella, Citrobacter, Pantoea,* and *Serratia*. A possible phylogenetic artifact included *Kluyvera cochleae, Kluyvera intermedia,* and *Buttiauxella agrestis* in the *Citrobacter* cluster.

The groups formed by *E. sakazakii* and other *Enterobacter* species are compatible with their presently accepted classification. One cluster included sequences of the *E. cloacae* complex, except *E. nimipressuralis,* associated with *E. amnigenus* in a separate branch. The sequences from *E. sakazakii,* one partial of 506 bp (GenBank accession no. AY370844) and one complete, sharing 89.82% similarity, formed a branch paraphyletic of the other *Enterobacter*.

hsp60 *sequence analysis*

The phylogenetic tree based on *hsp60* helps to clarify relationships between *E. sakazakii,* the other *Enterobacter* species, and neighboring species (Fig. 3).

The *E. cloacae* complex corresponded to a well-delimited group of sequences from 90.7 to 100% similarity. Sequences from *E. sakazakii* formed three distinct clusters of strains validated by highly significant bootstrap values. In contrast to the findings of the 16S rRNA gene analysis, the sequence from strain NTU84 of cluster 4 and the sequences of cluster 2 (sequence similarities, 98.54%) were included in a common phylogenetic group on the *hsp60* phylogenetic tree. Robust genetic cluster 1 (sequence similarities, 98.52%) and cluster 3 (sequence similarities, 100%) were clearly individualized. The three

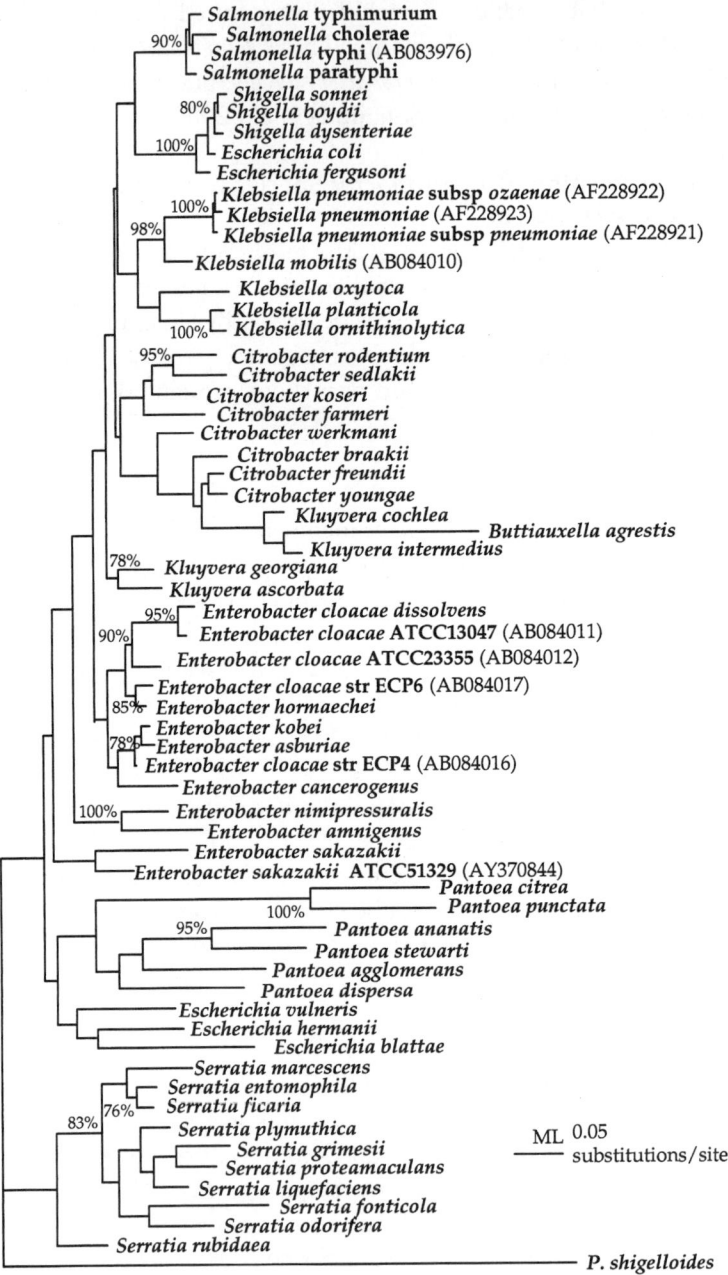

Figure 2 Maximum-likelihood tree of the *Enterobacteriaceae,* including strains of *E. sakazakii* and related *Enterobacter* species based on *gyrB* (1,192 bp) sequences. The phylogenetic tree was built using the K81 evolutionary model, a gamma parameter at 0.935, and a heuristic strategy with tree bisection reconnection (TBR) branch swapping.

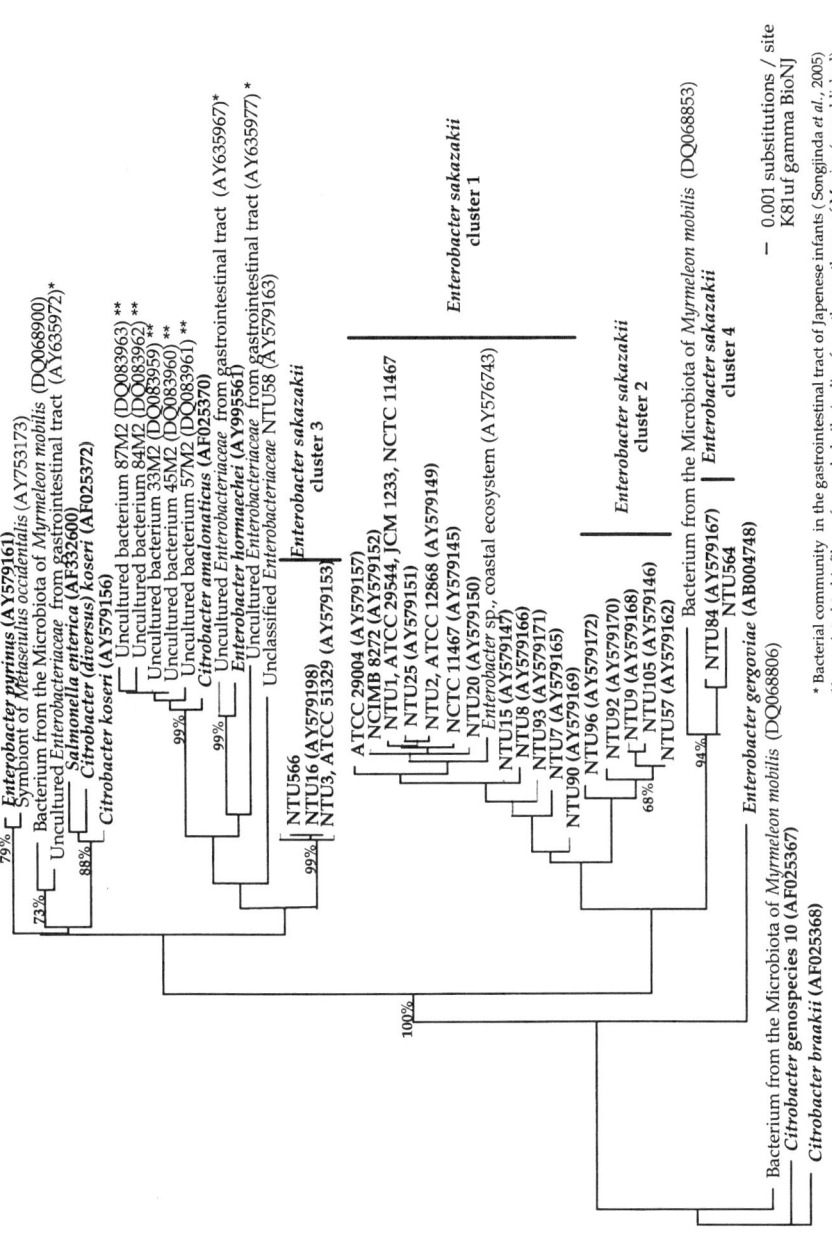

Figure 3 Neighbor-joining tree of *E. sakazakii* and related species based on partial *hsp60* (343 bp) sequences. The phylogenetic tree was built using the Tamura–Nei (TrN) evolutionary model, a gamma parameter at 0.6, and the bioNJ algorithm.

groups of strains shared close relationships with the *E. cloacae* complex and the other *Enterobacter* spp.

EVOLUTIONARY SYSTEMATIC STUDIES REVEALED UNCERTAINTIES OF CLASSIFICATION

Species Delineation

Phylogenetic trees based on both 16S rRNA gene and *hsp60* analyses divided *E. sakazakii* strains into several clusters, suggesting they may belong to more than one species.

These clusters of strains differed slightly according to the phylogenetic marker used. Cluster 1, containing many reference strains, including the type strain (JCM1233), formed a validated cluster on the *hsp60* phylogenetic tree. Cluster 3, including strains NTU3, NTU16, NTU566, and the Perceptrol strain ATCC 51329, was individualized on both 16S rRNA gene and *hsp60* phylogenetic trees. The strains NTU84 and NTU564 formed a separate group (cluster 4) on the 16S rRNA gene phylogenetic tree. Strains from cluster 2 were not clearly individualized from strains of cluster 1 on the 16S rRNA gene phylogenetic tree. To circumscribe the taxon species, DNA-DNA reassociations were performed by Forsythe and VanDamme (unpublished data) (Table 1). Table 1 was kindly provided by S. Forsythe. The laboratory work was by C. Loc-Carrillo and T. Coenye from the laboratories of S. Forsythe and P. Vandamme, respectively. Cross-hybridization of strains within each cluster showed high within-group homogeneities, from 77 to 94% DNA hybridization values. The between-cluster hybridization experiments yielded hybridization values that were above the postulated 70% value for the distinction of genetic species, indicating that strains from the four clusters belong to the *E. sakazakii* species.

Genus Determination

gyrB and *hsp60* sequence analyses led to uncertainties for genus affiliation, because they produced phylogenetic trees (Fig. 2 and 3) different from the 16S rRNA tree (Fig. 1), which is usually used to infer the classification of species.

The mixed relationships between *Citrobacter* and *Enterobacter* found by 16S rRNA gene phylogenetic analysis were not confirmed on the *gyrB* phylogenetic tree. On the contrary, *Citrobacter* and *Enterobacter* genera appeared as distinct monophyletic groups on the *gyrB* phylogenetic tree. Although not supported by significant bootstrap values, this latter analysis was preferred because it was in agreement with phenotypic characteristics.

The position of *E. sakazakii* appeared uncertain on the 16S rRNA gene phylogenetic tree, and this was supported by the low level of phylogenetic

Table 1 DNA relatedness of *E. sakazakii* strains[a]

| Species and strain | % Relatedness of *E. sakazakii* strains in cluster: | | | | | | | | % Relatedness to: | |
| | 1 | | | 2 | | 3 | | 4 | | |
	NTU1	NTU2	NTU7	NTU9	NTU3	NTU566	NTU84	NTU564	*Citrobacter koseri*[T]	*E. cloacae*[T]
E. sakazakii										
NTU1	100									
NTU2	90	100								
NTU7	94	91	100							
NTU9	80	77	77	100						
NTU3	84	79	ND	77	100					
NTU566	85	ND[b]	ND	ND	94	100				
NTU84	82	ND	ND	73	86	84	100			
NTU564	80	78	ND	ND	89	92	94	100		
Citrobacter koseri[T]	33	ND	ND	ND	ND	ND	ND	ND	100	
E. cloacae[T]	34	ND	ND	37	37	ND	31	ND	37	100

[a]Adapted from S. Forsythe (unpublished data).
[b]ND, not determined.

Table 2 Percent similarity between the *hsp60* sequences of *E. sakaza-kii* strains and other *Enterobacter* species and *Citrobacter koseri*

Species	% Similarity to *E. sakazakii* strain (cluster)[a]:			
	NTU1 (1)	NTU9 (2)	NTU3 (3)	NTU84 (4)
E. amnigenus	86.56	*88.03*	87.15	*88.61*
E. cloacae	89.48	83.2	89.49	83.21
E. gergoviae	86.58	86.75	87.45	87.81
E. hormaechei	*89.73*	84.13	89.31	84.48
E. pyrinus	88.17	88.58	*89.72*	86.72
Enterobacter mean	88.1	86.14	88.63	86.17
Citrobacter koseri	85.95	85.08	87.13	84.5

[a]The highest similarity values observed for each strain are indicated in italics.

information on the *gyrB* phylogenetic tree. Ambiguities were due to the heterogeneous level of similarities according to the cluster of strains taken into account. 16S rRNA gene sequences from cluster 1 (NTU1) showed 98.1% similarity with sequences from *C. koseri* (the closest relative) and 97.34% with sequences from *E. pyrinus*, while sequences from cluster 3 (NTU3) showed 97.72 and 97.52% similarity with *E. pyrinus* and *C. koseri* sequences, respectively. In addition, DNA-DNA hybridization data showed no clear genus assignment for *E. sakazakii* to *Enterobacter* or *Citrobacter*, with 31 to 37% relatedness to *E. cloacae* and 33% relatedness to *C. koseri* (Table 1). The assignment of *E. sakazakii* to the *Enterobacter* genus was supported by data from *hsp60* sequences showing the highest level of similarities between the four clusters and *Enterobacter* species than for *C. koseri* (Table 2) and by phenotypic characteristics (see below).

Concluding Remarks

The branching orders for *Enterobacteriaceae*, *Enterobacter* species, and *E. sakazakii* strains were difficult to determine with the phylogenetic markers presently available. Further studies and many more markers are needed to resolve more accurately the phylogenetic tree of the *Enterobacteriaceae*. We hope that this will lead in the future to a classification of *Enterobacteriaceae* that is in obvious agreement with the evolutionary histories of its members.

SOME KEYS FOR IDENTIFICATION

This section covers key phenotypic tests for the *Enterobacter* species. For details on methods for the detection and identification of *E. sakazakii*, the reader should refer to chapter 2.

Biochemical Differentiation of *Enterobacter* and Neighboring Genera

Close relationships were observed between *Enterobacter*, *Citrobacter*, *Kluyvera*, and *Pantoea* genera on the 16S rRNA gene and the *gyrB* phylogenetic trees. Fortunately, the four genera are easily distinguishable based on phenotypic criteria.

All the genera belong to the family *Enterobacteriaceae* and are gram-negative, facultative anaerobic rods or coccobacilli. Most (~75%) *E. sakazakii* strains and some strains of *Pantoea agglomerans* form yellow-pigmented colonies which range from bright to pale yellow. Most strains of *Citrobacter* produce a distinctive odor.

The biochemical tests most useful for separating the species of each genera are given in Table 3. *Citrobacter* and *Kluyvera* are easily distinguishable from *Enterobacter* by being Voges-Proskauer (VP) negative. *Pantoea* differs from *Enterobacter* by having an ornithine decarboxylase enzyme.

Enterobacter species are straight rods, are motile by peritrichous flagella, are facultatively anaerobic, grow readily on laboratory media, and ferment glucose with the production of acid and gas. They also give positive VP results, positive citrate assimilation and *o*-nitrophenyl-β-D-galactopyranoside reactions, and a negative methyl red test. Most *Enterobacter* species are characterized by a positive test for ornithine decarboxylase and fermentation of sucrose, arabinose, mannitol, melibiose, raffinose, rhamnose, trehalose, and xylose. Most members of the genus are negative for DNase, H_2S, indole, and phenylalanine deaminase.

Biochemical Characteristics To Distinguish *Enterobacter* Species

Tests useful in the differentiation of *Enterobacter* species include adonitol and sorbitol fermentation as well as growth in KCN, arginine dihydrolase, lysine, and ornithine decarboxylase (Table 3). *E. cloacae* strains are arginine, ornithine, KCN, and sorbitol positive but are lysine decarboxylase negative (Table 4). Most isolates are also adonitol negative.

Table 3 Biochemical characteristics of *Citrobacter*, *Pantoea*, *Kluyvera*, and *Enterobacter*[a]

Characteristic	Reaction for genus:			
	Citrobacter	*Pantoea*	*Kluyvera*	*Enterobacter*
Citrate (Simmons)	+	v	+	+
LDC	−	−	v	v
ODC	v	−	+	+
VP	−	+	−	+
D-Sorbitol fermentation	+	v	+	v
Resistance to cephalothin			+	+

[a]Symbols: −, <10% of strains positive; +, >80% of strains positive; v, 10 to 79% of strains positive.

Table 4 Biochemical differentiation of *Enterobacter* species[a]

Substance or procedure	Reaction for species[b]:									
	E. cloacae subsp. *cloacae*	*E. cloacae* subsp. *dissolvens*	*E. asburiae*	*E. kobei*	*E. ludwigii*	*E. hormaechei* subsp. *hormaechei*	*E. cancerogenus*	*E. nimipressuralis*	*E. gergoviae*	*E. sakazakii*
Yellow pigmentation	–	–	–	–	–	–	–	–	–	+
ADH	+	+	+	+	+	+	+	+	–	+
LDC	–	–	–	–	d	–	–	–	+	–
ODC	+	+	+	+	+	+	+	+	+	+
Growth in KCN	–	d	d	d	d	+	+	+	–	+
VP test	+	+	d	+[c]	d	+	+	+	+	+
Esculin hydrolysis	–	+	+	–	(d)	–	+	+	+	(+)
DNase test	–	–	–	–	–	–	–	–	–	–
Acid from										
Adonitol	d	–	–	ND	ND	–	–	–	–	–
Lactose	–	+	d	+	ND	(+)	(+)	+	d	+
D-Sorbitol	+	+	+	+	ND	–	–	+	–	–[d]
Utilization of										
D-Arabinose	+	+	+	+	+	+	+	–	ND	ND
Dulcitol	–	–	–	d	–	+	–	–	–	d
Lactulose	–	–	d	d	d	d	–	–	(+)	+
Malonate	–	(d)	–	(d)	(d)	+	(d)	–	d	(d)
Melibiose	+	+	d	+	+	–	–	+	+	+
Phenylacetate	(+)	+	(+)	(d)	+	+	+	+	+	–
Me-gluco-pyranose	–	–	–	–	+	–	+	–	–	–

L-Rhamnose	+	+	d	+	+	+	+	+	+
D-Sorbitol	+	(+)	+	+	−	+	+	−	−
Sucrose	+	+	+	+	+	+	−	+	d
Xylitol	−	−	−	−	−	−	−	d	−

[a]Data are from Dickey and Zumoff (1988); O'Hara et al. (1989); Schonheyder et al. (1994); Brenner and Farmer (2005); and Hoffmann et al. (2005b, 2005c).
[b]Symbols: +, positive for 90 to 100% of strains; −, negative for 90 to 100% of strains; d, positive or negative; (), after 7 days; ND, not determined.
[c]Initially described as VP negative in Kosako et al. (1996).
[d]Some strains ferment D-sorbitol as reported in Gurtler et al. (2005).

E. asburiae differed from *E. cloacae* by being esculin positive, by being VP, malonate, melibiose, and L-rhamnose negative, and potentially by its utilization of lactulose, melibiose, and L-rhamnose.

E. cancerogenus is a lactose-fermenting rod, differing from *E. cloacae* by growing in KCN and being esculin positive.

E. cloacae subsp. *dissolvens* is phenotypically indistinguishable from *E. cloacae*, except by esculin test results: positive for *E. dissolvens* and negative for *E. cloacae* (Hoffmann et al., 2005b).

E. gergoviae differs from *E. cloacae* by being both lysine and ornithine positive but negative for adonitol, arginine, KCN, and sorbitol, while *E. hormaechei* differs from *E. cloacae* by being D-sorbitol, melibiose, and esculin negative and dulcitol and malonate positive.

E. kobei has the same biochemical characteristics as *E. cloacae*; the only difference initially described was that *E. kobei* was negative by the VP test (Kosako et al., 1996). However, newly described isolates were VP positive (Hoffmann et al., 2005b). After analyzing 120 biochemical tests included in the API 20E and the Biotype100 systems (BioMerieux Inc., Marcy l'Etoile, France), 4 biochemical tests (growth on xylitol, dulcitol, and phenyl acetate but not on lactulose) were identified as potentially differentiating this new biotype from *E. cloacae*.

E. ludwigii differs from *E. cloacae* by potentially having a lysine decarboxylase, by hydrolyzing esculin, and by growth in KCN. It also differs from the other *Enterobacter* species in using 3-*O*-methyl-D-glucopyranose.

E. nimipressuralis differs from *E. cloacae* by being sucrose and D-arabinose negative.

E. sakazakii has biochemical reactions very similar to those of *E. cloacae*, but it is D-sorbitol negative, and most strains produce yellow-pigmented colonies. Unfortunately, some sorbitol-positive *E. sakazakii* strains as well as nonpigmented *E. sakazakii* strains exist, which led to their misidentification as *E. cloacae* (Iversen and Forsythe, 2004; Iversen et al., 2004b). In addition to the tests commonly used, *E. sakazakii*, unlike *E. cloacae*, is arginine, ornithine, and KCN positive and gives a delayed-positive DNase test at 2 to 7 days. Tween 80 esterase and α-glucosidase activities but a lack of phosphoamidase activity also create major differences between *E. sakazakii* and the other *Enterobacter* species (Nazarowec-White and Farber, 1997c). An artificial neural network has been constructed to distinguish *E. sakazakii* from related organisms (Iversen et al., 2006a). The original 15 biotypes described by Farmer et al. (1980) have been correlated with the 16S cluster groups, and a new biotype 16 has been described (Iversen et al., 2006b). This topic is covered in more detail in chapter 2 in the section on genotyping methods.

PHYSIOLOGY OF *E. SAKAZAKII*

Survival and Growth Characteristics

Survival of *E. sakazakii* in powdered infant formulas and its potential to multiply in reconstituted infant formulas are critical factors with respect to the infection risk of infants. One of the indications that *E. sakazakii* is capable of surviving well in dry environments is its frequent isolation from powdered infant formulas (Muytjens et al., 1988; Iversen and Forsythe, 2004). Furthermore, it has been reported that the level of *E. sakazakii* in artificially contaminated powdered infant formulas declined only by 3 log units in 1.5 years (Edelson-Mammel et al., 2005). A similar study also showed survival in infant formula stored for more than 2 years (Caubilla-Barron and Forsythe, 2006). Other studies showed that *E. sakazakii* can better survive osmotic stress and air drying than, e.g., *Escherichia coli*, *Salmonella*, and *Citrobacter* strains (Breeuwer et al., 2003). Further indications of good survival in dry conditions are that *E. sakazakii* was found in more than 30% of household vacuum cleaners ($n = 16$) in one investigation (Kandhai et al., 2004) and in six out of seven vacuum cleaners in a second study (Breeuwer, 2005). Altogether, these findings clearly show that *E. sakazakii* is relatively resistant to dry stress, and it can be hypothesized that this characteristic contributes to the prevalence of *E. sakazakii* in powdered infant formulas.

The mechanism by which *E. sakazakii* survives better in dry environments is not clear. A cDNA-amplified fragment length polymorphism study on the RNA extracted after desiccation indicated that the response of *E. sakazakii* to dry stress involved a genome-wide expression of functionally different groups of genes. However, no specific resistance mechanism could be identified for *E. sakazakii* (Breeuwer et al., 2004). Most likely, trehalose plays a role in the resistance, because this compatible solute is highly accumulated upon dry stress (Breeuwer et al., 2003). There are at least three pathways for trehalose biosynthesis, but the most common in bacteria is the OtsA-OtsB pathway. Trehalose is synthesized from UDP-glucose and glucose 6-phosphate. In a study by Diep and Breeuwer (2006), the *otsA* gene, which encodes a trehalose synthase, was inactivated in *E. sakazakii* and *Escherichia coli,* and the survival to desiccation of the mutant strains and the wild-type strains was compared. Contrary to what was expected, the impaired synthesis of trehalose in the *otsA* mutant did not influence the survival of *E. sakazakii* upon desiccation. However, the same mutation in *Escherichia coli* resulted in hypersensitivity to desiccation. This suggests that *E. sakazakii* can somehow compensate for the loss of trehalose (Diep and Breeuwer, 2006), but further work is necessary to verify this.

Another hypothesis is that extracellular polysaccharides (EPS) play a role in the resistance against desiccation stress. In the cyanobacterium *Nostoc commune,* colonies with EPS were highly water absorbent and were rapidly hydrated by atmospheric moisture. The cells embedded in EPS in *Nostoc* colonies were highly desiccation tolerant (Tamaru et al., 2005). Furthermore, treatment to remove EPS resulted in desiccation-sensitive cells. *E. sakazakii* is capable of producing EPS, such as cellulose and colanic acid-like structures (Lehner et al., 2005), but the importance of this for the resistance to dry stress is not clear at this moment.

E. sakazakii can grow in the presence of up to 1 M NaCl at mesophilic growth temperatures. At 45°C, the bacteria still showed good growth in the presence of 0.5 M NaCl (Guillaume-Gentil et al., 2005). Consequently, this feature was used in the development of a selective enrichment medium, because the most relevant competitive flora, such as *Citrobacter* spp., *Serratia* spp., *Pantoea agglomerans*, *Escherichia vulneris*, and others, was generally not able to grow well under these conditions.

The growth of *E. sakazakii* has been quantified with various growth media, including reconstituted infant formulas. In general, it has been shown that *E. sakazakii* can grow between approximately 5.5 and 47°C, and the doubling time in rich media such as brain heart infusion broth and infant formula at 37°C is around 20 min, depending on the strains (Nazarowec-White and Farber, 1997a; Breeuwer et al., 2003; Iversen et al., 2004a; Kandhai et al., 2006). The effect of temperature on the growth rate was described with the expanded square root model of Ratkowsky and with the secondary Rosso equation. The resultant minimum and maximum growth temperatures estimated with these equations were 3.6 and 47.6°C, respectively (Kandhai et al., 2006). The 3.6°C is a theoretical limit, and growth at this temperature has not been demonstrated practically in time-limited experiments. In the same study, the lag time of *E. sakazakii* in infant formulas was investigated in detail, and it was demonstrated that at optimal growth temperatures, the bacteria will rapidly start growing after reconstitution; the lag time was approximately 1.7 h at 37°C. Furthermore, the lag time in this rich medium was apparently not dependent on the growth history of the isolates. The importance of the growth characteristics was quickly recognized by institutions such as the European Society for Paediatric Gastroenterology Hepatology and Nutrition, the United Kingdom Food Standards Agency, the U.S. Food and Drug Administration, and the World Health Organization. These organizations made specific recommendations to avoid growth after reconstitution such that powdered infant formula, if possible, should be freshly prepared for each feeding (Agostoni et al., 2004; Food Standards Agency, 2005; Food and Agriculture Organization-World Health Organization, 2004).

Cooling after reconstitution is an obvious approach to limit the growth of *E. sakazakii.* To obtain information on the potential for growth of *E. sakazakii* during cooling and storage, the rate by which the temperature of infant formulas decreased in refrigerators was measured (Kandhai, 2005). The results from this study indicated that reconstitution at 40°C and subsequent storage in standard household refrigerators still allowed an increase of almost 1 log unit of the *E. sakazakii* cell count within 24 h. Furthermore, the use of larger reconstitution volumes considerably increased the time to reach the refrigerator temperature and increased the risk for outgrowth of *E. sakazakii*. In another study, the survival of various *E. sakazakii* isolates was investigated at 4°C (Breeuwer, 2005). On agar plates at 4°C, *E. sakazakii* colonies on agar plates lost their viability earlier than *Escherichia coli*, *Klebsiella* spp., and *Salmonella* spp., but they nevertheless remained viable for at least 140 days. Whereas in a liquid medium stationary-phase *E. sakazakii* remained viable at 4°C over the complete length of the study, i.e., for more than 210 days (Fig. 1), *E. sakazakii* cells stored at 4°C in the exponential phase slowly decreased in viable count during the study and became undetectable after 210 days. Interestingly, at 8°C, the exponential-phase cells were able to grow and reach stationary phase, which subsequently allowed them to survive for more than 210 days.

The heat resistance of *E. sakazakii* has been investigated by several groups (Nazarowec-White and Farber, 1997b; Breeuwer et al., 2003; Edelson-Mammel and Buchanan, 2004; Iversen et al., 2004a). The most important conclusions from these studies is that the heat resistance among the various *E. sakazakii* isolates is quite diverse and that there exists, as it seems, a subgroup of isolates with a higher thermotolerance (average D value at 58°C of 9.9 min) than most of the *Enterobacteriaceae* (Edelson-Mammel and Buchanan, 2004). In an elegant follow-up study, it was shown that the strains in this thermotolerant subgroup produced a protein homologous to a protein in the thermotolerant bacterium *Methylobacillus flagellatus* (Williams et al., 2005). It remains to be confirmed whether this protein does indeed confer the elevated heat resistance. It should be noted, however, that even the most resistant *E. sakazakii* strain will be reduced by more than 8 log units by standard (15 s at 72°C) pasteurization practices. Similarly, a 5-log reduction was calculated in a study where *E. sakazakii* was treated at 68°C for 16 s (Nazarowec-White et al., 1999). From these data, the World Health Organization concluded that inactivation of *E. sakazakii* will occur very quickly at temperatures above 70°C, and they proposed that this method should be considered as a strategy to reduce of risk of *E. sakazakii* in reconstituted powdered infant formula (Food and Agriculture Organization-World Health Organization, 2004, 2006). Although this treatment will efficiently reduce the number of bacteria, it is worth mentioning that this method also has shortcomings, such as the risk of scalding of the mother and babies and the environmental consequences of cooling down the formulas with, e.g., cold running water.

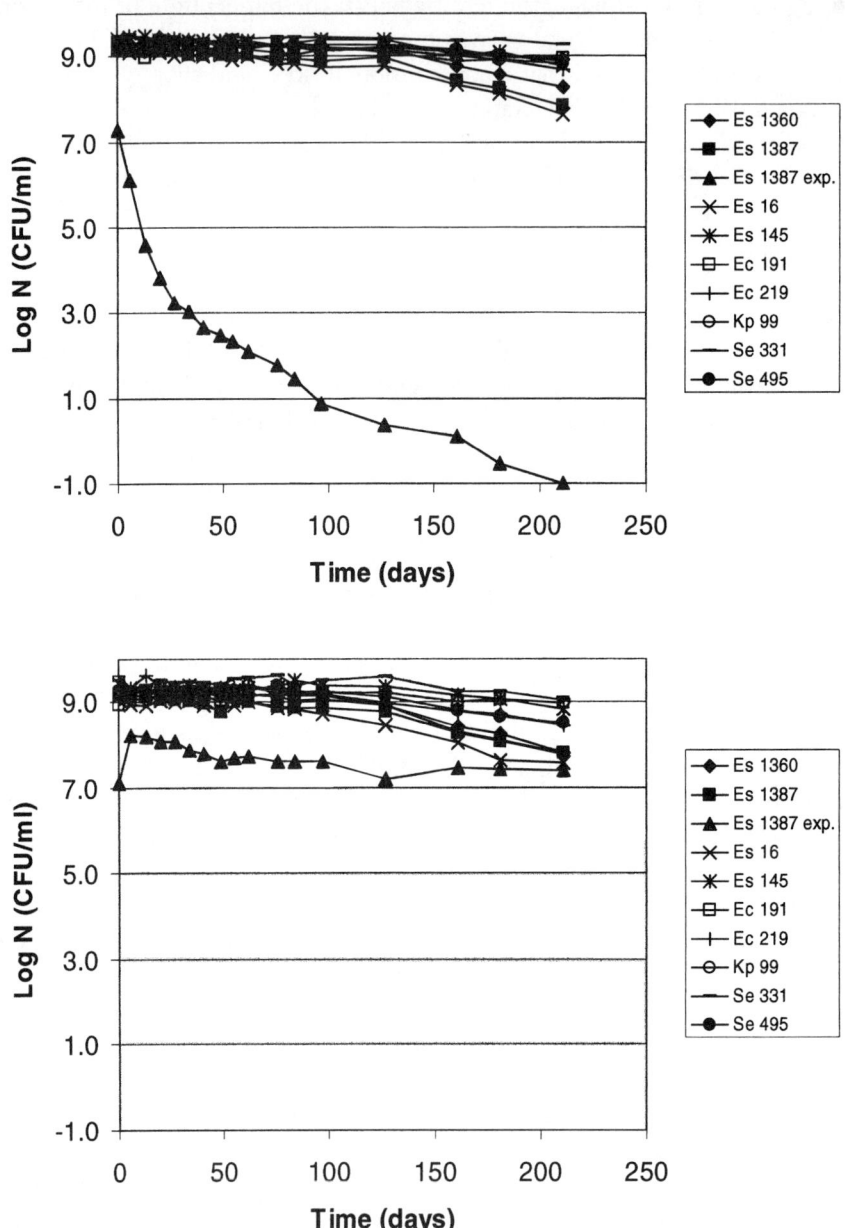

Figure 4 Survival of *Enterobacteriaceae* at 4°C (top) and 8°C (bottom) in brain heart infusion broth. Abbreviations: Es, *E. sakazakii*; Ec, *Escherichia coli*; Kp, *Klebsiella pneumoniae*; Se, *Salmonella enterica* serovar Enteritidis. All strains were in stationary phase, except for the Es 1387 exp strain, which was in exponentional phase.

With respect to production of infant formula powders, the heat resistance data demonstrate that *E. sakazakii* will not survive the first processing steps, including the pasteurization. Consequently, efforts should be focused on avoiding postpasteurization contamination via either the factory environment or the dry-mix ingredients added after the pasteurization process.

Not much information has been published on the acid resistance of *E. sakazakii.* Modeling efforts at the Nestlé Research Center suggested that the minimum pH for growth in unbuffered tryptone soya broth acidified with HCl at 30°C was around 3.9 (R. W. J. Lambert, personal communication). However, at this pH, the time to detection of growth was very long (>5 days to reach an optical density at 600 nm of 0.2, starting from an inoculation level of ×10^5 CFU/ml). At pH 4.3, the time to detection was approximately 20 h. Nevertheless, in infant rice cereal reconstituted with apple juice at pH 4.3 at 30°C, growth of *E. sakazakii* could not be detected within 72 h (Richards et al., 2005). This is probably due to the presence of other inhibiting factors in the reconstituted cereal mix, e.g., malic acid. In practice, acidification is a well-proven concept for limiting growth of enteric pathogens. This is illustrated in a study on the behavior of pathogens, including *E. sakazakii*, in infant formulas showing that acidification by addition of lactic acid (or by fermentation) to a pH below 5 has strong bacteriostatic properties (Joosten and Lardeau, 2004).

One of the particularities of *E. sakazakii* is the constitutive expression of its α-glucosidase gene. This property is, in fact, exploited for the identification of *E. sakazakii* (see chapter 2, "Isolation of *E. sakazakii* Using Chromogenic and Fluorogenic Media"). Supposedly, the function of the α-glucosidase enzyme (also called maltase) is the breakdown of maltose as a source of glucose. However, the relevance of this enzyme to the metabolism of *E. sakazakii* and why it is constitutively expressed have never been investigated.

Antimicrobial Resistance

Antibiotics

The susceptibility of *E. sakazakii* to antibiotics is of interest, because it has a direct impact on the effectiveness of the treatment of patients in the hospitals. Stock and Wiedemann (2002) investigated the natural antibiotic susceptibilities of 35 clinical isolates of *E. sakazakii* to 69 different antimicrobial agents and compared them to the relevant clinical breakpoints. The results showed that the *E. sakazakii* isolates were susceptible to most β-lactam and cephalosporin antibiotics, although the bacterium showed a relatively high natural resistance to benzylpenicillin and oxacillin. These results were in agreement with earlier published data (Farmer et al., 1980; Muytjens and van der Ros-van de Repe, 1986), which showed further that only cephalothin (which was not tested by Stock and Wiedemann) and sulfamethoxazoles showed a relatively high MIC.

One of the major concerns with respect to the treatment of *Enterobacter* infections is the emergence of antibiotic-resistant clinical strains. A recent survey indicated that about 14% of *Enterobacter* isolates produce extended-spectrum β-lactamases, conferring resistance against β-lactam antibiotics (Paterson et al., 2005). This confirmed earlier work in which extended-spectrum beta-lactamases were found in various species of *Enterobacter*, including *E. sakazakii* (Pitout et al., 1997; Girlich et al., 2001). Furthermore, the results of Stock and Wiedemann (2002) showed that some strains were resistant to amoxicillin, cefalozin, cefoxitin, and cefixim, suggesting that these isolates may have acquired antibiotic resistance genes. Other reports mention unusual resistance of a clinical *E. sakazakii* isolate from a wound infection in an adult to ampicillin, gentamycin, and cefotaxime (Dennison and Morris, 2002) and resistance of β-lactamase-positive clinical isolates from a hospital in Jerusalem to cefazolin (Block et al., 2002). In summary, these data classify *E. sakazakii* among those gram-negative pathogens that may express β-lactamases, thus conferring resistance against a range of penicillins and cephalosporins. Interestingly, it was observed in 1983 that the treatment of *E. sakazakii* meningitis by ampicillin and gentamicin was not very effective (Muytjens et al., 1983), but this was attributed to the low level of penetration of these antibiotics into the central nervous system (Willis and Robinson, 1988) and was not due to antibiotic resistance. In either case, to circumvent antibiotic resistance and achieve better penetration, the treatment of *E. sakazakii* meningitis has shifted to extended-spectrum cephalosporins, sometimes in combination with a second agent, e.g., trimethoprim (Lai, 2001; Block et al., 2002).

Other antimicrobials

In recent years, there has been an increasing interest in the use of natural antimicrobials, such as free fatty acids, for food preservation purposes. It has been shown that such compounds added to infant formula were able to inactivate a range of microorganisms (Isaacs et al., 1995). In a recent study, the antimicrobial effect of one of these antimicrobials, monocaprylin (monoglyceride of octanoic acid), on *E. sakazakii* was investigated, and it was shown that at 50 mM in reconstituted infant formula, the bacteria were efficiently killed by this monoglyceride (Nair et al., 2004). However, it should be taken into account that at this elevated concentration, these types of compounds may give rise to strong off-flavors.

Chitosan and chitosan oligomers are other natural antimicrobial compounds reported to decrease the viable cell count of *E. sakazakii* (No et al., 2002). The MICs for the higher-molecular-weight chitosans were on the order of 0.01% (wt/vol). It has been suggested that the antimicrobial effect may be attributed to the interaction between the positively charged chitosan molecules and the negatively charged bacteria. This also implies that the pH of the medium plays an important role in the efficiency of these compounds.

ADDENDUM

Since the completion of this chapter, a further publication on the taxonomy of *Enterobacter sakazakii* has been released. The Editors are aware of the apparent discrepancy between the DNA-DNA hybridization values of the two articles, and this matter is presently under further investigation. This shows the need for the research community to collaborate widely as we endeavor to control this opportunistic pathogen.

REFERENCES

Agogué, H., E. O. Casamayor, M. Bourrain, I. Obernosterer, F. Joux, G. J. Herndl, and P. Lebaron. 2005. A survey on bacteria inhabiting the sea surface microlayer of coastal ecosystems. *FEMS Microbiol. Ecol.* **54**:269–280.

Agostoni, C., I. Axelsson, C. Braegger, O. Goulet, B. Koletzko, K. F. Michaelsen, J. Rigo, R. Shamir, H. Szajewska, D. Turck, and L. T. Weaver. 2004. Probiotic bacteria in dietetic products for infants: a commentary by the ESPGHAN Committee on Nutrition. *J. Pediatr. Gastroenterol. Nutr.* **38**:365–374.

Block, C., O. Peleg, N. Minster, B. Bar-Oz, A. Simhon, I. Arad, and M. Shapiro. 2002. Cluster of neonatal infections in Jerusalem due to unusual biochemical variant of *Enterobacter sakazakii. Euro. J. Clin. Microbiol. Infect. Dis.* **21**:613–616.

Boy, K., and D. S. Hansen. 2003. Sequencing of 16S rDNA of *Klebsiella*: taxonomic relations within the genus and to other *Enterobacteriaceae. Int. J. Med. Microbiol.* **292**:495–503.

Breeuwer, P. 2005. Unpublished data.

Breeuwer, P., A. Lardeau, M. Peterz, and H. M. Joosten. 2003. Desiccation and heat tolerance of *Enterobacter sakazakii. J. Appl. Microbiol.* **95**:967–973.

Breeuwer, P., L. Michot, and H. Joosten. 2004. Genetic basis of dry stress resistance on *Enterobacter sakazakii*, abstr. 175. *Abstr. 91st Ann. Meet. Int. Assn. Food Protect.* IAFP, Des Moines, Iowa.

Brenner, D. J., C. Richard, A. G. Steigerwalt, M. A. Asbury, and M. Mandel. 1980. *Enterobacter gergoviae* sp. nov.: a new species of *Enterobacteriaceae* found in clinical specimens and the environment. *Int. J. Syst. Bacteriol.* **30**:1–6.

Brenner, D. J., A. C. McWhorter, A. Kai, A. G. Steigerwalt, and J. J. Farmer III. 1986. *Enterobacter asburiae* sp. nov., a new species found in clinical specimens, and reassignment of *Erwinia dissolvens* and *Erwinia nimipressuralis* to the genus *Enterobacter* as *Enterobacter dissolvens* comb. nov. and *Enterobacter nimipressuralis* comb. nov. *J. Clin. Microbiol.* **23**:1114–1120.

Brenner, D. J., and J. J. Farmer III. 2005. Order XIII "Enterobacteriales" family I. Enterobacteriaceae, vol. 2, part B, p. 587–607. *In* D. J. Brenner, N. R. Krieg, J. T. Staley, and G. M. Garrity (ed.), *Bergey's Manual of Systematic Bacteriology*, 2nd ed. The Williams & Wilkins Co., Baltimore, Md.

Caubilla-Barron, J., and S. J. Forsythe. 2006. Long-term persistence and recovery of *Enterobacter sakazakii* and other *Enterobacteriaceae* from powdered infant milk formula, abstr. P-019, p. 444. *Abstr. 106th Gen. Meet. Am. Soc. Microbiol.* American Society for Microbiology, Washington, DC.

Clark, N. C., B. C. Hill, C. M. O'Hara, O. Steingrimsson, and R. C. Cooksey. 1990. Epidemiologic typing of *Enterobacter sakazakii* in two neonatal nosocomial outbreaks. *Diagn. Microbiol. Infect. Dis.* **13:**467–472.

Davin-Regli, A., C. Bosi, R. Charrel, E. Ageron, L. Papazian, P. A. D. Grimont, A. Cremieux, and C. Bollet. 1997. A nosocomial outbreak due to *Enterobacter cloacae* strains with the *E. hormaechei* genotype in patients treated with fluoroquinolones. *J. Clin. Microbiol.* **35:**1008–1010.

Dennison, S. K., and J. Morris. 2002. Multiresistant *Enterobacter sakazakii* wound infection in an adult. *Infect. Med.* **19:**533–535.

Dickey, R. S., and C. H. Zumoff. 1988. Emended description of *Enterobacter cancerogenus* comb. nov. (formerly *Erwinia cancerogena*). *Int. J. Syst. Bacteriol.* **38:**371–374.

Diep, B., and P. Breeuwer. 2006. Resistance of *Enterobacter sakazakii* to desiccation, abstr. 089. *10th Int. Symp. Genet. Indust. Microorg.*, Prague, Czech Republic, 24 to 28 June 2006.

Dunn, A. K., and E. V. Stabb. 2005. Culture-independent characterization of the microbiota of the ant lion *Myrmeleon mobilis* (Neuroptera: Myrmeleontidae). *Appl. Environ. Microbiol.* **71:**8784–8794.

Edelson-Mammel, S. G., and R. L. Buchanan. 2004. Thermal inactivation of *Enterobacter sakazakii* in rehydrated infant formula. *J. Food Protect.* **67:**60–63.

Edelson-Mammel, S. G., M. K. Porteous, and R. L. Buchanan. 2005. Survival of *Enterobacter sakazakii* in a dehydrated powdered infant formula. *J. Food Protect.* **68:**1900–1902.

Farmer, J. J. I., M. A. Asbury, F. W. Hickman, D. J. Brenner, and The Enterobacteriaceae Study Group. 1980. *Enterobacter sakazakii*: a new species of "*Enterobacteriaceae*" isolated from clinical specimens. *Int. J. System. Bacteriol.* **30:**569–584.

Farmer, J. J., III, G. R. Fanning, B. R. Davis, C. M. O'Hara, C. Riddle, F. W. Hickmann-Brenner, M. A. Asbury, V. A. Lowery III, and D. J. Brenner. 1985. *Escherichia fergusonii* and *Enterobacter taylorae*, two new species of *Enterobacteriaceae* isolated from clinical specimens. *J. Clin. Microbiol.* **21:**77–81.

Fernandez-Baca, V., F. Ballestros, J. A. Hervas, P. Villalon, M. A. Dominguez, V. J. Benedi, and S. Alberti. 2001. Molecular epidemiological typing of *Enterobacter cloacae* isolates from a neonatal intensive care unit: three years of prospective study. *J. Hosp. Infect.* **49:**173–182.

Food and Agriculture Organization-World Health Organization (FAO-WHO). 2004. *Enterobacter sakazakii* and other microorganisms in powdered infant formula: meeting report. *Microbiological Risk Assessment Series 6.* World Health Organization-Food and Agriculture Organization of the United Nations, Geneva and Rome. WHO Press, Geneva, Switzerland. http://www.who.int/foodsafety/publications/micro/mra6/en/index.html.

Food and Agriculture Organization-World Health Organization (FAO-WHO). 2006. *Enterobacter sakazakii* and *Salmonella* in powdered infant formula: meeting report. *Microbiological Risk Assessment Series 10*. World Health Organization-Food and Agriculture Organization of the United Nations, Geneva and Rome. WHO Press, Geneva, Switzerland. http://www.who.int/foodsafety/publications/micro/mra10/en/index.html.

Food Standards Agency. 2005. Guidance on preparing infant formula. http://www.food.gov. uk/news/newsarchive/2005/nov/infantformulastatementnov05.

Ganeswire, R., K. L. Thong, and S. D. Puthucheary. 2003. Nosocomial outbreak of *Enterobacter gergoviae* bacteraemia in a neonatal intensive care unit. *J. Hosp. Infect.* **53**:292–296.

Garazzino, S., A. Apreto, A. Maiello, A. Massé, A. Biasibetti, F. G. de Rosa, and G. Di Perri. 2005. Osteomyelitis caused by *Enterobacter cancerogenus* infection following a traumatic injury: case report and review of the literature. *J. Clin. Microbiol.* **43**:1459–1461.

Girlich, D., L. Poirel, A. Leelaporn, A. Karim, C. Tribuddharat, M. Fennewald, and P. Nordmann. 2001. Molecular epidemiology of the integron-located VEB-1 extended-spectrum beta-lactamase in nosocomial enterobacterial isolates in Bangkok, Thailand. *J. Clin. Microbiol.* **39**:175–182.

Guillaume-Gentil, O., V. Sonnard, M. C. Kandhai, J. D. Marugg, and H. Joosten. 2005. A simple and rapid cultural method for detection of *Enterobacter sakazakii* in environmental samples. *J. Food Protect.* **68**:64–69.

Gurtler, J. B., J. L. Kornacki, and L. R. Beuchat. 2005. *Enterobacter sakazakii*: a coliform of increased concern to infant health. *Int. J. Food. Microbiol.* **104**:1–34.

Hoffmann, H., and A. Roggenkamp. 2003. Population genetics of the nomenspecies *Enterobacter cloacae*. *Appl. Environ. Microbiol.* **69**:5306–5318.

Hoffmann, H., S. Schmoldt, K. Trülzsch, A. Stumpf, S. Bengsch, T. Blankenstein, J. Heesemann, and A. Roggenkamp. 2005a. Nosocomial urosepsis caused by *Enterobacter kobei* with aberrant phenotype. *Diagn. Microbiol. Infect. Dis.* **53**:143–147.

Hoffmann, H., S. Stindl, W. Ludwig, A. Stumpf, A. Mehlen, J. Heesemann, D. Monget, K. H. Schleifer, and A. Roggenkamp. 2005b. Reassignment of *Enterobacter dissolvens* to *Enterobacter cloacae* as *E. cloacae* subspecies *dissolvens* comb. nov. and emended description of *Enterobacter asburiae* and *Enterobacter kobei*. *Syst. Appl. Microbiol.* **28**:196–205.

Hoffmann, H., S. Stindl, W. Ludwig, A. Stumpf, A. Mehlen, D. Monget, D. Pierard, S. Ziesing, J. Heesemann, A. Roggenkamp, and K. H. Schleifer. 2005c. *E. hormaechei* subsp. *oharae* subsp. nov., *E. hormaechei* subsp. *hormaechei* comb. nov., and *E. hormaechei* subsp. *steigerwaltii* subsp. nov., three new subspecies of clinical importance. *J. Clin. Microbiol.* **43**:3297–3303.

Hoffmann, H., S. Stindl, S. Stumpf, A. Mehlen, D. Monget, J. Heesemann, K. H. Schleifer, and A. Roggenkamp. 2004. Description of *Enterobacter ludwigii* sp. nov. A novel *Enterobacter* species of clinical relevance. *Syst. Appl. Microbiol.* **28**:206–212.

Isaacs, C. E., R. E. Litov, and H. Thormar. 1995. Antimicrobial activity of lipids added to human milk, infant formula, and bovine milk. *J. Nutr. Biochem.* **6**:362–366.

Iversen, C., and S. Forsythe. 2004. Isolation of *Enterobacter sakazakii* and other *Enterobacteriaceae* from powdered infant formula milk and related products. *Food Microbiol.* **21**:771–777.

Iversen, C., M. Lane, and S. J. Forsythe. 2004a. The growth profile, thermotolerance and biofilm formation of *Enterobacter sakazakii* grown in infant formula milk. *Lett. Appl. Microbiol.* **38**:378–382.

Iversen, C., M. Waddington, S. L.W. On, and S. Forsythe. 2004b. Identification and phylogeny of *Enterobacter sakazakii* relative to *Enterobacter* and *Citrobacter*. *J. Clin. Microbiol.* **42**:5368–5370.

Iversen, C., L. Lancashire, M. Waddington, S. Forsythe, and G. Ball. 2006a. Identification of *Enterobacter sakazakii* from closely related species: the use of artificial neural networks in the analysis of biochemical and 16S rDNA data. *BMC Microbiol.* **6**:28.

Iversen, C., M. Waddington, J. J. Farmer III, and S. J. Forsythe. 2006b. The biochemical differentiation of *Enterobacter sakazakii* genotypes. *BMC Microbiol.* **6**:94.

Joosten, H. M., and A. Lardeau. 2004. Enhanced microbiological safety of acidified infant formulas tested in vitro. *S. Afr. J. Clin. Nutr.* **17**:87–92.

Kämpfer, P., S. Ruppel, and R. Remus. 2005. *Enterobacter radicincitans* sp. nov., a plant growth promoting species of the family Enterobacteriaceae. *Syst. Appl. Microbiol.* **28**:213–221.

Kandhai, M. C. 2005. Personal communication.

Kandhai, M. C., M. W. Reij, L. G. Gorris, O. Guillaume-Gentil, and M. van Schothorst. 2004. Occurrence of *Enterobacter sakazakii* in food production environments and households. *Lancet* **363**:39–40.

Kandhai, M. C., M. W. Reij, C. Grognou, M. van Schothorst, and M. H. Zwietering. 2006. The effect of pre-culturing conditions on the lag time and the specific growth rate of *Enterobacter sakazakii* in reconstituted powdered infant formula. *Appl. Environ. Microbiol.* **72**:2721–2729.

Kosako, Y., K. Tamura, R. Sakazaki, and K. Miki. 1996. *Enterobacter kobei* sp. nov., a new species of the family Enterobacteriaceae resembling *Enterobacter cloacae*. *Curr. Microbiol.* **33**:261–265.

Lai, K. K. 2001. *Enterobacter sakazakii* infections among neonates, infants, children, and adults. Case reports and a review of the literature. *Medicine* (Baltimore) **80**:113–122.

Lehner, A., K. Riedel, L. Eberl, P. Breeuwer, B. Diep, and R. Stephan. 2005. Biofilm formation, extracellular polysaccharide production, and cell-to-cell signaling in various *Enterobacter sakazakii* strains: aspects promoting environmental persistence. *J. Food Protect.* **68**:2287–2294.

Muytjens, H. L., H. Roelofs-Willemse, and G. H. Jaspar. 1988. Quality of powdered substitutes for breast milk with regard to members of the family Enterobacteriaceae. *J. Clin. Microbiol.* **26**:743–746.

Muytjens, H. L., and J. van der Ros-van de Repe. 1986. Comparative in vitro susceptibilities of eight *Enterobacter* species, with special reference to *Enterobacter sakazakii*. *Antimicrob. Agents Chemother.* **29**:367–370.

Muytjens, H. L., H. C. Zanen, H. J. Sonderkamp, L. A. Kollee, I. K.Wachsmuth, and J. J. Farmer III. 1983. Analysis of eight cases of neonatal meningitis and sepsis due to *Enterobacter sakazakii*. *J. Clin. Microbiol.* **18**:115–120.

Nair, M. K., J. Joy, and K. S. Venkitanarayanan. 2004. Inactivation of *Enterobacter sakazakii* in reconstituted infant formula by monocaprylin. *J. Food Protect.* **67**:2815–2819.

Nazarowec-White, M., and J. M. Farber. 1997a. Incidence, survival, and growth of *Enterobacter sakazakii* in infant formula. *J. Food Protect.* **60**:226–230.

Nazarowec-White, M., and J. M. Farber. 1997b. Thermal resistance of *Enterobacter sakazakii* in reconstituted dried-infant formula. *Lett. Appl. Microbiol.* **24**:9–13.

Nazarowec-White, M., and J. M. Farber. 1997c. *Enterobacter sakazakii*: a review. *Int. J. Food Microbiol.* **34**:103–113.

Nazarowec-White, M., R. C. McKellar, and P. Piyasena. 1999. Predictive modelling of *Enterobacter sakazakii* inactivation in bovine milk during high-temperature short-time pasteurization. *Food Res. Int.* **32**:375–379.

No, H. K., N. Y. Park, S. H. Lee, H. J. Hwang, and S. P. Meyers. 2002. Antibacterial activities of chitosans and chitosan oligomers with different molecular weights on spoilage bacteria isolated from tofu. *J. Food Sci.* **67**:1511–1514.

O'Hara, C. M., A. G. Steigerwalt, B. C. Hill, J. J. Farmer III, G. R. Fanning, and D. J. Brenner. 1989. *Enterobacter hormaechei*, a new species of the family *Enterobacteriaceae* formerly known as enteric group 75. *J. Clin. Microbiol.* **27**:2046–2049.

Paterson, D. L., F. Rossi, F. Baquero, P. R. Hsueh, G. L.Woods, V. Satishchandran, T. A. Snyder, C. M. Harvey, H. Teppler, M. J. Dinubile, and J. W. Chow. 2005. In vitro susceptibilities of aerobic and facultative gram-negative bacilli isolated from patients with intra-abdominal infections worldwide: the 2003 Study for Monitoring Antimicrobial Resistance Trends (SMART). *J. Antimicrob. Chemother.* **55**:965–973.

Pitout, J. D., E. S. Moland, C. C. Sanders, K. S. Thomson, and S. R. Fitzsimmons. 1997. Beta-lactamases and detection of beta-lactam resistance in *Enterobacter* spp. *Antimicrob. Agents Chemother.* **41**:35–39.

Richards, G. M., J. B. Gurtler, and L. R. Beuchat. 2005. Survival and growth of *Enterobacter sakazakii* in infant rice cereal reconstituted with water, milk, liquid infant formula, or apple juice. *J. Appl. Microbiol.* **99**:844–850.

Schonheyder, H. C., K. T. Jensen, and W. Frederiksen. 1994. Taxonomic notes: synonymy of *Enterobacter cancerogenus* (Urosevic 1966) Dickey and Zumoff 1988 and *Enterobacter taylorae* Farmer et al. 1985 and resolution of an ambiguity in the biochemical profile. *Int. J. Syst. Bacteriol.* **44**:586–587.

Songjinda, P., J. Nakayama, Y. Kuroki, S. Tanaka, S. Fukuda, C. Kiyohara, T. Yamamoto, K. Izuchi, T. Shirakawa, and K Sonomoto. 2005. Molecular monitoring of the developmental bacterial community in the gastrointestinal tract of Japanese infants. *Biosci. Biotechnol. Biochem.* **69**:638–641.

Stewart, J. M., and J. R. Quirk. 2001. Community-acquired pneumonia caused by *Enterobacter asburiae*. *Am. J. Med.* **111**:82–83.

Stock, I., and B. Wiedemann. 2002. Natural antibiotic susceptibility of *Enterobacter amnigenus*, *Enterobacter cancerogenus*, *Enterobacter gergoviae* and *Enterobacter sakazakii* strains. *Clin. Microbiol. Infect.* **8**:564–578.

Swofford, D. L. 1998. *PAUP. Phylogenetic Analysis Using Parsimony, Version 3.1.1.* Illinois National History Survey, University of Illinois, Champaign.

Tamaru, Y., Y. Takani, T. Yoshida, and T. Sakamoto. 2005. Crucial role of extracellular polysaccharides in desiccation and freezing tolerance in the terrestrial cyanobacterium *Nostoc commune*. *Appl. Environ. Microbiol.* **71**:7327–7333.

Urmenyi, A. M. C., and A.W. Franklin. 1961. Neonatal death from pigmented coliform infection. *Lancet* **i**:313–315.

Williams, T. L., S. R. Monday, S. Edelson-Mammel, R. Buchanan, and S. M. Musser. 2005. A top-down proteomics approach for differentiating thermal resistant strains of *Enterobacter sakazakii*. *Proteomics* **5**:4161–4169.

Willis, J., and J. E. Robinson. 1988. *Enterobacter sakazakii* meningitis in neonates. *Pediatr. Infect. Dis. J.* **7**:196–199.

Enterobacter sakazakii
Edited by Jeffrey M. Farber and Stephen J. Forsythe
© 2008 ASM Press, Washington, D.C.

Isolation and Identification of *Enterobacter sakazakii*

2

Séamus Fanning and Steve Forsythe

DISTRIBUTION OF *ENTEROBACTER SAKAZAKII* IN RAW FOOD, READY-TO-EAT FOODS, SOIL AND WATER, HUMAN AND ANIMAL CARRIAGE, AND OTHER HABITATS

Occurrence of *Enterobacter sakazakii* in Foods and the Environment

Due to the association of *Enterobacter sakazakii* with neonatal infections, the organism has primarily been associated with powdered infant formula. However, *E. sakazakii* is widely distributed in the environment and in foods, with the most probable sources of the organism being water, soil, and vegetables. Table 1 summarizes the various published papers reporting the isolation of *E. sakazakii* from clinical samples, foodstuffs, and the environment. The organism has been isolated from a range of dairy (cheese, ultrahigh-temperature pasteurized milk, milk powder, and powdered infant milk formula) and nondairy products (fermented bread, tofu, and sour tea), including meat (cured meats, minced beef, and sausage meat). Gassem (1999) found *E. sakazakii* in Khamir bread due to being part of the sorghum seed surface flora. The organism has also been isolated from rice seeds, lettuce, alfalfa sprouts, and tomatoes. In a large survey of over 500 foodstuffs and ingredients, a large proportion (~25%) of herbs and spices were shown to contain *E. sakazakii* (Iversen and Forsythe, 2004).

Although *E. sakazakii* has been isolated from water, sediment, and soil (Table 1), Muytjens and Kollee (1990) were unable to isolate the organism from a number of environmental sources: raw cow's milk, cattle, rodents, grain, bird feces, domestic animals, surface water, soil, mud, and rotting

SÉAMUS FANNING, Centre for Food Safety, School of Agriculture, Food Science, and Veterinary Medicine, University College Dublin, Belfield, Dublin 4, Ireland. STEVE FORSYTHE, School of Biomedical and Natural Sciences, Nottingham Trent University, Nottingham, United Kingdom.

Table 1 Sources of *E. sakazakii*[a]

Source	Detail(s)	References
Neonates and infants	Meningitis	Urmenyi and Franklin (1961), Jøker et al. (1965), Kleiman et al. (1981), Muytjens et al. (1983), Naqvi et al. (1985), Arseni et al. (1987), Willis and Robinson (1988), Biering et al. (1989), Lecour et al. (1989), Simmons et al. (1989), Clark et al. (1990), Muytjens and Kollee (1990), Noriega et al. (1990), Gallagher and Ball (1991), Ries et al. (1994), Burdette and Santos (2000), Bar-Oz et al. (2001), Lai (2001), Coignard et al. (2006), Loc-Carrillo et al. (2006), Bowen and Braden (2006)
	Bacteremia	Monroe and Tift (1979), Clark et al. (1990), Bar-Oz et al. (2001), Coignard et al. (2006), Jarvis (2005), Stoll et al. (2004), Loc-Carrillo et al. (2006)
	Necrotizing enterocolitis	Van Acker et al. (2001), Coignard et al. (2006)
	Wound exudates, appendicitis, conjunctivitis	Reina et al. (1989)
Adults	Bacteremia	Jimenez and Gimenez (1982), Pribyl et al. (1985), Murray et al. (1990), Hawkins et al. (1991), Lai (2001)
	Throat	Dennison and Morris (2002), Gosney et al. (2006)
Food and drink	Powdered infant formula	Farmer et al. (1980), Postupa and Aldová (1984), Block et al. (1988), Biering et al. (1989), Simmons et al. (1989), Muytjens et al. (1988), Smeets et al. (1998), Bar-Oz et al. (2001), Heuvelink et al. (2001), Himelright et al. (2002), Iversen and Forsythe (2004), Estuningsih et al. (2006), Loc-Carrillo et al. (2006), Igimi et al. (2006), Shaker et al. (2007)
	Preparation equipment: blender, spoons	Block et al. (1988), Clark et al. (1990), Smeets et al. (1998), Bar-Oz et al. (2001)
	Milk powder	Postupa and Aldová (1984), Heuvelink et al. (2001), Iversen and Forsythe (2004)
	Water, pipelines, biofilm	Bartolucci et al. (1995), Al-Hadithi et al. (1995), Oliver (1997)
	Hydrothermal springs	
	Rice seed	Mosso et al. (1994)
	Beer mugs	Cottyn et al. (2001)
	Cured meat	Schindler and Metz (1990)
	Fermented bread	Watanabe and Esaki (1994)
	Lettuce	Gassem (1999)
	Tofu	Soriano et al. (2001)

Table 1 (*continued*)

Source	Detail(s)	References
	Sour tea	No et al. (2002)
	Cheese, minced beef, sausage meat, vegetables	Tamura et al. (1995)
	Dried infant food	Leclercq et al. (2002)
	Cheese, fresh and dried foods, herbs and spices	Iversen and Forsythe (2004)
	Soybean	Iversen and Forsythe (2004)
	Ready-to-eat salads	Kuklinsky-Sobral et al. (2004)
	Artisanal cheese	Weiss et al. (2005)
	Fermented cassava	Chaves-López et al. (2006)
	Eggs	Coulin et al. (2006), Brito et al. (2006)
Environmental	Hospital air	Masaki et al. (2001)
	Hospital infant formula preparation area	Palcich et al. (2006)
	Clinical material	Tuncer and Ozsan (1988), Janicka et al. (1999)
	Milk powder, chocolate, cereal, potato flour, pasta and spices factories	Kandhai et al. (2004a)
	Household dust	
	Flies	Kandhai et al. (2004a)
	Rats	Kuzina et al. (2001), Gakuya et al. (2001)
	Soil	Neelam et al. (1987)
	Rhizosphere	Emilani et al. (2001)
	Sediment, wetlands	Espeland and Wetzel (2001)
	Crude oil	Assadi and Mathur (1991)
	Cutting fluids	Suliman et al. (1997)
	Paper sludge	Beuchamp et al. (2006)

[a]Data are from Iversen and Forsythe (2003).

wood. However, this may have been due to the lack of positive selection in their isolation method, which would have resulted in the overgrowth by other (non-*E. sakazakii*) *Enterobacteriaceae*.

The natural habitat of *E. sakazakii* is presently unknown. However, based on some of the organism's physiological features, it is likely that it may have an environmental niche on plant material (Iversen et al., 2004b). These physiological traits aid environmental survival and include the ability to produce a yellow pigment that protects the cell against UV rays in sunlight, capsular and fimbriae formation to aid in adherence to surfaces, and its ability to resist desiccation during long dry periods of over 2 years (Caubilla-Barron and Forsythe, 2007). This may also explain why the organism is isolated from various plant-related products and ingredients, including cereals, potato flour, pasta, herbs, and spices (Table 1). These characteristics may also lend themselves to aid the organism's persistence in the formula manufacturing and domestic environments.

Kandhai et al. (2004a) isolated *E. sakazakii* from almost all environments examined, including milk powder manufacturing facilities and household vacuum cleaners. Eight of the nine facilities samples were positive, and five of the sixteen households were also found to be positive for *E. sakazakii*. This evidence confirms the ubiquitous nature of this pathogenic organism. Moreover, contaminated utensils used for the preparation of infant feed, blenders and spoons, as well as prolonged storage of reconstituted infant formula in bottle warmers, have been linked to *E. sakazakii* infection of infants (Jaspar et al., 1990; Noriega et al., 1990; Bar-Oz et al., 2001; Block et al., 2002). In a German study of the bacterial flora of rinsed beer mugs, *E. sakazakii* was identified as a common organism, which indicated that unhygienic dishwashing procedures may play an important role in the epidemiology of clinically relevant microorganisms (Schindler and Metz, 1990).

Two reports have documented the isolation of *E. sakazakii* from rats in Kenya (Gakuya et al., 2001) and from Mexican fruit flies in California (Kuzina et al., 2001). Despite the list of contaminated foods and other sources, the frequency of contamination is unknown, and therefore the risk of exposure cannot be reliably quantified.

Contamination of Powdered Infant Formula

Although the mode of transmission is not always clear, contaminated infant formula is a recognized source of *E. sakazakii*. Such infection poses a serious risk to newborn infants due to its high morbidity rates and occasional mortality rates (see chapter 4) (Forsythe, 2005; Bowen and Braden, 2006). Additionally, Postupa and Aldová (1984) isolated two strains of *E. sakazakii* from infant formula in former Czechoslovakia. Clark et al. (1990) investigated two unrelated cases involving meningitis and bacteremia, and they found that each victim could be linked to a particular infant formula. Unlike ready-to-feed liquid infant formula, dried formula powders are not sterile. It is of interest that one of the strains used to define *E. sakazakii* by Farmer et al. (1980) was a strain that was originally isolated from dried milk by Thornley (1960). Therefore, *E. sakazakii* has probably been present in dried milk products for many decades, including 1958, when the first *E. sakazakii* meningitis case was reported (Urmenyi and Franklin, 1961).

Table 2 Summary of *Enterobacteriaceae* cultured from different infant formulations[a]

Organism	Frequency	Concn (CFU/100 g)
Pantoea spp. 1 to 4, *Escherichia vulneris* (*Enterobacter agglomerans*)	35	0.3–14.94
Enterobacter cloacae	30	0.36–91.78
Enterobacter sakazakii	20	0.36–66
Klebsiella pneumoniae	13	0.36–46.22
Citrobacter freundii	5	~0.36
Escherichia coli	4	~4.6
Klebsiella oxytoca	4	0.36–0.92
Citrobacter koseri (formerly *C. diversus*)	3	0.19–0.36
Hafnia alvei	1	0.15

[a]Data are from Muytjens et al. (1988). The original paper gave the *"Entrobacter agglomerans"* designation to some isolates. However, since Muytjens et al. (1988), this organism has been reclassified and subsequently it is not possible to determine whether the orginal isolates were *Pantoea* species or *Escherichia vulneris*.

Two main routes by which *E. sakazakii* can enter reconstituted infant formula are known:

1. intrinsic contamination, either through contaminated ingredients added after drying or from the processing environment following drying and before packing, and/or

2. through external contamination of the formula during reconstitution and handling (e.g., through poorly cleaned utensils).

Muytjens et al. (1988) conducted a survey on commercially acquired infant formulas for *E. sakazakii*. *Enterobacter* species were cultured from over half of the 141 infant formulas examined from 35 different countries. In addition, *Pantoea agglomerans*, *E. cloacae*, *E. sakazakii*, and *Klebsiella pneumoniae* were the most frequently isolated organisms (Table 2). Twenty of the 141 samples were from 13 of the 35 countries. Further surveys of *Enterobacteriaceae* in powdered infant formula and milk powders confirmed this range of organisms (Iversen and Forsythe, 2004).

As the bacterial numbers determined by Muytjens et al. (1988) were under the required limit of 1 CFU/g of dry powder, all samples were in accordance with the microbiological guidelines of that time as set out by The International Commission on Microbiological Specifications for Foods (ICMSF) (http://www.nap.edu/books/0309034973/html/152.html). The levels of *E. sakazakii* determined ranged from 0.36 to 66 CFU/100 g. More recently, these microbiological criteria have been revised in the European

Union (EU) (European Commission, 2005) and are presently under revision by the Codex Alimentarius Commission. In an earlier study, Muytjens et al. (1983) isolated *E. sakazakii* from reconstituted infant formula but were unable to culture the organism from the dried infant formula or the water used to reconstitute the formula. A Canadian survey of the incidence of *E. sakazakii* in commercial dried infant formula showed that 8 of 120 cans (6.7%) tested positive for *E. sakazakii*. These eight positive cans represented five different manufacturers, with a level of 0.36 CFU/100 g (Nazarowec-White et al., 1999).

Following the increased awareness concerning *E. sakazakii* and powdered infant milk formula, a number of surveys have been published recently and are summarized in Table 3. It should be noted that the surveys used different detection methods and sample volumes. Nevertheless, the incidence of *E. sakazakii* reported ranged from 2 to 14%. Only three reports determined the concentrations of the organism, which were between 0.2 and 92 CFU/ 100 g. Importantly, *E. sakazakii* has never been reported at levels greater than 1 CFU/g. This is similar to the value of 8 cells/100 g estimated by Simmons et al. (1989) for an open can of powdered infant formula in use during a neonatal care unit outbreak.

It is important that *E. sakazakii* infection from rehydrated infant formula can arise from contaminated water, introduction by the caregiver, or the environment. Therefore, considerable effort has been put into improving hygiene during the preparation and storage of infant formula. There is a misconception that bottled water is sterile. Schindler (1994) surveyed 54

Table 3 Surveys of powdered infant formula or milk powders for *E. sakazakii*

Reference	Method[a]	Volume sampled (g)	Sample no.	No. positive for *E. sakazakii*	Enumeration
Muytjens et al. (1988)	BPW, EE, VRBGA	333–555	141	20 (14.2)	0.19–91.78
Nazarowec-White and Farber (1997)	DW, EE, VRBGA	333	120	8 (6.7)	0.36
Heuvelink et al. (2002, 2003)	BPW, EE, VRBLA	25	101	2 (2)	ND[b]
Kandhai et al. (2004)	BPW, mLST, VRBLA	10	68	14 (20.6)	ND
Kress et al. (2005)	Various	333, 300	74, 20	10 (17.6), 2 (10)	ND
Iversen et al. (2004)	BPW, EE, DFI	25	102	3 (2.9)	ND
Santos (personal communication)	FDA, BAX	100 (5×)	98	12 (12.2)	0.22–1.61/ 100 g

[a]See Tables 6 to 8 for further details on isolation methods.
[b]ND, not determined; DW, distilled water; BPW, buffered peptone water.

manufacturers of mineral, spring, and table water and found *E. cloacae* (17 samples out of 54 positive) as well as *E. amnigenus* (11 samples out of 54 positive). The clinical occurrence of *E. sakazakii* is covered in chapter 4.

ICMSF AND EU SAMPLING PLANS FOR FINISHED PRODUCTS

As previously stated, powdered infant milk is not manufactured as a sterile product but must conform to international microbiological criteria and internal quality standards. The present Codex Alimentarius Commission (CAC) code of hygiene practice for foods for infants and children (CAC, 1979) requires a minimum of four to five samples with <3 coliforms/g and a maximum of one out of five control samples with >3 but ≤ 20 coliforms/g. Other criteria include the absence of *Salmonella* ($n = 60$; $m = 0/25$ g, where m is the maximum number of bacteria) and a maximum aerobic plate count of 10^4 CFU/g. A survey by Townsend et al. (2006) showed that while most samples meet these criteria, 3 and 6% of samples analyzed exceeded the upper limits for coliforms and aerobic plate counts, respectively (Table 4). This coliform count has not been exceeded by the numbers of *E. sakazakii* present in infant formula associated with outbreaks. Nevertheless, after the Alfaré outbreak (van Acker et al., 2001), Nestlé upgraded its facilities and applied more stringent release criteria of <0.3 coliforms/g and 0 *E. sakazakii* isolates/10 g.

Despite the rare occurrence of reported *E. sakazakii* outbreaks, the international criteria are being revised. The criteria proposed by the Codex Commission for Food Hygiene (CCFH), along with the ICMSF, for finished products are given in Table 5. These criteria, especially the choice of a two-class plan ($n = 30$; $m = 0/10$ g), have not yet been accepted, and presently three-class plans are being considered as an alternative. In addition to specific criteria for *E. sakazakii*, the replacement of the original indicator organism group, coliforms with *Enterobacteriaceae*, has been proposed. Although *E. sakazakii* is a member of the *Enterobacteriaceae*, the specific *Enterobacteriaceae* test is a

Table 4 General microbial flora cultured from powdered infant formula[a]

Organism or plate count	% of sample[b] at viable count (CFU/g) of:				
	$<10^2$	10^2	$>10^2–10^3$	$>10^3–10^4$	$>10^4$
Aerobic plate count	59	23	13	4	1
Enterobacteriaceae	99	0	1	0	0
Bacillus species	53	23	23	1	0
Staphylococcus species	96	4	0	0	0
Clostridium species	100	0	0	0	0

[a]Data are from Townsend et al. (2007).
[b]$n = 75$.

Table 5 Proposed CCFH and ICMSF microbiological criteria[a]

Microorganism(s)	n	c	m	M	Class plan
Mesophilic aerobic bacteria[b]	5	2	$5 \times 10^2/g$	$5 \times 10^3/g$	3
Enterobacteriaceae spp.	10	2	0/10 g	NA[b]	2
E. sakazakii	30	0	0/10 g	NA	2
Salmonella enterica serovars	60	0	0/25 g	NA	2

[a]NA, not applicable; *n*, number of sample units from a lot that must be examined; *c*, maximum acceptable number of sample units that may exceed the value of *m*; *m*, maximum number of relevant bacteria (values greater than this are either marginally acceptable or unacceptable); *M*, maximum number of relevant bacteria (values greater than this are unacceptable).
[b]These proposed criteria for mesophilic aerobic bacteria are reflective of good manufacturing practices and do not include nonpathogenic microorganisms that may be intentionally added, such as probiotics.

Table 6 Presence of *E. sakazakii* in samples in which *Enterobacteriaceae* were present in or absent from powdered infant formula samples

Source of data	Presence (+) or absence (−) of *Enterobacteriaceae*	*E. sakazakii* negative	*E. sakazakii* positive	Total
Muytjens et al. (1988)	−	69	5	74
	+	52	15	67
	Total	121	20	141
Iversen et al. (2004)	−	73	2	75
	+	7	0	7
	Total	80	2	82

general microbial hygiene indicator and is particularly sensitive to stressed cells (as occurs during powdered infant formula manufacture) due to the direct plating of the sample on selective agar. The *Enterobacteriaceae* method is not as sensitive as the *E. sakazakii* method. As shown in Table 6, *E. sakazakii* has been isolated from *Enterobacteriaceae*-negative samples (Muytjens et al., 1988; Iversen and Forsythe, 2004). Meanwhile, from 1 January 2006, the EU implemented the same criteria (*n* = 30; m = 0/10 g) for *E. sakazakii* and *Enterobacteriaceae* (European Commission, 2005). These criteria are to be applied to direct infant formula and dried dietary foods for special medical purposes intended for infants under 6 months. Note that, as of the end of 2006, the EU microbiological criteria for these products were under revision.

CULTURAL ISOLATION METHODS: RESUSCITATION, ENRICHMENT, AND SELECTIVE-DIFFERENTIAL MEDIA

Isolation of *E. sakazakii* after Selection for *Enterobacteriaceae*

The initial method for the detection of *E. sakazakii* in powdered infant formula was developed by Muytjens et al. (1988) and was used with minor

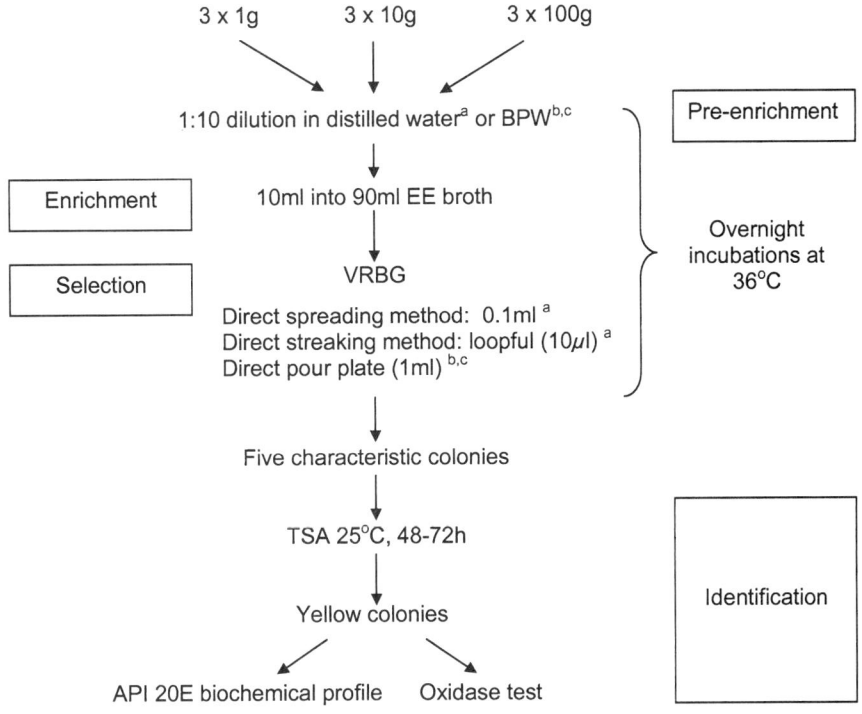

3 x 1g   3 x 10g   3 x 100g

1:10 dilution in distilled water[a] or BPW[b,c] | Pre-enrichment |

Enrichment   10ml into 90ml EE broth

Selection   VRBG

Direct spreading method: 0.1ml [a]
Direct streaking method: loopful (10μl) [a]
Direct pour plate (1ml) [b,c]

Overnight
incubations at
36°C

Five characteristic colonies

TSA 25°C, 48-72h

Yellow colonies   Identification

API 20E biochemical profile   Oxidase test

a FDA method   b Muytjens *et al.* (1988) c Nazarowec-White & Farber (1997)

Figure 1 Conventional cultural detection of *E. sakazakii* (from Forsythe, 2005).

modifications by Nazarowec-White and Farber (1997), and it is the U.S. Food and Drug Administration (FDA)-recommended method (Fig. 1; Table 7).

These methods have three key steps, beginning with a preenrichment step, wherein the infant formula is rehydrated in sterile distilled water overnight at 36°C. After preenrichment, the sample is enriched in *Enterobacteriaceae* enrichment (EE) broth. After overnight incubation at 36°C, the sample is surface plated and streaked onto violet red bile glucose agar (VRBGA). The plates are incubated overnight (36°C) before subculturing presumptive-positive colonies onto tryptic soy agar (TSA). These TSA plates are incubated for 48 to 72 h at 25°C. Only yellow-pigmented colonies are selected, and their identities are confirmed using the API 20E biochemical identification system. This test requires a further 18 to 24 h to complete. In total, the FDA method requires a minimum of 5 days to complete (www.cfsan.fda .gov/~comm/mmesakaz.html). The initial overnight preenrichment step is

Table 7 Culture methods for the detection of *E. sakazakii* in powdered infant formula

Method	Preenrichment	Enrichment	Primary isolation	Presumptive identification
FDA[a]	DW (37°C)	EE broth (37°C)	VRBGA	Yellow pigment on TSA at 25°C for 48–72 h
DFI	BPW (37°C)	EE broth (37°C)	DFI agar (37°C)[b]	Blue-green colonies
NES[c]	BPW (37°C)	mLST + vancomycin (45°C)	TSA (37°C with enhanced light exposure)	Yellow, α-glucosidase-positive colonies
AES[d]	ESSB (37°C)	ESSB (37°C)	ESIA (44°C)	Blue-green colonies
R&F[e]	DW	EE broth (37°C)	ESPM	Blue-back or blue-gray colonies

[a]U.S. Food and Drug Administration (2002). DW, distilled water; BPW, buffered peptone water.
[b]Iversen et al. (2004a).
[c]Guillaume-Gentil et al. (2005).
[d]AES Laboratorie (http://www.aes-lab.com).
[e]Restaino et al. (2006).

Table 8 Composition of *E. sakazakii* enrichment broths

Broth	Principal nutrient (g/liter)	Selective ingredients (g/liter[a])	Incubation conditions	Reference or source
EE	Peptone (10), glucose (5)	Ox bile (20), brilliant green (0.0135)	37°C, 18–24 h	Oxoid Ltd.
Lauryl sulfate tryptose broth		Sodium lauryl sulfate (0.1), sodium chloride (0.5 M)	45°C, 22–24 h	Kandhai et al. (2004)
Modified lauryl sulfate tryptose–vancomycin medium	Enzymatic digest of animal and plant tissue (20), lactose (5)	Sodium chloride (0.5 M), sodium lauryl sulfate (0.1), vancomycin (0.01)	45°C, 22–24 h	Guillaume-Gentil et al. (2005)
Modified lauryl sulfate tryptose–vancomycin medium	Enzymatic digest of animal and plant tissue (20), lactose (5)	Sodium chloride (34), sodium lauryl sulfate (0.1), vancomycin (0.01)	45°C, 22–24 h	ISO/DTS 22964

[a]Unless otherwise indicated.

necessary, as both EE broth and VRBGA contain selective and differential ingredients (Oxgall and brilliant green in the EE broth and bile salts no. 3 and crystal violet in VRBGA) that may prevent resuscitation of stressed *E. sakazakii* cells (Tables 8 and 9). Therefore, VRBGA cannot be used for direct detection in powdered infant formula and other foods.

In addition to this being a time-consuming procedure, the FDA method will not detect nonpigmented *E. sakazakii* strains (Iversen et al., 2004a; Iversen et al., 2004b; Besse et al., 2006) and uses only one biochemical

Table 9 Composition of selective and differential agars for *E. sakazakii*

Medium	Principal component(s) (g/liter)	Indicator(s) (g/liter)	Selective agent(s) (g/liter)	Incubation conditions	Reference
DFI	Tryptone soya agar, enzymatic digest of casein (15), soya (5)	5-Bromo-4-chloro-3-indolyl-α, D-glucopyranoside (0.1), sodium thiosulfate (1), ammonium iron (III) citrate (1)	Sodium deoxycholate (1)	37°C, 24 h	Iversen et al. (2004)
OK	Tryptone (20)	4-Methylumbelliferyl-α-D-glucoside (0.05), sodium thiosulfate (1), ferric citrate (1)	Bile salt no. 3 (1.5)	37°C, 24 h	Oh and Kang (2004)
TSBA	Tryptone soya agar, enzymatic digest of casein (15), soya (5)	Exposure to artificial white light (ca. 600 lux) to induce pigment production	Bile salt no. 3 (1.5)	37°C, 24 h	Guillaume-Gentil et al. (2005)
ESIA	Peptone (7), yeast extract (3)	5-Bromo-4-chloro-3-indolyl-α, D-glucopyranoside (0.15)	Sodium desoxycholate (0.6), crystal violet (0.002)	44°C, 24 h	ISO/DTS 22964
ESPM	Peptone (4.3), tryptone (4), proteose peptone (3), yeast extract (6), sugars (sorbitol, D-arabitol, adonitol) (23)	X-α-D-glucopyranoside (0.15), X-α-D-cellobioside (0.15), phenol red (0.1)	Bile salts (1.25), vancomycin (not given), cefsulodin (not given)	35°C, 24 h	Restaino et al. (2006)

Table 10 Percentage of strains showing increase in optical density after 24 h of incubation in enrichment media[a]

Organism (*n*)	Incubation medium[b]			
	BPW (37°C)	EE (37°C)	ESSB (37°C)	mLSB (44°C)
E. sakazakii (160)	100	98	96	94
Non-*E. sakazakii* Enterobacteriaceae spp. (74)	100	100	95	83

[a]Data are from Iversen and Forsythe (2007).
[b]BPW, buffered peptone water; mLSB, modified lauryl sulfate broth.

Table 11 Percent growth of *E. sakazakii* strains and other *Enterobacteriaceae* on various selective and differential agars[a]

Organism (*n*)	Medium[b]				
	DFI[c] (37°C)	VRBGA[d] (37°C)	VRBL[d] (37°C)	TSA (44°C)	ESIA (44°C)
E. sakazakii (160)	100	100	100	99	97
Non-*E. sakazakii*, Enterobacteriaceae spp. (74)	54	100	93	97	34

[a]Data are from Iversen and Forsythe (2007).
[b]Data are given as the percentage of strains.
[c]Blue-green colonies.
[d]Red colonies.

test kit, whereas other end tests (including ID32E) have been shown to be more reliable (Iversen and Forsythe, 2004). In addition, approximately 2% of *E. sakazakii* strains are sensitive to EE broth and will be missed by this method (Iversen and Forsythe, 2007) (Table 10). Comparisons of various enrichment broths and selective agars for *E. sakazakii* are given in Tables 10 and 11. Although the International Organization for Standardization (ISO) method (DTS 22964) includes the identification criterion of yellow pigmentation, it does acknowledge that some *E. sakazakii* strains are nonpigmented. As of the end of 2006, a horizontal method is under consideration, and it is anticipated that the method will not use the raised temperature of 44°C or yellow pigment production as discriminatory features; see Besse et al. (2006).

The FDA-approved method is a presumptive, most-probable-number test based on the three-tube enrichment format, so that low levels of the organism, if present in the product, can be detected and quantified. Triplicate samples of 1, 10, and 100 g of powder are analyzed as described above. However, the protocol is only selective for *Enterobacteriaceae* and is not specific for *E. sakazakii*. The final step requires the biochemical identification of any presumptive-positive *E. sakazakii* colonies.

Isolation of *E. sakazakii* with Chromogenic and Fluorogenic Media

Several media have been developed recently for specifically detecting *E. sakazakii*, rather than general *Enterobacteriaceae*, in powdered infant formula. A summary of the main features of *E. sakazakii* on these media is given in Table 9.

These media take advantage of the key biochemical characteristic of this bacterium, the production of α-glucosidase. Muytjens et al. (1984) first identified the α-glucosidase enzyme activity with specific reference to *E. sakazakii*. Furthermore, this enzymatic reaction was not found in any other *Enterobacter* species. These authors showed that α-glucosidase activity was present in 129 (100%) isolates of *E. sakazakii* and absent in other *Enterobacter* spp., including *E. cloacae* (60 strains), *Pantoea agglomerans* (18 strains), and *E. aerogenes* (19 strains).

Oh and Kang (2004) described a differential selective medium (designated OK) containing a fluorogenic substrate, 4-methyl-umbelliferyl-α-D-glucopyranoside, which similarly detects α-glucosidase activity (Table 9). The fluorogen serves as an indicator of the production of α-glucosidase by *E. sakazakii*. Bile salts no. 3 selects for enteric bacteria, and ferric citrate and sodium thiosulfate differentiate H_2S-producing *Enterobacteriaceae*.

The same fluorogen is also included in a differential nonselective medium, described by Leuschner et al. (2004), for the presumptive detection of *E. sakazakii* in infant formula. However, this paper was later retracted by the journal editor and will not be considered further (Tortorello, 2004).

A perceived problem with fluorogens, as opposed to chromogens, is that the resultant fluorescence is diffuse and can be difficult to associate with a particular colony in a mixed culture. In contrast, chromogens remain within the bacterial biomass, resulting in a distinctly colored colony.

A chromogenic agar was developed, designated Druggan-Forsythe-Iversen (DFI) agar, where a chromogenic substrate, 5-bromo-4-chloro-3-indolyl-α-D-glucopyranoside, is used as an indicator of α-glucosidase activity (Iversen et al., 2004b) (Table 8). This moiety is cleaved by α-glucosidase, creating blue-green *E. sakazakii* colonies. The medium was developed as a differential, selective medium for recovering *E. sakazakii* in powdered infant formula after preenrichment and enrichment in preference to VRGBA (see Table 7). A selective agent, sodium desoxycholate, along with sodium thiosulfate and ferric ammonium citrate, is also present in the medium. The latter two act as indicators of hydrogen sulfide production to discriminate *E. sakazakii* from *Proteus* and *Salmonella*, which also can grow on the agar and appear as black-colored colonies (*E. sakazakii* is H_2S negative). Potentially, DFI agar could be used as a dual-detection system for *E. sakazakii* and *Salmonella*, but this has never been fully investigated.

The R&F *Enterobacter sakazakii* chromogenic plating medium (ESPM) contains chromogens that cause *E. sakazakii* colonies to appear blue-black or blue-gray in color (Restaino et al., 2006) (Tables 7 and 9). The medium contains three sugars (sorbitol, D-arabitol, and adonitol) and two chromogens, X-α-D-glucopyranoside and X-α-D-cellobioside. Selection is achieved by the inclusion of bile salts, vancomycin, and cefsulodin. Presumptive isolates are confirmed by the lack of acid production on a second (biplate) agar (ESPM) that detects acid production from sucrose or melibiose after 6 h at 35°C.

Guillaume-Gentil et al. (2005) described a method based on selective enrichment for the isolation of *E. sakazakii* from environmental samples. With this strategy, a modified enrichment broth was tested, containing 0.5 M NaCl and 10 mg of vancomycin/liter in lauryl sulfate broth (mLSB) (Table 8). Enrichment was carried out at 45°C for 22 to 24 h, followed by streaking onto TSA and a further incubation at 37°C for 24 h. To enhance pigment production, the plates were exposed to artificial white light (ca. 600 lux) to induce pigment production. As per the FDA method, yellow colonies are selected for further identification using API 20E as well as ribotyping. The presence of α-glucosidase is determined by homogenizing a presumptive *E. sakazakii* colony in saline (0.25 ml) with a ready-to-use diagnostic tablet (Diatab 50421; Rosco, Taastrup, Denmark) and incubation for 4 h before observing the production of a yellow color due to *p*-nitrophenol release. The method is a development from Kandhai et al. (2004b). Vancomycin (10 mg/liter) is added to mLSB in the enrichment stage to suppress the growth of gram-positive bacteria. Violet red bile lactose agar (VRBLA) then replaces VRBGA, and finally artificial light exposure is used during TSA incubation.

The vertical ISO method (DTS 22964) modifies the protocol of Guillaume-Gentil et al. (2005) by the use of the chromogenic agar *Enterobacter sakazakii* isolation agar (ESIA) after enrichment. ESIA is incubated at 44°C for 24 h. Presumptive *E. sakazakii* colonies are confirmed by pigment production on TSA at 25°C, not 37°C, with enhanced light exposure. ESIA contains the chromogen 5-bromo-4-chloro-3-indolyl-α-D-glucopyranoside (similar to the DFI method) as well as crystal violet. Colonies obtained after incubation appear green to blue-green in color. The selective agents included are sodium desoxycholate and crystal violet (Table 8). The agar is manufactured prepoured by AES Laboratoire, which also produces an *E. sakazakii* selective broth (ESSB). This broth is recommended for direct enrichment at 37°C without initial preenrichment (Table 11). Table 12 compares the present vertical ISO method (DTS 22964) with the FDA-recommended method and shows the greater sensitivity and specificity of the former. However, present (2007) development of a new horizontal ISO method will not use the raised temperature of

Table 12 Comparison of sensitivity and specificity of the FDA method with ISO (DTS 22964) for the isolation of *E. sakazakii*[a]

Result by FDA method	No. positive or negative by mLST-ESIA (ISO)	
	Positive	Negative
Positive	126	7
Negative	36	196

[a]Total number of strains tested was 365.

44°C and yellow pigment production as discriminatory features due to the number of *E. sakazakii* strains that are not detected under such conditions.

Despite the welcome development and availability of selective agars for detection of *E. sakazakii*, there still remains considerable variation in the recovery of this organism by a range of protocols. First, Farmer et al. (1980) reported that not all strains of *E. sakazakii* are yellow pigmented. Additionally, pigment production in some strains is temperature dependent. Subsequently, the FDA method incubates TSA at 25°C. Pigment production is enhanced by light exposure, and hence the use of artificial white light by Guillaume-Gentil et al. (2005), albeit at 37°C. A number of selection methods use raised incubation temperatures of 44 to 45°C. However, Nazarowec-White and Farber (1997) reported that only 3 of 11 *E. sakazakii* strains grew at 44 to 45°C. The maximum growth temperature for 4 of 11 strains was 41°C. This included the type strain ATCC 29544. Finally, the correct identification of *E. sakazakii* based on biochemical profiles is problematic, and discrepancies between kits have been reported (Iversen and Forsythe, 2004; Restaino et al., 2006). See the section "Phenotypic Methods: Biochemical Identification Kits" (next page) for more details on phenotyping.

Other Culture Methods
Bead capture and detection of **E. sakazakii**
Detection by culture is labor-intensive and requires long incubation times. Recently, a 24-h capture and detection protocol, based on charged separation using cationic beads, was developed (Mullane et al., 2006). These positively charged magnetic beads electrostatically attract the negatively charged lipopolysaccharide found on the surface of gram-negative bacteria. Briefly, the protocol consists of a 6-h enrichment in buffered peptone water at 42°C, followed by a 30-min cationic bead capture and then direct plating onto selective DFI agar. Results have shown that this method can reliably detect 2 CFU/500 g of powder and that other potential gram-negative organisms, including *Salmonella* spp., do not significantly interfere with the recovery of

E. sakazakii. A same-day method combining cationic capture and real-time PCR technology is also being assessed. This approach will further reduce the time to detection.

IDENTIFICATION OF *E. SAKAZAKII*

Phenotypic Methods: Biochemical Identification Kits

The *Enterobacter* genus includes 14 species of highly motile bacteria. The genus is biochemically similar to *Klebsiella*, but unlike *Klebsiella*, *Enterobacter* is ornithine positive. The main biochemical features distinguishing the *Enterobacter* species are summarized in chapter 1, Table 4. Biotyping strategies are based on the phenotypic characteristics that are unique to strains and that can be used for identification. Biochemical profiling systems, such as API 20E, ID32E, and Microbact, are convenient and simple methods to identify many bacteria. Biolog Microlog uses 96 biochemical tests and hence is expected to be more discriminatory than the above-mentioned biochemical kits.

Iversen and Forsythe (2004) reported the apparent disagreement between test kits for *E. sakazakii* and related organisms. A significant number of isolates identified as *Pantoea* spp. (nonpathogenic) using API 20E were characterized as *E. sakazakii* using the ID32E test kit. Further studies using strains reliably identified using 16S rRNA gene sequence analysis showed that Biolog GN2 had better sensitivity and specificity than API 20E and ID32 (Table 13). Weiss et al. (2005) isolated 19 strains of *E. sakazakii* from ready-to-eat salads, identified using API 20E. However, none could be confirmed by other methods, including the BAX detection system. Restaino et al. (2006) reported that one-third of API 20E-identified *E. sakazakii* isolates were not confirmed by the Biolog system. The key test was gelatinase activity, whereby gelatinase-positive isolates were identified as *E. sakazakii* with Biolog and as non-*E. sakazakii* with API 20E. Gelatinase-negative isolates also would be identified as *E. sakazakii* with API 20E, indicating a problem with the API 20E system. These authors concluded that *E. sakazakii* confirmation should be based on more than one system and recommended API 20E and Biolog Microlog 3 (release 4.20) systems together. Unfortunately,

Table 13 Sensitivity and specificity of three biochemical profiling methods[a]

Profiling method	Sensitivity (%)	Specificity (%)
API 20E	98.3	88.9
ID32E	100	69.4
Biolog GN2	100	88.9

[a]Data are from Iversen et al. (unpublished).

isolates were not subjected to any DNA sequence analysis for confirmation of their correct identification.

In 1980, Farmer et al. described 15 biotypes of *E. sakazakii*. These were principally based on metabolism (inositol, ornithine, malonate, indole, and nitrate reduction) and motility. More recently, these biotypes have been compared with 16S rRNA gene sequences (Iversen et al., 2006). Biotypes 1 to 5, 7 to 9, and 11 to 14 corresponded to the major 16S cluster group 1. Biotype 15 corresponded to cluster group 3, and cluster 4 included biotypes 6, 10, and 12. Cluster group 2 corresponded with a new biotype, 16 (Table 14).

Genotypic Methods

Laboratory detection and recognition methods for infectious agents have developed at a remarkable pace in recent years. This has been aided by the rapid and innovative developments in nucleic acid amplification technologies (reviewed in Singh et al., 2006). Some of these approaches, as applied to *E. sakazakii* and related organisms, are presented below.

Comparative DNA sequence analysis of housekeeping genes

Studies based on the 16S rRNA gene and *hsp60* indicate that culture methods and biotyping lack the necessary discriminatory power required to confidently identify distinct phylogenetic lineages among *E. sakazakii* isolates (see chapter 1). Sequence analysis of both partial 16S rRNA genes and *hsp60* (*groEL*) showed four distinct clusters (see chapter 1, Fig. 1 and 3). The majority of strains were located in cluster 1, including the type strain, whereas the Preceptrol strain ATCC 51329 was in the smaller cluster 3. An artificial neural network has been constructed that distinguishes between *E. sakazakii* and related organisms (Iversen et al., 2006). In addition to biochemical traits, the artificial neural network also demonstrated the sequence positions on the 16S rRNA gene that had the greatest predictive values for identification (see chapter 1).

Citrobacter koseri was 97.8% similar to *E. sakazakii*. This organism causes neonatal meningitis with a pathophysiology similar to that of *E. sakazakii*. This organism also had a greater DNA-DNA hybridization value compared to that of *E. sakazakii* than did *Citrobacter freundii* (Farmer et al. 1980) (see chapter 1, Table 1). *Citrobacter freundii* and *E. cloacae* were 96.0 and 97.0% similar, respectively. Similarly, Lehner et al. (2004) analyzed the full-length 16S rRNA gene sequences from 13 *E. sakazakii* isolates cultured from food, environmental, and human sources. Probably due to the smaller number of strains analyzed, these authors only distinguished two lineages, the second corresponding to cluster 3 of Iversen et al. (2004c). More recent full-length 16S rRNA sequence analyses and DNA-DNA hybridization determination

Table 14 *E. sakazakii* biotypes and 16S rRNA gene sequence cluster groups[a]

Farmer et al. (1988) biogroup[b]	Presence (+) or absence (−) of phenotype[c]										16S rRNA gene cluster group (Iversen et al., 2004c)
	VP	MR	Nit	Orn	Mot	Ino	Dul	Ind	Malo	Gas	
1	+	−	+	+	+	+	−	−	−	+	1
2	+	−	+	+	(+)[d]	−	−	−	−	+	
3	+	−	+	+	−	+	−	−	−	+	
4	+	−	+	−	(+)	+	−	−	+	+	
5	+	−	+	+	(+)	+	−	−	−	+	
7	+	−	+	+	+	+	−	−	−	−	
8	+	−	−	+	+	(+)	−	−	(+)	+	
9	(+)	−	+	+	+	−	−	−	+	+	
11	+	−	+	+	+	+	+	−	−	+	
13	+	(+)	+	(+)	(+)	−	−	−	−	+	
14	+	−	+	−	+	−	−	−	(+)	+	
New 16	+	−	+	(+)	(+)	+	+	−	(+)	+	2
15	+	−	+	+	+	+	+	+	+	+	3
6	+	−	+	+	+	+	−	+	−	+	4
10	+	−	+	+	+	−	−	+	−	+	4
12	+	−	+	+	+	+	−	+	+	+	4

[a]Modified from Iverson et al. (2006b).

[b]VP, Voges–Proskauer; MR, methyl red; Nit, nitrate reduction; Orn, ornithine utilization; Mot, motility at 37°C; Ino, acid production from inositol; Dul, acid production from dulcitol; Ind, indole production; Malo, malonate utilization; Gas, gas production from glucose.

[c]Readers should consult the original paper for details regarding number of strains within each biotype and subgroupings.

[d]Parentheses indicate that not all strains exhibited the trait.

by Forsythe and VanDamme (personal communication) (see chapter 1, Table 1) have confirmed the four lineages of the *E. sakazakii* species. This diversity of *E. sakazakii* should be considered with respect to the implementation of the DNA-based detection methods referred to below.

End detection of E. sakazakii *using PCR*

Several conventional PCR assay formats have been developed to detect *E. sakazakii*. These take advantage of unique sequences. Previously identified targets include the 16S rRNA gene, a 16S-23S internal-transcribed spacer (ITS), and *ompA*.

A summary of the *E. sakazakii* genome-specific targets, along with the corresponding primers (and probes) used for amplification, is shown in Table 15.

Seo et al. (2003), along with Seo and Brackett (2005), developed and evaluated a 5′-nuclease real-time PCR assay for the detection of *E. sakazakii* in infant formula. This method used a TaqMan approach to specifically amplify part of the macromolecular synthesis operon, the *rpsU* gene 3′ end and the primase (*dnaG*) gene 5′ end. The assay was specific for differentiating *E. sakazakii* and *E. cloacae* and almost 50 other genera of *Enterobacteriacae*, allowing detection of as few as 100 CFU of *E. sakazakii* per ml of infant formula without enrichment. Consequently, enrichment procedures would be necessary to increase the population to about 100 CFU/ml to enable detection. Given that the protocol for analyzing dry infant formula requires a 1:10 dilution in enrichment broth, about 15 generations or approximately 4 to 5 h at an optimal growth temperature would be required to reach this population.

In a modified version of the original real-time PCR assay design, Drudy et al. (2006a and unpublished data) applied it to purified template genomic DNA from a large range of isolates. A study collection of 56 *E. sakazakii* isolates (cultured from food, environmental, and clinical sources) and 25 non-*E. sakazakii* isolates (comprising *E. cloacae, Pantoea agglomerans, E. aerogenes,* and *E. gergoviae*) were included. Sixty non-*Enterobacter* spp., including *Citrobacter freundii, Carnobacterium divergens, K. pneumoniae, K. oxytoca, Salmonella enterica* serovar Typhimurium, *S. enterica* serovar Infantis, *Escherichia coli, Campylobacter coli, Campylobacter jejuni, Staphylococcus aureus,* and *Clostridium difficile,* were also evaluated by this method.

Species-specific PCR methods have been designed to detect *E. sakazakii* using the 16S-23S rRNA gene ITS region, which is located between both loci. Briefly, following the sequence characterization of the ITS regions from six *E. sakazakii* strains, two primers and 10 oligonucleotide probes were designed and tested against an oligonucleotide array of 88 non-*E. sakazakii* strains (Liu et al., 2006). The sensitivity of this method was reported as 1.3 CFU in 100 g of infant formula following selective enrichment. Combined

Table 15 PCR primers and their thermodynamic characteristics, genome targets, and sequences used for the detection of *E. sakazakii*

Primer	Primer sequence[a]	% G+C	T_m	Target site	Reference
Esak2	5' CCC GCA TCT CTG CAG GAT TCT C 3'	59	64	16S rRNA gene	Keyser et al. (2003)
Esak3	5' CTA ATA CCG CAT AAC GTC TAC G 3'	45	58		
Esakf	5' GCT YTG CTG ACG AGT GGC GG 3'	68	65	16S rRNA gene	Lehner et al. (2004)
Esakr	5' ATC TCT GCA GGA TTC TCT GG 3'	50	57		
Es-16S-f	5' CAA GTC GAA CGG TAA CAG GG 3'	55	59	16S rRNA gene	Malorny and Wagner (2005)
Es-16S-r	5' GTC CCC CAC TTT GGT CCG 3'	67	61		
SG-F	5' GGG TTG TCT GCG AAA GCG AA 3'	55	59	ITS-G, ITS-IA	Liu et al. (2006)
SG-R	5' GTC TTC GTG CTG CGA GTT TG 3'	55	59		
SI-F	5' CAG GAG TTG AAG AGG TTT AAC T 3'	41	57		Liu et al. (2006)
SI-R	5' GTG CTG CGA GTT TGA GAG ACT C 3'	55	62		
FP	5' CCG GAA CAA GCT GAA AAT TGA 3'	43	56	ITS tRNA[Glu]	Liu et al. (2005)
RP	5' TCT TCG TGC TGC GAG TTT G 3'	53	57		
TaqMan probe	5'-(FAM) ACT CTG ACA CAC CGC GCA TTC CTG 3' (TAMRA)	58	66	ITS tRNA[Giu] tRNA[Ile] tRNA[Ala]	Liu et al. (2005)
FP	5' TAT AGG TTG TCT GCG AAA GCG 3'	48	58	ITS	Liu et al. (2005)
RP	5' GTC TTC GTG CTG CGA GTT TG 3'	55	59		
ESSF	5' GGA TTT AAC CGT GAA CTT TTC C 3'	41		*ompA*	Nair and Ventikanaraynan (2006)
ESSR	5' CGC CAG CGA TGT TAG AAG A 3'	53			

[a]FAM, 6-carboxyfluorescein; TAMRA, 6-carboxytetramethylrhodamine.

with the enrichment stage, PCR, and oligonucleotide array, this procedure takes 48 h. Further modifications of this approach included the use of SYBR Green dye and TaqMan probes, which were tested against 35 *E. sakazakii* strains, and it was capable of detecting 1.1 CFU in 100 g of powdered formula (Liu et al., 2006).

MOLECULAR SUBTYPING OF *E. SAKAZAKII*

Molecular subtyping of bacteria by profiling either proteins or nucleic acids is a useful approach to investigate the epidemiological relationship(s) of isolates involved in outbreaks of food-borne disease. Several approaches have been applied. Generally, methods based on phenotype analysis are acknowledged to be unreliable due to the unstable expression of the corresponding marker(s). For this reason, DNA-based protocols offer an attractive alternative. Furthermore, provided that the protocols applied are standardized, DNA fingerprinting allows for a direct comparison of isolates in outbreaks (Singh et al., 2006).

Previously, DNA fingerprinting protocols have been applied to *E. sakazakii* (Clark et al., 1990; Nazarowec-White and Farber, 1997). Reported methods used include ribotyping, pulsed-field gel electrophoresis (PFGE), and random amplification of polymorphic DNA (RAPD) (Clark et al., 1990; Nazarowec-White and Farber, 1997). These molecular tools facilitate strain surveillance, providing for the trace-back of outbreak isolates from clinical sources to the contaminated batch(es) of powdered infant milk formula and/or the manufacturing environment. In addition, these protocols are useful tools as part of a microbiological risk-reduction management plan to control and reduce the risk of transmission. A comparison of several molecular subtyping protocols applied to *E. sakazakii* has been reported in the literature by Nazarowec-White and Farber (1997). These authors characterized 18 *E. sakazakii* isolates using three molecular typing methods, including ribotyping, PFGE, and RAPD.

RAPD

In the limited study by Nazarowec-White and Farber (1997), 18 isolates were compared using two random 10-mer primers. These primers were selected following trial experiments with various designs and were aimed at providing a reproducible RAPD profile. Data obtained demonstrated that this approach under these laboratory conditions could provide stable and reproducible DNA patterns. In a more recent study using a similar approach, Drudy et al. (2006a) analyzed the RAPD profiles of 56 *E. sakazakii* isolates cultured from environmental, food, and clinical sources, along with a number of type strains. Each isolate produced 3 to 11 amplified products, ranging in size from approximately 350 to 2,600 bp. Three clusters were identified based on a quantitative analysis of the banding patterns obtained. One contained six isolates from the Australasia region that were highly related, clustering at 100%. Three additional isolates were highly related to these six, clustering at 88%. In a second cluster, there were eight isolates that were 84% similar. All isolates in this group were cultured from a common European environmental source. The remaining cluster contained six isolates that were 84% identical, despite the fact that two isolates were obtained from Australasia, along with four isolates from Europe. These data demonstrated that individual strains can have indistinguishable RAPD patterns but originate from geographically dispersed regions, in this case Europe and Australasia. Moreover, there was no known epidemiological connection between the sources of these two isolates. The remaining 33 isolates had diverse RAPD banding profiles, with similarities of less than 80%, and these were considered to be genetically unrelated. Interestingly, one isolate (confirmed as *E. asburiae*) was among the most diverse of the isolates, demonstrating only 35% similarity to the latter clusters.

Subtyping by PFGE

The PFGE molecular subtyping method is regarded as a "gold-standard" technique. PFGE is a highly discriminatory method based on the whole genome. It is a modification of the more traditional restriction enzyme analysis. Rare-cutting restriction enzymes such as XbaI are used to cleave the target DNA into relatively few, comparatively large fragments. The DNA fragments are then separated on the basis of size using special electrophoretic conditions. A strain-specific genetic profile (also called a macrorestriction profile) is obtained that results from variations in the presence of restriction sites. PulseNet is a standardized protocol developed at the Centers for Disease Control and Prevention that facilitates the direct comparison of DNA fingerprints generated in laboratories in different geographical regions under agreed standardized conditions (Swaminathan et al., 2001). Following normalization of banding patterns, gel profiles can be directly compared after the transfer of gel images over the Internet as TIF files. This system has facilitated the global traceability of a number of enteric pathogens. It is expected that *E. sakazakii* will become a designated PulseNet-listed organism in the future, following the development of an agreed approach to typing using PFGE.

Nazarowec-White and Farber (1997) reported the PFGE profiles for 18 isolates of *E. sakazakii*. Purified genomic DNA was digested with XbaI and SpeI in separate reactions. These enzymes were selected based on the fact that they produced banding patterns that could be easily interpreted. In

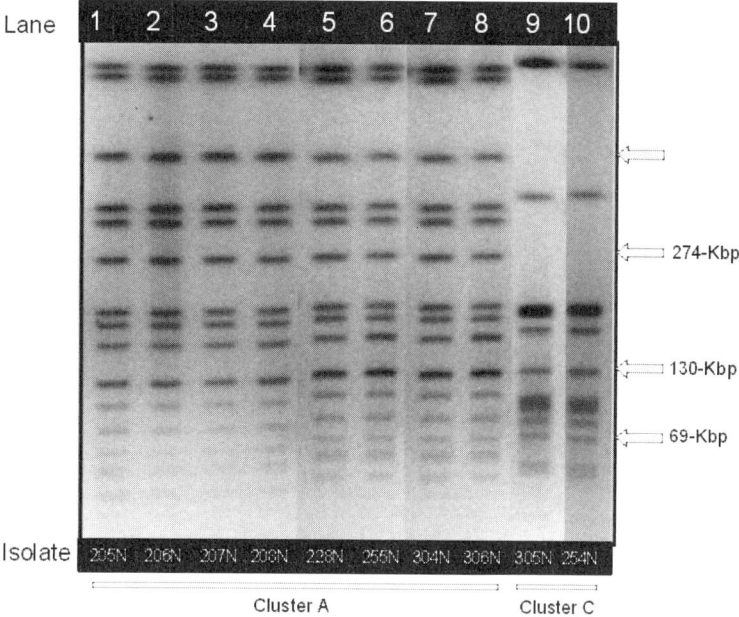

Figure 2 XbaI-generated PFGE profiles for a collection of *E. sakazakii* isolates.

comparison to the RAPD data described above, the discriminatory power of PFGE was similar. Much larger studies (Fanning and Forsythe, personal communication) have analyzed >500 *E. sakazakii* isolates by XbaI PFGE. Banding patterns consisted of between 9 and 19 DNA fragments, ranging in size from 50 to 1,000 kbp. All of these profiles were reproduced on at least three independent occasions. Quantitative analysis of the DNA profiles established the existence of 19 clusters at greater than 95% similarity using the software default settings; see Fig. 2 for an example.

ERIC

Several methods based on the use of specific primers designed for the outer regions of repetitive sequences found in bacterial genomes have been reported (Versalovic et al., 1994). One application, focusing on the enterobacterial repetitive intergenomic consensus (ERIC) repeat, has been applied to *E. sakazakii*. In this approach, a primer pair originally described following alignments of various ERIC elements was applied to the purified genomic DNA preparations of the organism. These primers contained between 32 and 34 bases. Mullane (unpublished data) assessed the application of ERIC-PCR subtyping on a subset of a larger collection. Figure 3 shows a 1.5%

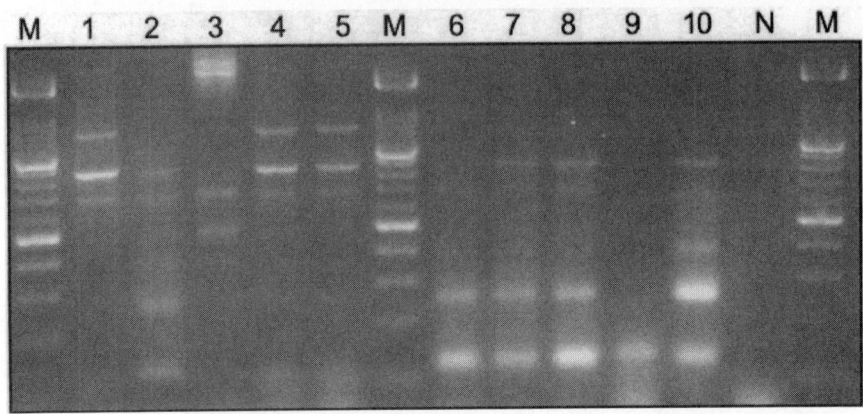

Figure 3 ERIC-PCR DNA profiles for a collection of *E. sakazakii* isolates and a *Pantoea agglomerans* strain.

agarose gel stained with ethidium bromide containing nine *E. sakazakii* isolates and one *Pantoea agglomerans* isolate. The two typed *E. sakazakii* strains, NCTC 10467 and NCTC 8155 (lanes 1 and 2), and *Pantoea agglomerans* NCTC 9831 (lane 3) were compared to eight randomly selected isolates from a larger database consisting of clinical, food, and environmental strains. DNA profiles in lanes 1, 4, and 5, containing the former typed strain and two environmental isolates, were indistinguishable. Similarly, lanes 6, 7, 8, and 10 contained isolates recovered from clinical sources (lanes 6 and 7) and the environment. These appeared to be indistinguishable from each other, with only minor differences in comparison to the previous three strains. Lanes 1 and 9 contained *E. sakazakii* isolates that were very different. *Pantoea agglomerans* produced a unique ERIC DNA profile.

When the banding profiles of these isolates were further compared using RAPD and PFGE, the isolates in lanes 4 and 5 were indistinguishable and were determined to be 85% similar by PFGE. Furthermore, the NCTC 10467 isolate was also similar to those in the latter lanes by RAPD but were less similar by PFGE (~75%). The *E. sakazakii* isolates in lanes 6 and 7 (Fig. 3) were indistinguishable by both PFGE and RAPD. Those in lanes 8 and 10 were considered very different based on their corresponding RAPD and PFGE profiles.

Ribotyping

DNA polymorphism refers to nucleotide sequence variation between members of the same species. These can be detected by restriction enzyme analysis and separation of resultant fragments by gel electrophoresis. The banding patterns are referred to as restriction fragment length polymorphisms. The different banding patterns between strains are the result of nucleotide differences causing a loss or gain in a restriction enzyme site. The method can result in large numbers of bands, ranging from 1,000 to 20,000 bp, which can cause resolution problems. This can be solved by the use of PFGE and, alternatively, by using a specific probe, and the method can be refined to detect certain DNA fragments used in ribotyping.

Ribotyping is based on DNA polymorphism in the chromosomal regions containing the highly conserved rRNA genes (*rrn*) for 5S, 16S, and 23S rRNA molecules in the ribosome (Bingen et al., 1994). The corresponding genes are cotranscribed as a large 30S precursor RNA in the order 5′ 16S spacer tRNA-23S-5S-3′. In *Escherichia coli* the *rrn* operon spans 6 to 7 kbp of DNA, and there are seven copies. The higher the number of copies, the more discriminatory the technique. In *Escherichia coli*, the spacer tRNA (ITS) is either for glutamic acid or alanine and isoleucine. Presently the number of copies of *rrn* in *E. sakazakii* is unknown, but it is reasonable to expect a number comparable to that of *Escherichia coli*. Liu et al. (2006) used real-time PCR with SYBR Green-labeled probes for both ITS regions (see below).

Most ribotyping probes use sequences from *Escherichia coli* 5S, 16S, and 23S rRNA genes. Clark et al. (1990) used ribotyping, along with a number of other typing methods, to link two outbreak isolates with strains from contaminated powdered infant milk formula. *E. sakazakii* ribotyping has been commercialized by DuPont.

Multilocus Sequence Typing

Recent 16S rRNA gene sequencing studies suggested that there are at least four genetically and biochemically distinct subgroups of *E. sakazakii*, valuable in their ability to differentiate between other *Enterobacteriaceae* and non-*E. sakazakii* α-glucosidase-positive *Enterobacteriaceae* (Iversen et al., 2006; also see chapter 1). rRNA genes, though eminent in their ubiquitous distribution and relatively slow rate of evolution, have several shortcomings with respect to assessment of microbial diversity as well as phylogenetic analysis, which impede the differentiation of closely related strains as well as resolution of evolutionary trees (Santos and Ochman, 2004). Other factors that may cause distortion of relationships between organisms include the occurrence of insertions and deletions (indels) as well as resistance to lateral gene transfer (Santos and Ochman, 2004).

In order to overcome problems related to 16S rRNA gene analyses as well as issues of pathogen-specific loci, 10 housekeeping genes have been identified as potential universal candidates for phylogenetic analyses of bacterial species (Santos and Ochman, 2004). Housekeeping genes are defined as constitutive genes, the products of which are necessary for maintenance of the cell. Furthermore, their expression is virtually identical between different genetic strains. Such genes are typically present in single copies within genomes, are subject to low rates of insertions and deletions, and can be readily partitioned into synonymous and nonsynonymous sites, which undergo diverse rates and patterns of evolution (Santos and Ochman, 2004).

A multilocus sequence typing method using 10 housekeeping genes is presently under development for *E. sakazakii* and looks very promising as a molecular typing tool for the epidemiology of *E. sakazakii* (Pagotto et al., unpublished results).

REFERENCES

Al-Hadithi, H. T., and T. A. A. Al-Edani. 1995. A comparative study on the antibiotic susceptibility of six species of faecal *Enterobacter* isolated from aquatic and clinical sources. *Dirasat (Pure and Applied Sciences)* 22B:35–41.

Arseni, A., E. Malamou-Ladas, C. Koutsia, M. Xanthou, and E. Trikka. 1987. Outbreak of colonization of neonates with *Enterobacter sakazakii J. Hosp. Infect.* 9:143–150.

Assadi, M. M., and R. P. Mathur. 1991. Application of an HPLC system in the analysis of biodegraded crude oil compounds. *J. Liq. Chromatogr.* 14:3623–3629.

Barbour, M. February 2004. Patent WO 2004/104550 A2: Rapid and specific detection of *Enterobacter sakazakii* (BAX).

Bar-Oz, B., A. Preminger, O. Peleg, C. Block, and I. Arad. 2001. *Enterobacter sakazakii* infection in the newborn. *Acta Paediatr.* 90:356–358.

Bartolucci, L., A. Pariani, F. Westall, F. Gardini, and M. E. Guerzoni. 1996. Interaction by microbiological processes between water, biofilm and pipe materials in water distribution systems. A proposed method for determining bacterial colonization in drinking water pipe networks. *Water Supply* 14:457–463.

Beauchamp, C. J., A.-M. Simao-Beaunoir, C. Beaulieu, and F.-P. Chalifour. 2006. Confirmation of *E. coli* among other thermotolerant coliform bacteria in paper mill effluents, wood chips screening rejects and paper sludges. *Water Res.* 40:2452–2462.

Besse, N. G., A. Leclercq, Y. Maladen, C. Tyburski, and B. Lombard. 2006. Evaluation of the International Organization for Standardization-International Dairy Federation (ISO-IDF) draft standard method for detection of *Enterobacter sakazakii* in powdered infant food formulas. *J. AOAC Int.* 89:1309–1316.

Biering, G., S. Karlsson, N. V. C. Clark, K. E. Jonsdottir, P. Ludvigsson, and O. Steingrimsson. 1989. Three cases of neonatal meningitis caused by *Enterobacter sakazakii* in powdered milk. *J. Clin. Microbiol.* 27:2054–2056.

Bingen, E. H., E. Denamur, and J. Elion. 1994. Use of ribotyping in epidemiological surveillance of nosocomial outbreaks. *Clin. Microbiol. Rev.* 7:311–327.

Block, C., O. Peleg, N. Minster, B. Bar-Oz, B. A. Simhon, I. Arad, and M. Shapiro. 2002. Cluster of neonatal infections in Jerusalem due to unusual biochemical variant of *Enterobacter sakazakii. Eur. J. Clin. Microbiol. Infect. Dis.* 21:613–616.

Bowen, A. B., and C. R. Braden. 2006. Invasive *Enterobacter sakazakii* disease in infants. *Emerg. Infect. Dis.* **12**:1185–1189.

Brito, J., S. E. Gilbreth, M. T. Musgrove, J. E. Call, and J. B. Luchansky. 2006. Characterization of *Enterobacter* spp. isolated from shell eggs using pulsed-field gel electrophoresis. Poster P2–38. International Association of Food Protection, Calgary, Canada.

Burdette, J. H., and C. Santos. 2000. *Enterobacter sakazakii* brain abscess in the neonate: the importance of neuroradiologic imaging. *Pediatr. Radiol.* **30**:33–34.

Caubilla-Barron, J., and S. J. Forsythe. Dry stress and survival time of *Enterobacter sakazakii* and other *Enterobacteriaceae* in dehydrated powdered infant milk formula. *J. Food Protect.,* in press.

Chaves-López, C., M. De Angelis, M. Martuscelli, A. Serio, A. Paparella, and G. Suzzi. 2006. Characterization of the *Enterobacteriaceae* isolated from an artisanal Italian ewe's cheese (Pecorino Abruzzese). *J. Appl. Microbiol.* **101**:353–360.

Clark, N. C., B. C. Hill, C. M. O'Hara, O. Steingrimsson, and R. C. Cooksey. 1990. Epidemiologic typing of *Enterobacter sakazakii* in two neonatal nosocomial outbreaks. *Diagn. Microbiol. Infect. Dis.* **13**:467–472.

Coignard, B., V. Vaillant, J.-P. Vincent, A. Leflèche, P. Mariani-Kurkdjian, C. Bernet, F. L'Hériteau, H. Sénéchal, P. Grimont, E. Bingen, and J.-C. Desenclos. 2006. Infections sévères à *Enterobacter sakazakii* chez des nouveau-nés ayant consommé une préparation en poudre pour nourrissons. *Bull. Épidémiol. Hebdomadaire* **2–3**:10–13.

Coulin, P., Z. Farah, J. Assanvo, H. Spillmann, and Z. Puhan. 2006. Characterisation of the microflora of attieke, a fermented cassava product, during traditional small-scale preparation. *Int. J. Food Microbiol.* **106**:131–136.

Dennison, S. K., and J. Morris. 2002. Multiresistant *Enterobacter sakazakii* wound infection in an adult. *Infect. Med.* **19**:533–535.

Drudy, D., M. O'Rourke, M. O'Mahony, M. Murphy, N. R. Mullane, R. O'Mahony, L. Kelly, M. Fischer, S. Sanjaq, P. Shannon, P. Wall, P. Whyte, and S. Fanning. 2006a. Comparison of *Enterobacter sakazakii* isolates by phenotypic and genotypic methods. *Int. J. Food Microbiol.* **1110**:127–134.

Emiliani, F., R. Lajmanovich, and S. M Gonzalez. 2001. *Escherichia coli*: biochemical phenotype diversity in fresh waters (Santa Fe Province, Argentina). *Rev. Argentina Microbiol.* **33**:65–74.

Espeland, E. M., and R. G. Wetzel. 2001. Complexation, stabilization, and UV photolysis of extracellular and surface-bound glucosidase and alkaline phosphatase: implications for biofilm microbiota. *Microb. Ecol.* **42**:572–585.

Estuningsih, E., C. Kress, A. A. Hassan, O. Akineden, E. Schneider, H. Becker, and E. Usleber. 2006. Enterobacteriaceae in dehydrated powdered infant formula manufactured in Indonesia and Malaysia. *J. Food Protect.* **69**:3013–3017.

European Commission. 2005. Commission regulation (EC) number 2073/2005 of 15 November 2005 on microbiological criteria for foodstuffs. *Official J. Eur. Union* **L338**:1–26.

Farmer, J. J., II, M. A. Asbury, F. W. Hickman, D. J. Brenner, and The *Enterobacteriaceae* Study Group. 1980. *Enterobacter sakazakii*: a new species of "*Enterobacteriaceae*" isolated from clinical specimens. *Int. J. Syst. Bacteriol.* **30**:569–584.

Forsythe, S. J. 2005. *Enterobacter sakazakii* and other bacteria in powdered infant milk formula. *Matern. Child Nutr.* **1**:44–50.

Gakuya, F. M., M. N. Kyule, P. B. Gathura, and S. Kariuki. 2001. Antimicrobial resistance of bacterial organisms isolated from rats. *East Afr. Med. J.* **78**:646–649.

Gallagher, P. G., and W. S. Ball. 1991. Cerebral infarctions due to CNS infection with *Enterobacter sakazakii. Pediatr. Radiol.* **21**:135–136.

Gassem, M. A. A. 1999. Study of the microorganisms associated with the fermented bread (khamir) produced from sorghum in Gizan region, Saudi Arabia. *J. Appl. Microbiol.* **86**:221–225.

Gosney, M. A., M. V. Martin, A. E. Wright, and M. Gallagher. 2006. *Enterobacter sakazakii* in the mouths of stroke patients and its association with aspiration pneumonia. *Eur. J. Intern. Med.* **17**:185–188.

Guillaume-Gentil, O., V. Sonnard, M. C. Kandhai, J. D. Marugg, and H. Joosten. 2005. A simple and rapid cultural method for detection of *Enterobacter sakazakii* in environmental samples. *J. Food Protect.* **68**:64–69.

Hawkins, R. E., C. R. Lissner, and J. P Sanford. 1991. *Enterobacter sakazakii* bacteremia in an adult. *South. Med. J.* **84**:793–795.

Heuvelink, A. E., M. Ahmed, F. D. Kodde, J. T. M. Zwartkruis-Nahuis, and E. de Boer. 2001. *Enterobacter sakazakii* in melkpoeder. Keuringsdienst van Waren Oost. Project number OT 0110.

Heuvelink, A. E., M. Ahmed, F. D. Kodde, J. T. M. Zwartkruis-Nahuis, and E. de Boer. 2002. *Enterobacter sakazakii* in melkpoeder. *De Ware Chemicus* **32**:17–30.

Heuvelink, A. E., J. T. M. Zwartkruis-Nahuis, A. H. van der Wit, B. van Oosterom, and E. de Boer. 2003. Handhavingsactie *Enterobacter sakazakii* in zuigelingenvoeding. Keuringsdienst van Waren Oost. Project number OT 02108p.

Himelright, I., E. Harris, V. Lorch, and M. Anderson. 2002. *Enterobacter sakazakii* infections associated with the use of powdered infant formula—Tennessee, 2001. *JAMA* **287**:2204–2205.

Igimi, S., H. Asakura, T. Morita-Ishihara, A. Ishiwa, Y. Okada, and S. Yamamoto. 2006. Isolation and genetical characterization of *Enterobacter sakazakii* in Japan. *Abstr. Food Micro 2006: 20th Int. ICFMH Symp.*, Bologna, Italy.

Iversen, C., and S. Forsythe. 2003. Risk profile of *Enterobacter sakazakii*, an emergent pathogen associated with infant milk formula. *Trends Food Sci. Technol.* **14**:443–454.

Iversen, C., and S. Forsythe. 2007. Comparison of media for the isolation of *Enterobacter sakazakii. Appl. Environ. Microbiol.* **73**:48–52.

Iversen, C., and S. Forsythe. 2004. Isolation of *Enterobacter sakazakii* and other *Enterobacteriaceae* from powdered infant formula milk and related products. *Food Microbiol.* **21**:771–777.

Iversen, C., P. Druggan, and S. Forsythe. 2004a. A selective differential medium for *Enterobacter sakazakii*, a preliminary study. *Int. J. Food Microbiol.* **96**:133–139.

Iversen, C., M. Lane, and S. J. Forsythe. 2004b. The growth profile, thermotolerance and biofilm formation of *Enterobacter sakazakii* grown in infant formula milk. *Lett. Appl. Microbiol.* **38**:378–382.

Iversen, C., M. Waddington, S. L.W. On, and S. Forsythe. 2004c. Identification and phylogeny of *Enterobacter sakazakii* relative to *Enterobacter* and *Citrobacter. J. Clin. Microbiol.* **42**:5368–5370.

Iversen, C., L. Lancashire, M. Waddington, S. Forsythe, and G. Ball. 2006. Identification of *Enterobacter sakazakii* from closely related species: the use of artificial neural networks in the analysis of biochemical and 16S rDNA data. *BMC Microbiol.* **6**:28.

Janicka, G., I. Kania, B. Ulatowska, E. Krusznska, and M. Wojda. 1999. The occurrence of the *Enterobacter* genus rods in the clinical materials and materials taken from a hospital environment. *Wiadomosci Lekaarskie* **52**:554–558.

Jarvis, C. 2005. Fatal *Enterobacter sakazakii* infection associated with powdered infant formula in a neonatal intensive care unit in New Zealand. *Am. J. Infect. Control* **33**:e19.

Jaspar, A. H. J., H. L. Mutyjens, and L. A. A. Kolee. 1990. Neonatale meningitis door *Enterobacter sakazakii*: melkpoeder is niet sterile in bacterien lusten ook melk. (Neonatal meningitis caused by *E. sakazakii*: milk powder is not sterile and bacteria like milk too). *Tijdschr. Kindergenee Skd.* **58**:151–155.

Jimenez, E. B., and C. Gimenez. 1982. Septic shock due to *Enterobacter sakazakii. Clin. Microbiol. Newsl.* **4**:30.

Jöker, R. N., T. Norholm, and K. E. Siboni. 1965. A case of neonatal meningitis caused by a yellow *Enterobacter. Dan. Med. Bull.* **12**:128–130.

Kandhai, M. C., M. W. Reij, L. G. M. Gorris, O. Guillaume-Gentil, and M. van Schothorst. 2004a. Occurrence of *Enterobacter sakazakii* in food production environments and households. *Lancet* **363**:39–40.

Kandhai, M. C., M. W. Reij, K. van Puyvelde, O. Guillaume-Gentil, R. R. Beumer, and M. van Schothorst. 2004b. A new protocol for the detection of *Enterobacter sakazakii* applied to environmental samples. *J. Food Protect.* **67**:1267–1270.

Keyser, M., R. C. Witthuhn, L.-C. Ronquest, and T. J. Britz. 2003. Treatment of winery effluent with upflow anaerobic sludge blanket (UASB) – granular sludges enriched with *Enterobacter sakazakii. Biotechnol. Lett.* **25**:1893–1898.

Kleiman, M. B., S. D. Allen, P. Neal, and J. Reynolds. 1981. Meningoencephalitis and compartmentalization of the cerebral ventricles caused by *Enterobacter sakazakii. J. Clin. Microbiol.* **14**:352–354.

Kress, C., A. A. Hassan, Ö. Akineden, E. Schneider, S. Estuningsih, H. Becker, and E. Usleber. 2005. *Enterobacter sakazakii* in Trockenerzeugnissen auf Milchbasis, p. 190–195. *In* Arbeitstagung des Arbeitsgebietes Lebensmittelhygiene der Deutschen Veterinärmedizinischen Gesellschaft e.V., 27.-30.09.2005, vol. 46. Deutsche Veterinärmedizinische Gesellschaft, Garmisch-Partenkirchen, Germany.

Kuklinsky-Sobral, J., W. L. Araujo, R. Mendes, I. O. Geraldi, A. A. Pizzirani-Kleiner, and J. L. Azevedo. 2004. Isolation and characterisation of soybean-associated bacteria and their potential for plant growth promotion. *Environ. Microbiol.* **6**:1244–1251.

Kuzina, L. V., J. J. Peloquin, D. C. Vacek, and T. A. Miller. 2001. Isolation and identification of bacteria associated with adult laboratory Mexican fruit flies, *Anastrepha ludens* (Diptera: Tephritidae). *Curr. Microbiol.* **42**:290–294.

Leclercq, A., C. Wanegue, and P. Baylac. 2002. Comparison of fecal coliform agar and violet red bile lactose agar for fecal coliform enumeration in foods. *Appl. Environ. Microbiol.* 68:1631–1638.

Lecour, H., A. Seara, J. Cordeiro, and M. Miranda. 1989. Treatment of childhood bacterial meningitis. *Infection* 17:343–346.

Lehner, A., T. Tasara, and R. Stephan. 2004. 16S rRNA gene based analysis of *Enterobacter sakazakii* strains from different sources and development of a PCR assay for identification. *BMC Microbiol.* 4:43.

Leuschner, R. G. K., F. Baird, B. Donald, and L. J. Cox. 2004. A medium for the presumptive detection of *Enterobacter sakazakii* in infant formula. *Food Microbiol.* 21:527–533. Retracted by the journal editor. See Tortorello (2004).

Liu, Y., Q. Gao, X. Zhang, Y. Hou, J. Yang, and X. Huang. 2006. PCR and oligonucleotide array for detection of *Enterobacter sakazakii* in infant formula. *Mol. Cell. Probes* 20:11–17.

Liu, Y., X. Cai, X. Zhang, Q. Gao, X. Yang, Z. Zheng, M. Luo, and X. Huang. 2005. Real time PCR using TaqMan and SYBR Green for detection of *Enterobacter sakazakii* in infant formula. *J. Microbiol. Methods* 65:21–31.

Loc-Carrillo, C., O. Fayet, J. Caubilla Barron, M.-F. Prere, and S. Forsythe. 2006. Fatal outbreak of *Enterobacter sakazakii* in a neonatal intensive care unit in 1994: a retrospective study, P-019. *Abstr. 106th Gen. Meet. Am. Soc. Microbiol.* American Society for Microbiology, Washington, DC.

Malorny, B., and M. Wagner. 2005. Detection of *Enterobacter sakazakii* strains by real-time PCR. *J. Food Protect.* 68:1623–1627.

Masaki, H., N. Asoh, M. Tao, H. Ikeda, S. Degawa, K. Matsumoto, K. Inokuchi, K. Watanabe, H. Watanabe, K. Oishi, and T. Nagatake. 2001. Detection of gram-negative bacteria in patients and hospital environments at a room in geriatric wards under the infection control against MRSA. *J. Jpn. Assoc. Infect. Dis.* 75:144–150.

Monroe, P. W., and W. L. Tift. 1979. Bacteremia associated with *Enterobacter sakazakii* (yellow-pigmented *Enterobacter cloacae*). *J. Clin. Microbiol.* 10:850–885.

Mosso, M. A., M. C. de la Rosa, C. Vivar, and M. R. Medina. 1994. Heterotrophic bacterial populations in the mineral waters of thermal springs in Spain. *J. Appl. Bacteriol.* 77:370–381.

Mullane, N. R., D. Drudy, P. Whyte, M. O'Mahony, A. G. M. Scannell, P. G. Wall, and S. Fanning. 2006. *Enterobacter sakazakii*: biological properties and significance in dried infant milk formula (IMF). *Int. J. Dairy Technol.* 59:102–111.

Murray, M. M., D. F. Welch, and T. L. Kuhls. 1990. *Serratia osteochondritis* after puncture wounds of the foot. *Pediatr. Infect. Dis.* 9:523–524.

Muytjens, H. L., and L. A Kolee. 1990. *Enterobacter sakazakii* meningitis in neonates: causative role of formula? *Pediatr. Infect. Dis.* 9:372–373.

Muytjens, H. L., H. C. Zanen, H. J. Sonderkamp, L. A. Kollee, I. K. Wachsmuth, and J. J. Farmer III. 1983. Analysis of eight cases of neonatal meningitis and sepsis due to *Enterobacter sakazakii*. *J. Clin. Microbiol.* 18:115–120.

Muytjens, H. L., J. van der Ros-van de Repe, and H. A. M. van Druten. 1984. Enzymatic profiles of *Enterobacter sakazakii* and related species with special reference to the alpha glucosidase reaction and reproducibility of the test system. *J. Clin. Microbiol.* 20:684–686.

Muytjens, H. L., H. Roelofs-Willemse, and G. H. Jaspar. 1988. Quality of powdered substitutes for breast milk with regard to members of the family *Enterobacteriaceae. J. Clin. Microbiol.* 26:743–746.

Nair, M. K. M., and K. S. Ventikanaraynan. 2006. Cloning and sequencing of *ompA* gene from *Enterobacter sakazakii* and the development of an *ompA*-targeted PCR for rapid detection of *Enterobacter sakazakii* in infant formula. *Appl. Environ. Microbiol.* 72:2539–2546.

Naqvi, S. H., M. A. Maxwell, and L. M. Dunkle. 1985. Cefotaxime therapy of neonatal gram-negative bacillary meningitis. *Pediatr. Infect. Dis.* 4:499–502.

Nazarowec-White, M., and J. M. Farber. 1997. Incidence, survival, and growth of *Enterobacter sakazakii* in infant formula. *J. Food Protect.* 60:226–230.

Nazarowec-White, M., R. C. McKellar, and P. Piyasena. 1999. Predictive modelling of *Enterobacter sakazakii* inactivation in bovine milk during high-temperature short-time pasteurization. *Food Res. Int.* 32:375–379.

Neelam, M., Z. Nawaz, and S. Riazuddin. 1987. Hydrocarbon biodegradation biochemical characterization of bacteria isolated from local soils. *Pak. J. Sci. Ind. Res.* 30:382–385.

No, H. K., N. Y. Park, S. H. Lee, H. J. Hwang, and S. P. Meyers. 2002. Antibacterial activities of chitosans and chitosan oligomers with different molecular weights on spoilage bacteria isolated from tofu. *J. Food Sci.* 67:1511–1514.

Noriega, F. R., K. L. Kotloff, M. A. Martin, and R. S. Schwalbe. 1990. Nosocomial bacteremia caused by *Enterobacter sakazakii* and *Leuconostoc mesenteroides* resulting from extrinsic contamination of infant formula. *Pediatr. Infect. Dis.* 9:447–449.

Oh, S., and D. Kang. 2004. Fluorogenic selective and differential medium for isolation of *Enterobacter sakazakii. Appl. Environ. Microbiol.* 70:5692–5694.

Oliver, E. D. 1997. Atypical, non-lactose fermenting isolates shown to be total coliforms by the β-galactosidase (ONPG) reaction, p. 225–231. Proceedings of the Water Quality Technology Conference. American Water Works Association, Denver, CO.

Palcich, G., C. Gillio, M. Langraf, B. D. G. M. Franco, and M. T. Destro. 2006. *Enterobacter sakazakii* in milk kitchens of maternities in São Paulo, Brazil. Poster P5-02. International Association of Food Protection, Calgary, Canada.

Postupa, R., and E. Aldová. 1984. *Enterobacter sakazakii*: a Tween 80 esterase-positive representative of the genus *Enterobacter* isolated from powdered milk specimens. *J. Hyg. Epidemiol. Microbiol. Immunol.* 28:435–440.

Pribyl, C., R. Salzer, J. Beskin, R. J. Haddad, B. Pollock, R. Beville, B. Holmes, and W. J. Mogabgab. 1985. Azteonam in the treatment of serious orthopaedic infections. *Am. J. Med.* 78:51–56.

Reina, J., F. Parras, S. Gil, F. Salva, and P. Alomar. 1989. Human infections caused by *Enterobacter sakazakii.* Microbiologic considerations. *Enferm. Infecc. Microbiol. Clin.* 7:147–150.

Reis, M., D. Harms, and J. Scharf. 1994. Multiple cerebrale infarzierungen mit resultierender multizystischer encephalomalazie bei einem fru(r) hgeborenen mit *Enterobacter sakazakii*-meningitis. *Klinische Pädiatrie* 206:184–186.

Restaino, L., E. W. Frampton, W. C. Lionberg, and R. J. Becker. 2006. A chromogenic plating medium for the isolation and identification of *Enterobacter sakazakii* from foods, food ingredients, and environmental sources. *J. Food Protect.* 69:315–322.

Santos, S. R., and H. Ochman. 2004. Identification and phylogenetic sorting of bacterial lineages with universally conserved genes and proteins. *Environ. Microbiol.* 6:754–759.

Schindler, P. R. 1994. Enterobacteria in mineral, spring and table water. *Gesundheitswesen* 56:690–693.

Schindler, P. R., and H. Metz. 1990. Coliform bacteria in rinsed mugs—identification with the API 20E system and resistance behavior. *Offentliche Gesundheitswesen* 52:592–597.

Seo, K. H., G. Thammasuvimol, R. E. Brackett, and S. G. Edelson-Mammel. 2003. 2003 FDA Science Forum Poster Abstract 190. Center for Food Safety and Applied Nutrition, Food and Drug Administration, College Park, MD. [Online] http://www.cfsan.fda.gov/~frf/forum03/F-01.htm.

Seo, K. H., and R. E. Brackett. 2005. Rapid, specific detection of *Enterobacter sakazakii* in infant formula using a real-time PCR assay. *J. Food Protect.* 68:59–63.

Shaker, R., T. Osaili, W. Al-Omary, Z. Jaradat, and M. Al-Zuby. 2007. Isolation of *Enterobacter sakazakii* and other *Enterobacter* sp. from food and food production environments. *Food Control* 18:1241–1245.

Simmons, B. P., M. S. Gelfand, M. Haas, L. Metts, and J. Ferguson. 1989. *Enterobacter sakazakii* infections in neonates associated with intrinsic contamination of a powdered infant formula. *Infect. Control Hosp. Epidemiol.* 10:398–401.

Singh, A., R.V. Goering, S. Simjee, S. L. Foley, and M. J. Zervos. 2006. Application of molecular techniques to the study of hospital infection. *Clin. Microbiol. Rev.* 19:512–530.

Smeets, L. C., A. Voss, H. L. Muytjens, J. F. G. M. Meis, and W. J. G. Melchers. 1998. Genetische karakterisatie van *Enterobacter sakazakii*-isolaten van Nederlandse patiënten met neonatale meningitis. *Nederlands Tijdschrift Med. Microbiol.* 6:113–115.

Soriano, J. M., H. Rico, J. C. Molto, and J. Manes. 2001. Incidence of microbial flora in lettuce, meat and Spanish potato omelette from restaurants. *Food Microbiol.* (London) 18:159–163.

Stoll, B. J., N. Hansen, A. A. Fanaroff, and J. A. Lemons. 2004. *Enterobacter sakazakii* is a rare cause of neonatal septicemia or meningitis in VLBW infants. *J. Pediatr.* 144:821–823.

Suliman, S. M. A., M. I. Abubakr, and E. F. Mirghani. 1988. Microbial contamination of cutting fluids and associated hazards. *Tribiol. Int.* 30:753–757.

Swaminathan, B., T. J. Barrett, S. B. Hunter, R. V. Tauxe, and the Centers for Disease Control and Prevention PulseNet Task Force. 2001. PulseNet: the molecular subtyping network for foodborne bacterial disease surveillance, United States. *Emerg. Infect. Dis.* 7:382–389.

Tamura, A., M. Kato, M. Omori, A. Nanba, K. Miyagawa, C. R. Wang, and W. H. Zhou. 1995. Flavor componenets and microorganisms isolated from Suancha (sour tea, Takeutsusancha in Japanese). *Nippon Kasei Gakkaishi* 46:759–764.

Thornley, M. J. 1960. The differentiation of *Pseudomonas* from other gram-negative bacteria on the basis of arginine metabolism. *J. Appl. Bacteriol.* 23:37–50.

Tortorello, M. L. 2006. Retraction notice to "A medium for the presumptive detection of *Enterobacter sakazakii* in infant formula" [*Food Microbiology* 21(5) (2004) 527–533]. *Food Microbiol.* 23:409.

Townsend, S., J. Caubilla-Barron, C. Loc-Carrillo, and S. Forsythe. 2006. The presence of endotoxin in powdered infant formula milk and the influence of endotoxin and *Enterobacter sakazakii* on bacterial translocation in the infant rat. *Food Microbiol.* **24**:67–74.

Tuncer, I., and K Ozsan. 1988. Biochemical typing of *Enterobacter* isolated from several clinical materials. *Mikrobiyoloji Bulteni* **22**:105–112.

Urmenyi, A. M. C., and A. W. Franklin. 1961. Neonatal death from pigmented coliform infection. *Lancet* **i**:313–315.

U.S. Food and Drug Administration. 2002. Isolation and enumeration of *Enterobacter sakazakii* from dehydrated powdered infant formula. [Online] http://www.cfsan.fda.gov/~comm./mmesakaz.html.

Van Acker, J., F. De Smet, G. Muyldermans, A. Bougatef, A. Naessens, and S. Lauwers. 2001. Outbreak of necrotizing enterocolitis associated with *Enterobacter sakazakii* in powdered milk formula. *J. Clin. Microbiol.* **39**:293–297.

Versalovic, J. S. M., F. J. De Bruijn, and J. R. Lupski. 1994. Genomic fingerprinting of bacteria using repetitive sequence-based polymerase chain reaction *Methods Mol. Cell. Biol.* **5**:25–40.

Wantanabe, I., and E. Makoto. 1994. Studies in an unusual case of fermentation of meatproducts during the curing process. *Bokin Bobai* **22**:9–14.

Weiss, C., B. Becker, and W. Holzapfel. 2005. Einsatz und Eignung dreier kommerzieller Systeme zum Nachweis von Enterobacter sakazakii in verzehrsfertigem Mischsalat. *Archiv für Lebensmittelhygiene* **56**:25–48.

Willis, J., and J. E. Robinson. 1988. *Enterobacter sakazakii* meningitis in neonates. *Pediatr. Infect. Dis. J.* **7**:196–199.

Enterobacter sakazakii
Edited by Jeffrey M. Farber and Stephen J. Forsythe
© 2008 ASM Press, Washington, D.C.

The Neonatal Intestinal Microbial Flora, Immunity, and Infections

3

Stacy Townsend and Stephen J. Forsythe

INTRODUCTION

It is well known that, during development in the womb, the intestinal tract of the human fetus is microbiologically sterile and that the neonate rapidly acquires its microbial flora during passage through the vagina at birth, through contact with the environment, and through feeding. Consequently, a dense, complex bacterial community is established in the human intestinal tract. However, the diversity of the gut flora, the influence of the diet on that flora, and the influence of the gut flora on the host immune system are still poorly understood. The microbial flora of the human intestine is a complex ecosystem and may have significant effects on the health of the host through involvement in the nutrition, pathological process, and immune function (Kien et al., 1987; Kirjavainen and Gibson, 1999; Wold and Adlerberth, 2000). Gnotobiotic animals have shown that the gut microbial flora contributes to the nutritional status of the host by providing vitamins and amino acids (Hooper et al., 2001) and by recycling bile acids, bilirubin, and cholesterol (Neish et al., 2000). The maturation of the immune system requires continual stimulation from developing gut flora (Duffy, 2000). It has been proposed that the increase in prevalence of atopic diseases is associated with improvements in public health and hygiene, leading to reduced exposure to microorganisms in early life and therefore failure to develop tolerance to innocuous antigens (Bach, 2002). There has been considerable interest in the supplementation of formula feeds with pre- and probiotics. However, the scope of this topic is too large to be given adequate coverage in this chapter.

STACY TOWNSEND and STEPHEN J. FORSYTHE, School of Biomedical and Natural Sciences, Nottingham Trent University, Clifton Lane, Nottingham NG11 8NS, United Kingdom.

Readers are directed to the work of Burns and Rowland (2000), Reuter (2001), and Rosberg-Cody et al. (2004) for further information.

This chapter summarizes our understanding in this area, with particular emphasis on the preterm intestinal flora and immunity development. We aim to widen the general microbiologist's knowledge of the issue of *Enterobacter sakazakii* and other *Enterobacteriaceae* as infectious agents of neonates in the context of the neonate having a complex intestinal flora, which includes *Enterobacter* species and where endogenous infections occur. Additionally, the preterm, low-birth-weight neonate has a greatly reduced ability to resist bacterial infections.

IMMUNOLOGICAL STATUS OF PREMATURE NEONATES, FULL-TERM NEONATES, AND INFANTS UP TO 1 YEAR

The mechanisms of the immune system of neonates are undermined by immaturity and function relative to gestational age. Thus, neonates and infants up to 12 months have a transitory yet considerable immunodeficiency that affects a broad number of immune functions, often aggravated by premature or traumatic delivery, maternal disease, and drugs. For example, premature and very-low-birth-weight (VLBW) infants occur in about 56,270 (1.4%) of all live births in the United States within a year (Arias et al., 2003). The incidence of sepsis in this group is 20%, compared to 0.1% in term infants (Stoll and Hansen, 2003). The improved ability to preserve the life of severely ill and premature neonates is an accomplishment that has expanded a host population that is at increased risk for bacterial infection. With the frequent use of invasive life support devices comes the increased opportunity for infection and complications evoking the immune response. The neonatal systemic and mucosal immune system exhibits a diminished capacity to prevent bacterial overgrowth, increasing the risk of disease and gastrointestinal immune disorders such as necrotizing enterocolitis (NEC) (Spencer et al., 1990). The mechanisms for this have not been completely elucidated. Also, great variation exists in what is considered normal healthy gut and immune function, especially between term infants and VLBW neonates. However, there are a number of common immunological aspects that put neonates at increased risk. Innate and adaptive immunity comprise the two branches of the immune system, and both are less robust in neonates than in adults in several different respects.

Innate Immune System

The innate immune response provides the rapid initial response against infection and is also responsible for orchestrating an effective adaptive immune response. Physiological barriers and secretions, peristaltic movements, serum complement proteins, and phagocytic cells are examples of innate immunity. The utility of the innate immune system is not dependent upon previous exposure to antigens. Thus, innate defense mechanisms are crucial to naïve infants with immature immune systems upon first contact with invading pathogens. This section will focus on deficiencies in the innate immune response.

Mechanical barriers. Infants, particularly premature infants, are made more susceptible to disease by reduced natural epidermal and epithelial barriers and secreted products that comprise part of the innate immune system. The outer epidermis layer or stratum corneum in premature infants at 26 weeks of gestation is only 3 layers thick (compared to 16 layers in a full-term infant) and is easily compromised via invasive procedures (Evans and Rutter, 1986). Levels of protective mucus and production of gastric acid are low in preterm infants (Udall, 1990). Proteolytic enzymes, essential for the degradation of toxins, exhibit lower levels of activity in premature infants (Udall et al., 1990). Less gut mobility, decreased numbers of B and T lymphocytes, as well as deficient secretory immunoglobulin A (IgA), mucin, and defensins allow bacteria to adhere more readily to the mucus membrane, reducing barrier function (Udall et al., 1990; Rognum et al., 1992; MacKendrick and Caplan, 1993; Mallow et al., 1996; Faix and Adams, 1994). Increased permeability exhibited by the infant gut wall also may facilitate bacterial invasion.

The carbohydrates exposed on the surface of neonatal host cells are developmentally regulated. Often, these molecules act as receptors for pathogens, and this may account for certain tissue tropisms and age specificity influencing host susceptibility. For example, the Gb3 receptor for Shiga toxin is expressed only in rat intestines postweaning. Also, the GM1 glycolipid exhibits specificity for cholera toxin and is located on the microvillous membrane (Fishman and Atikkan, 1980). Further, although the numbers of receptors expressed by mature and immature intestines were comparable, a difference in receptor affinity was observed. This resulted in an increase in cholera toxin-induced fluid secretion from the intestine of 2-week-old rats compared to that of 4-week-old rats, suggesting that cellular maturity is an important factor affecting the neonatal immune response (Chu and Walker, 1989).

Phagocytic cells. Cells responsible for phagocytosis are important for internalization and destruction of damaged tissue, microorganisms, and other cells, and they help orchestrate the inflammatory response during infections. Phagocytic cells such as macrophages, neutrophils, and natural killer (NK) cells are found at the site of infection before lymphocytes. Monocytes present in the circulation are the precursor cells for macrophages that enter tissues and differentiate. Neonatal monocytes are similar to those of adults in function and concentration. However, tissue macrophage chemotaxic, phagocytic, and bacteriocidal activities are diminished. Macrophages are not fully proficient in antigen presentation or cytokine production, which may be related to a decrease in T-cell production in the neonate.

Neutrophils (polymorphonuclear leukocytes) are a component of the innate immune system that responds to various sites of infection or damage, delivering an acute inflammatory reaction. Like macrophages, neutrophils nonspecifically ingest and degrade cellular debris and bacteria, aided by toxic cytoplasmic granules. Neutrophils populate the bone marrow, where progenitor cells are stimulated by colony-stimulating factors to differentiate into neutrophils. Populations also exist in specific tissues and in circulation. Neutrophils within the circulating group can be associated with the vascular endothelium and are called marginated. Marginated cells are induced to demarginate and enter the circulation by infection, corticosteroids, and epinephrine. The stress of birth is thought to account for the elevated white blood cell count during the first hours after birth due to demargination (Schelonka et al., 1994). The average life span of neutrophils is 8 h and is approximately 32 h for those that migrate into tissues. The pool of neutrophils that originates from the bone marrow of neonates is significantly smaller than that of adults and may be quickly depleted during sepsis (Christensen et al., 1982). Infants are unable to increase proliferation from the bone marrow, and a severe infection can exhaust the limited stores of neutrophils available, resulting in neutropenia (Christensen and Rothestein, 1980; Christensen et al., 1984).

The migration of cells along a concentration gradient toward a chemoattractant (chemotaxis) and adhesion are important functions of neutrophils. Neonate neutrophils also generate less actin, which is utilized during locomotion and cytoskeletal rearrangements that cause the formation of pseudopods (Wolach et al., 1992; Harris et al., 1993). The neonate neutrophils have more fluid (less rigid) membranes than adults and thus have less means to generate force that drives changing shape. However, by 10 days of age, healthy full-term neonates have chemotaxis function comparable to that of adults (Eisenfeld et al., 1990). L-Selectin is an adhesion molecule that facilitates the initial attachment and subsequent movements toward the infection or injury. In full-term infants, L-selectin is down-regulated, and expression comparable to that of adults is not reached until 15 years of age (Eisenfeld et al., 1990). It has been suggested that this is a consequence of immature neutrophils being released prematurely, with reduced L-selectin expression, from the bone marrow (Buhrer et al., 1994).

Several bactericidal deficiencies in neutrophils have been well documented. Once phagocytosis has occurred, primary or azurophilic granules fuse with the phagosome and expose the contents to toxic proteins and enzymes. Azurophilic granules are composed of defensins, lysozyme, azurocidin, bacterial permeability-increasing protein (BPI), elastase, cathepsin G, proteinase, and esterase N. BPI is required for oxygen-independent killing of gram-negative bacteria. Neonatal neutrophils have been shown to contain three- to fourfold less BPI per cell than adult neutrophils. Neonatal neutrophil acid extracts have lower levels of antibacterial activity than adult extracts. Forty percent of neonates showed up to a 9- to 10-fold reduction in BPI (Levy et al., 1989). This suggests the existence of another factor that may contribute to the variability in host susceptibility to infection.

Like neutrophils, monocytes and macrophages differentiate from common myeloid precursor cells in the bone marrow. Monocytes present in the circulation are precursor cells which can localize to specific tissues and differentiate into macrophages. However, there is a pool of undifferentiated monocytes within tissues that, upon activation, will differentiate into specific tissue macrophages. The level of monocytes present in neonates is comparable to the level found in adults. Monocytes circulate in the blood between 8 and 70 h and, once activated within tissues, will differentiate into tissue macrophages within 8 to 12 h. The life span of macrophages is long compared to that of neutrophils, and they can persist for several months (Thomas et al., 1976). Macrophage chemotaxis is generally believed to be non-antigen specific. Macrophages appear at sites of infection following neutrophil infiltration to clear necrotic debris. Macrophages also secrete proinflammatory and immunoregulatory cytokines that aid in the development of an appropriate adaptive immune response.

Opportunistic pathogens can take advantage of the weak immune response delivered by infants, resulting in more severe invasive infections. The longevity of macrophage cells is an attractive niche exploited by organisms, such as *Escherichia coli* K1 and *Citrobacter koseri,* that cause sepsis and meningitis in neonates. Studies have shown that intracellular survival of *Escherichia coli* K1 within macrophages is essential for efficient progression of systemic disease in the infant rat model (Hill et al., 2004). *C. koseri* has been shown to replicate within macrophages and to survive phagolysosomal fusion (Townsend et al., 2003). It has been shown that some pathogens can use macrophage survival to dysregulate cytokine expression and evade immune detection (Verreck et al., 2004). Clinical evidence indicates that relapse is common during these central nervous system infections, and it has been thought that persistence within macrophages may contribute to this complication. The affinity of these infections for the premature neonate indicates that inherent immune deficiency of neonates, particularly regarding macrophages, cannot be ruled out as a contributing factor to chronic infection.

Tissue macrophage chemotaxic (Pahwa et al., 1977), phagocytic (Schuit and Powell, 1980), metabolic (Das et al., 1977), and bactericidal activities are diminished in the neonate. Macrophages are not fully proficient in antigen presentation or cytokine and chemokine production, which may be related to a decrease in T-cell production (Taylor and Bryson, 1985; Wilson et al., 1986; Hariharan et al., 2000). Neonatal macrophages exhibit deficient production of interleukin-12 (IL-12)/IL-23, IL-18, and other proinflammatory cytokines and are hyporesponsive to gamma interferon (IFN-γ). This was not due to less IFN-γ receptor expression or ligand binding affinity (Taylor and Bryson, 1985; Marodi et al., 2000 and 2001). The impairment was shown to be associated with deficient phosphorylation of STAT-1, a signal transduction protein found to be associated with the IFN-γ receptor (Marodi et al., 2001). Together, these deficiencies are suggested to be a consequence of Th2 cytokine-biased expression in the neonate and will be discussed further below.

The innate immune response is also initiated through a special set of proteins located on the surface of macrophages called Toll-like receptors (TLR). TLRs are integral membrane proteins that activate signal transduction to induce specific inflammatory reactions in response to ligand-receptor coupling. There are 10 known TLRs, and each recognizes a specific antigen motif (Table 1). For example, TLR4 and TLR5 recognize lipopolysaccharide (LPS) and flagellin molecules, respectively (Poltorak et al., 2000; Tallant et al., 2004). Recent studies have shown that neonatal monocytes and macrophages exhibit a diminished response to several TLR ligands. Specifically, cord blood macrophages and monocyte TLR4 have a diminished response to LPS as a consequence of a decrease in phosphorylation of the cytoplasmic TLR4 adaptor protein MyD88, which resulted in a reduction in transcriptional activation mediated by NF-κB (Hasegawa et al., 2003; Yan et al., 2004). These deficiencies put the neonate at increased risk of infection by gram-negative bacteria and sepsis.

NK cells are also part of the innate response responsible for lymphocyte-mediated targeted cell killing and do not require previous exposure to an antigen. Host cells lacking major histocompatibility complex class I (MHC-I) receptors are immediately overwhelmed with cytolytic compounds delivered by NK cells when they fail to detect the "self" receptor. The concentration of these cells in the peripheral circulation of neonates is actually higher than that found in adults. Further, neonates have been shown to elicit a strong MHC-I-restricted cytotoxic lymphocyte response (Brown et al., 1980; Kaufmann et al., 1988). However, the immaturity of these cells is demonstrated by studies showing reduced cytolytic activity against herpesviruses that may be a consequence of inadequate expression of specific antigens on the cell surface.

Table 1 Toll-like receptors[a]

Receptor	Ligand	Origin of ligand
TLR1	Triacyl lipopeptides	Bacteria and mycobacteria
	Soluble factors	*Neisseria meningitidis*
TRL2	Lipoprotein/lipopeptides	Various pathogens
	Peptidoglycan	Gram-positive bacteria
	Lipoteichoic acid	Gram-positive bacteria
	Lipoarabinomannan	Mycobacteria
	Phenol-soluble modulin	*Staphylococcus epidermidis*
	Glycoinositolphospholipids	*Trypanosoma cruzi*
	Glycolipids	*Treponema maltophilum*
	Porins	*Neisseria*
	Atypical lipopolysaccharide	*Leptospira interrogans*
	Atypical lipopolysaccharide	*Porphyromonas gingivalis*
	Zymosan	Fungi
TLR3	Double-stranded RNA	Viruses
TLR4	LPS	Gram-negative bacteria
	Taxol	Plants
	Fusion protein	Respiratory syncytial virus
	Envelope protein	Mouse mammary-tumor virus
TLR5	Flagellin	Bacteria
TLR6	Diacyl lipopeptides	*Mycoplasma*
	Lipoteichoic acid	Gram-positive bacteria
	Zymosan	Fungi
TLR7	Single-stranded RNA	Viruses
TLR8	Single-stranded RNA	Viruses
TLR9	CpG-containing DNA	Bacteria and viruses
TLR10	Unknown	
TLR11	Unknown	Uropathogenic bacteria

[a]Adapted from Akira and Takeda (2004).

Opsonization. The complement system presently is known to contain at least 30 different proteins, which are primarily formed in the liver and circulate in their inactive form. These proteins, when activated, produce various complexes that identify (opsonize) and disrupt membranes of bacteria and red blood cells. Opsonization, or enhanced attachment, refers to the complement proteins and certain antibodies, coupling antigens to phagocytes facilitating efficient phagocytosis. Circulating complement proteins have anti-infective properties activated by different pathways. The "classical" pathway involves the binding of the C1 complement protein to an antibody coupled to an antigen. The "alternative" pathway consists of nonspecific binding of C1 to various foreign moieties. Both pathways subsequently form the membrane attack complex (MAC), which creates a pore in biological membranes inducing lysis. Another protein involved in MAC formation is mannose-binding lectin, which is secreted by hepatocytes and binds to mannose. Similar to the alternative complement pathway, activation is independent of antibodies, although specific proteases cleave C3, inducing MAC formation. The activation of complement also releases the chemoattractants that establish a concentration gradient essential for the chemotaxis of macrophages and acute inflammatory cells.

The alternate pathway is more prone to deficiencies than the classical pathway. The expression of Fc and C3 receptors is comparable to that of adults; however, C3 receptors delay up-regulation expression following activation (Griffioen et al., 1992). Concentrations of proteins (i.e., C3) important in early complement activation in the neonate reach between 60 and 80% of adult levels. Levels of terminal proteins (i.e., MAC proteins) are about 10% of what is expected in adults and are especially inefficient against gram-negative bacteria. Despite a reduced quantity of complement proteins, the function remains similar to that of adults as measured by total hemolytic complement activity (CH_{50}) (Schelonka and Infante, 1998). A further consequence of complement protein deficiency is the reduced production of molecular signals stimulating chemotaxis and opsonization. Premature infants have lower levels of complement protein and lower levels of function than term infants. However, complement levels increase with age, and most neonates have adult complement levels by 3 months of age (Miyano et al., 1996).

IgG and IgM antibodies are opsonization factors capable of crossing the placenta. Premature infants are disadvantaged by a reduced time for diffusion of maternal IgG. IgA secretion begins at approximately 2 to 5 weeks of age and also may be passed on to the neonate passively in breast milk, bolstering the mucosal immunity (Hayward, 1983). However, the infant mucosal immune system has fewer B cells, IgA-producing plasma cells, and T cells than fully mature immune system (Spencer et al., 1990). It has been noted that the immature intestine has an exaggerated secretory response (Chu and Walker, 1989).

Cytokines. Cytokines are small peptide molecules that provide intracellular signals influencing growth, proliferation, and activation of various host cells involved in the immune response. The balance between the Th1/Th2 cytokines is important to maintain a healthy immune status in adults. Various cell types are able to produce cytokines during an immune response. Important cell types, such as macrophages, NK cells, T cells, and mast cells, are pivotal in facilitating an effective immune response during infection. Cytokines produced by macrophages, NK cells, T cells, and mast cells are essential mediators of an effective immune response to infection and serve to bridge the gap between the innate and adaptive immune responses. These cytokines, classified as type 1 and type 2, cause the differentiation of CD4$^+$ T helper cells into either Th1 or Th2 cells, respectively. The enhancement of resistance to intracellular pathogens comprises the type 1 immune response. Characteristics of this response include induction of cellular immunity, cytotoxic T-lymphocyte responses, and macrophage activation. Type 2 responses aggressively attack parasitic worms during helminthic disease and aid in the establishment of humoral immunity. This response is characterized by the activation of the allergy response mediated by eosinophils and mast cells. The dichotomy of type 1 and type 2 cytokine responses is that activation of a type 1-specific response prevents the initiation of the type 2 response, just as induction of a type 2-specific response precludes the induction of a type 1 response (Schmitz et al., 1993; Trinchieri, 1997). The macrophage is known to express the immunoregulatory cytokine interleukin-12 (IL-12) and immunosuppressive cytokine IL-10. IL-12 expression favors IFN-γ production, which causes phagocytic cells to become activated and clonal expansion of cell types specific to a type 1 response, inhibiting the type 2 response (Sher et al., 1992). Type 2 cell proliferation is favored when IL-10 is induced, inhibiting IL-12 and IFN-γ expression (Flesch and Kaufmann, 1994; Hirsch et al., 1996).

Cytokine secretion in neonates is imbalanced. Neonates show a bias toward Th2 immune responses and are deficient in secretion of IFN-γ and IL-10 (Kotiranta-Ainamo et al., 2004). Hyporesponsive macrophages are largely responsible for deficient IFN-γ production in neonates. There was no detectable secretion of proinflammatory IL-1β, IL-12, and tumor necrosis factor-alpha (TNF-α), and only some secretion of IL-6 was detected when neonatal macrophages were treated with LPS. There was a three- to fivefold more immunosuppressive IL-10 expression in neonates compared to that of treated adult macrophages (Joyner et al., 2000; Chelvarajan et al., 2004). Secretion of IL-18, the IFN-γ-inducing factor, is significantly reduced in both neonatal lymphocytes and mixed mononuclear cells in cord blood compared to that of adults (Nomura et al., 2001; La Pine et al., 2003). Neutralizing antibodies specific to IL-10 restored the neonatal macrophage ability to induce B cells to produce antibodies. Placental up-regulation of the Th2 response may be a natural adaptation of mammalian evolution that consequently down-regulates the

Th1 response in order to protect the fetus from Th1-induced damage invoked by "foreign" (fetal) material (Marodi, 2006). This deficiency in Th1 cytokine production is known to persist in the neonate. Neonates show a diminished response to LPS, an endotoxin associated with gram-negative bacterial cell membranes (Marodi, 2005). This, coupled with the lack of a proper Th1 response due to the neonatal Th2 bias, could contribute to the significant increase in sepsis observed in VLBW and premature infants.

Adaptive Immunity

The adaptive or acquired immune response provides the targeted antibody response against specific antigens during infection and is also accountable for life-long protective immunity. T and B lymphocytes are key cells of adaptive immunity. Time (in terms of weeks) is required for antigen processing and antibody production. Further, long-term adaptive immunity is dependent upon previous exposure to antigens. However, the significance of the adaptive immune response is the specificity and memory of the immune response providing a precise reaction that limits destruction of healthy tissues that is recalled over the long term. Thus, adaptive defense mechanisms can provide infants with specific immunity following infection or vaccination.

T cells. T cells are a diverse set of lymphocytes, each contributing to a specialized and specific immune response comprising the cellular branch of the acquired immune response. Subpopulations of T cells are distinguished by factors such as developmental origin, cell surface markers, and cytokine production. For example, $CD4^+$ T cells are involved in antigen processing and presentation with MHC-II molecules. $CD4^+$ T cells are further differentiated into Th1 (inflammatory) and Th2 (antibody) cells as discussed earlier; $CD8^+$ T cells are largely responsible for cytotoxic activity against tumors and virally infected host cells. The thymus is essential for the development and maturation of peripheral lymphoid tissue and is highly active during fetal development. Humoral substances, such as cytokines, that influence T-cell differentiation are produced in the thymus, and mature T cells are detectable by 14 weeks of gestation. The number of neonatal circulating T cells increases until 6 months of age, and they are influenced by antigenic stimulation.

Neonates have an abnormal increase in the number of $CD4^+$ lymphocytes compared to the number in adults. $CD4^+$ T cells are abundant compared to the low number of $CD8^+$ T cells, as reflected in the $CD4^+/CD8^+$ ratio. The difference between the numbers gradually decreases with age (Gasparoni et al., 2003). Neonatal deficiency in antigen-specific antibodies is significant following infection and may be a consequence of an increased amount of antigenically naive T cells or increased T-cell suppressor activity observed in neonates under specific conditions. Adult circulating lymphocytes are largely primed CD4[+] memory T cells and are more able to respond to antigens and produce cytokines. In neonates, although IL-2 secretion is adequate, the expression of other important cytokines (IL-3 to IL-5) is reduced compared to that of adults (Ehlers and Smith, 1991). Also, IFN-γ expression was not observed from CD4[+] T cells from stimulated cord blood, where as it was observed in adult T cells (Cerbulo-Vazquez et al., 2003). In line with this observation are data showing that the level of IFN-γ-inducing cytokine IL-18 expression is less than that observed from adult lymphocytes (Nomura et al., 2001; La Pine et al., 2003). However, the immunosuppressive cytokines IL-10 and IL-13 are expressed from cord blood T cells at a higher level than that observed from adult T cells (Blanco-Quiros et al., 2000; Ribeiro-do-Couto et al., 2001). Furthermore, neonatal cells are not as proliferative as adult cells when activated (Porcu et al., 1998). Overall, neonates have a partial T-cell immunodeficiency that contributes to increased susceptibility to infection.

The inherent neonatal functional deficiencies in the adaptive immune cells can be compensated for under specific conditions in animal models. However, the restoration of function does not result in a fully protective immune response. Factors contributing to these undeveloped responses might include the reduced number of immune cells in neonates compared to the number in adults, the presence of circulating immune cells recently derived from fetal hematopoietic precursors in neonates but not adults, apoptosis of selective subsets of T helper cells, and the unique neonatal environment, which promotes homeostatic proliferation of both CD4[+] and CD8[+] T cells (Adkins et al., 2004).

B cells. The humoral adaptive immune response requires the secretion of antibodies or immunoglobulin molecules from B cells. Fetal development occurs in an immune-privileged site, considering the general lack of antigens present during gestation. Thus, only trace amounts of Ig molecules have been found. During development, the fetus acquires IgG via transplacental diffusion; however, IgA, IgM, IgD, and IgE do not cross the placenta (Garty et al., 1994). Neonatal catabolism of maternally acquired IgG following birth causes an IgG deficiency (hypogammaglobulinemia) between 2 and 6 months of age. Premature infants show an increased risk of severe hypogammaglobulinemia, because, due to the reduced gestational time, less IgG diffuses into the fetus. However, studies have shown that intravenous administration of IgG to premature infants prevents hypogammaglobulinemia (Ruderman et al., 1990). Neonatal IgG levels are adequate once IgG production increases at 6 months of age. IgG levels reach 70% of what is found in average adults by the age of 1 year. Adult levels of IgM expression are genetically constrained until 2 months of age, yet adult levels are achieved after 1 year (Shiokawa et al.,

1999). However, IgG and IgA levels do not match those of adults until late childhood (Adkins et al., 2004).

The passive immunity provided by antibodies passed from mother to infant in breast milk has proven effective in providing protection from serious diseases, such as those caused by *Bordetella pertussis, Haemophilus influenzae* type b, *Streptococcus pneumoniae,* and *Neisseria meningitidis* (Kassim et al., 1989). Complement proteins, lysozyme, lactoferrin, and secretory IgA are components of breast milk that protect the mucus membranes of the gastrointestinal and upper respiratory tracts of nursing infants from infectious pathogens (Hanson et al., 1985). Again, the modest levels of maternal antibodies found in VLBW infants are depleted between 2 and 4 months of age, and thus they are unable to benefit from this added protection.

Weak reactions to thymus-independent antigens, such as polysaccharides, increase neonatal susceptibility to infection by encapsulated bacteria (Chelvarajan et al., 2004). Vaccines utilizing polysaccharide antigens obtain a poor response unless administered with an adjuvant. For example, *Haemophilus* and pneumococcal polysaccharide vaccines are coupled to diphtheria toxin (Ahman et al., 1996). Premature infants are less responsive to vaccination than term infants. On the other hand, impaired immune function in neonates results in a significantly lower frequency of disease caused by immunological rejection of grafted tissues.

The immune status of neonates and infants is inherently compromised. Weaknesses in both the innate and adaptive immune responses show a direct relationship to gestational age as the infant makes the transition from the protective womb into an antigenically aggressive world. Reduced bactericidal properties, developmental immaturity, decreased proliferative responses, and dysregulation of cytokine networks are all factors that put infants at increased risk of infection. Physicians and scientists have made extraordinary progress in the field of neonatal immunity and continue to elucidate specific mechanisms that underlie the inherent immune deficiency in neonates.

INTRODUCTION TO THE HUMAN INTESTINAL MICROBIAL FLORA

The gastrointestinal tract is divided into the main regions of the stomach, the small intestine, and the large intestine. In the fully developed intestinal tract, each region has its own environmental conditions. The stomach is an extremely acidic environment with a high flow rate. The stomach pH is around 2.5. Gram-negative organisms, including pathogens such as *Salmonella* spp., cannot survive at this pH without the protection of a food matrix. The bacterial flora of the stomach is dominated by gram-positive and aerobic organisms at ca. 10^3 to 10^5 CFU/ml. There is a transition in the small intestine from the low-level stomach flora toward the more densely colonized large intestines. In the small intestine, the bacterial count varies from 10^4 to 10^8 CFU/ml, and there is a redox potential in the distal ileum of ca. -150 mV. This environment has a near-neutral pH due to alkaline biliary and pancreatic secretions and bicarbonate anions from Brunner's glands in the duodenum. The densest bacterial concentration is in the large intestines, at 10^{11} to 10^{12} CFU/ml. Due to the low redox potential in this environment, the majority of bacteria in the colon are strict anaerobes.

Metabolic Activities of the Intestinal Bacteria

The metabolic activity of the intestinal microbial flora has been studied for many years (Salyers, 1979; Smith and Bryant, 1979; Egert et al., 2006). These studies reflect the changes in techniques available, ranging from carbohydrate metabolism studies via assaying specific enzymes (i.e., glycosidases) by Salyers (1979) to metagenomics and isotope tracking (Egert et al., 2006). With respect to the neonatal flora, Walker et al. (1989) studied the urinary excretion of organic acids as products of bacterial carbohydrate fermentation in 52 preterm neonates. VLBW neonates born before 33 weeks of gestation excreted 2,3-butanediol more often and in higher concentrations than neonates of 33 to 36 weeks. In the majority of cases, neonates excreting 2,3-butanediol were colonized by *Klebsiella*, *Enterobacter*, or *Serratia* species (Edwards and Parrett, 2002). Bacterial fermentation of carbohydrates in the intestines produces short-chain fatty acids (SCFA). The concentration of these acids is initially low in breast-fed infants and increases in the first year of life. Acetic and lactic acids dominate in these infants, whereas the major SCFA in adults are acetic, propionic, and *n*-butyric acids. The SCFA profiles of formula-fed infants are similar to those of adults, probably due to the differences in intestinal flora between the two groups (see "Intestinal Flora of Breast-Fed and Formula-Fed Neonates," below). Butyric acid is an energy source for colonocytes (Edwards and Parrett, 2002).

Studying the Neonatal Gut Flora

Until recently, our understanding of the microbial gut flora of infants and adults was based largely on the use of traditional culture techniques. This involved selective and nonselective agar plates inoculated with a sample dilution series and incubated primarily under strictly anaerobic conditions. A considerable body of knowledge was gained from such studies. Early studies by Drasar et al. (1969), Haenel (1970), and Gall (1970) were very thorough in their analyses of the small and large intestinal tracts and are valuable sources of information, despite being "pre-PubMed." Some of Drasar's studies involved 132 individuals and included microbial flora studies of the oral cavity and stomach as well as the small and large intestines. Methods for the cultivation and identification of strict anaerobes were developed in the 1970s, notably at the Virginia Polytechnic Institute by Holdeman et al. (1977). Savage (2001) gives an excellent overview of the contribution of early workers to this subject area.

The cultivation method, however, may favor the growth of certain bacterial groups and prevent the growth of others, resulting in a biased view of the bacterial community. Additionally, certain bacteria remain nonculturable, and most media used for quantification of bacterial groups are nonspecific (Harmsen et al., 2000). In recent years, molecular methods have been applied to studying the diversity of the intestinal flora (Vaughan et al., 2000). One commonly used method is based on the gene for the 16S rRNA molecule.

The use of 16S rRNA gene sequences has greatly facilitated the study of the microbial flora of the intestinal tract, because it allows the culture-independent analysis of the microbiota (Favier et al., 2002). Phylogenetic analysis of the DNA sequences is used to identify the members of the microbial flora. The simplest and presently most widely used method is to amplify the 16S rRNA gene directly from total community DNA using rRNA-specific primers. Universal primers can be designed that anneal to rRNA genes from the *Archaea*, *Eubacteria*, and *Eucaryota*. Alternatively, primers can be designed to amplify rRNA genes from a specific group(s) of organisms (Muyzer and Smalla, 1998; Muyzer, 1999; Walter et al., 2001; Satokari et al., 2001; Heilig et al., 2002). Specific 16S rRNA gene-based oligonucleotide probes have been developed that detect different bacterial groups without cultivation directly by using fluorescent in situ hybridization (Harmsen et al., 2000). 16S rRNA gene sequence analysis has been used increasingly in clinical microbiology and has been very well reviewed by Clarridge (2004).

In order to separate 16S rRNA gene PCR products from different organisms in the same sample, the PCR-denaturing gradient gel electrophoresis (PCR-DGGE) technique is applied (Muyzer et al., 1993). This exploits the hypervariable regions of the 16S rRNA gene. The 16S rRNA genes are amplified from the community DNA using specific primers, one of which has a "G+C clamp" of approximately 39 nucleotides attached to the 5′ end. This G+C clamp prevents the two DNA strands from dissociating completely, even under highly denaturing conditions. The resultant mixture of 16S rRNA gene fragments (from different organisms) is separated along a denaturing gradient of urea and formamide. This results in a DNA banding profile of the microbiota present. The variation in distance travelled by the near-equal-length PCR products is because individual amplicons will stop migrating when the DNA denatures. This will vary according to their G+C contents. This method therefore enables the separation of individual sequences from a single sample of mixed species. The bands in the profile will represent most of the dominant microbial populations in the community. Changes in banding profile will therefore reflect changes in the microbial diversity. Although the technique does not enumerate the identified organisms (see the description of real-time PCR below), the intensity of each band gives a relative proportion of the target sequence in the sample.

However, this method can lead to some sampling bias, as some sequences may amplify better than others (Wintzingerode et al., 1997). Because the PCR products correspond to the relative $G+C$ content of the amplified V6-V8 region of the 16S rRNA gene, bacterial species with similar $G+C$ content in this amplified region may form assemblages and appear as a single band (Muyzer and Smalla, 1998). Therefore, the PCR-DGGE method may not show the full diversity of the microbial flora present in the sample. Additionally, the 16S rRNA gene exists in many copies in some bacteria, and the sequence variation may hinder interpretation of the results (Satokari et al., 2001).

Real-time PCR assays are being used to detect and quantify bacterial populations that are small in number and have been applied to the large-intestine microbial flora of adults and infants (Bartosch et al., 2004; Hopkins et al., 2005; Penders et al., 2005).

A second molecular method, which is more rapid than 16S rRNA gene analysis, is terminal restriction fragment length polymorphism (T-RFLP) analysis. T-RFLP is a robust and reproducible method that has been used to compare microbial communities in soil, sediment, and intestinal contents from various animals. It is very applicable to compare spatial and temporal changes in diverse bacterial communities and therefore is very relevant to studying the neonatal gut flora. The data sets generated by this method may require principal component analysis, a multivariate projection method, for interpretation (Wang et al., 2004).

Intestinal Flora of Breast-Fed and Formula-Fed Neonates

The majority of studies of the human intestinal gut flora have been fecal flora studies of the adult. However, the immune system and gastric secretions of the neonate differ significantly from those of the adults, and hence there are considerable differences in the colonization of the neonatal intestinal tract. Of particular importance with respect to *E. sakazakii* is that a better understanding of the neonatal flora may explain, in part, the reason why this ubiquitous organism has become associated with neonatal illnesses such as NEC, septicemia, and meningitis. Positive samples are often obtained from the throat, stomach, and intestines via swabs, gastric aspirates, and fecal collections, respectively. *E. sakazakii* has also been isolated from neonatal eye infections via swabbing (Reina et al., 1989; Block et al., 2002).

There have been a limited number of studies using cultivation and molecular techniques to monitor the changes in the neonatal gut flora in the first weeks of life. Some of these studies have distinguished neonates according to birth weight and method of nutrition (i.e., breast or formula fed). Unfortunately, with respect to nutrition information, a breast-fed neonate may nevertheless be fed breast milk with added fortifiers, and this important detail may be omitted in published studies. Similarly, the antibiotic history of the neonates is not always declared. The majority of studies have been of duodenal and stool samples, which represent the small and large intestines, respectively. Identification of cultivated organisms has primarily relied on phenotyping. However, even using DNA sequence-based techniques, it is not always possible to identify intestinal organisms (Favier et al., 2002).

In general, bacterial colonization starts soon after birth, and facultative anaerobic bacteria, such as *Enterobacter* species, can be isolated from the neonate's feces after the first hours of life. The facultative anaerobes lower the redox potential due to their use of oxygen, and this enables the growth of strict anaerobes, which normally appear at high levels during the first week of life.

In vaginally delivered neonates, these first bacteria are of maternal origin. In Cesarean section-delivered neonates, the flora is characterized by the lack of anaerobic bacteria. Instead, the flora is dominated by microaerophilic and facultative bacteria and spore formers such as clostridia. Subsequently, the environment and hospital staff are a major source of the colonizing bacteria (Bezirtzoglou, 1997). After this initial colonization, the diversity of the subsequent bacterial floral community is influenced by the diet (Harmsen et al., 2000; Wold and Alderberth, 2000).

While *E. sakazakii* has been associated with contaminated powdered infant milk formula and the authors agree that breast milk should be the source of infant nutrition where possible (Oddy, 2002), it should be noted that human milk often has small numbers of streptococci, micrococci, lactobacilli, staphylococci, diphtheroids, and bifidobacteria (Bezirtzoglou, 1997). Hence, colonization by staphylococci is more frequent in breast-fed infants, probably due to increased contact with the mother's skin during feeding. Expressed human breast milk can be a reservoir for staphylococci due to secondary contamination (Carneiro et al., 2004). There was one case in which the mother asymptomatically excreted *Salmonella enterica* serotype Panama for 2 weeks and caused a neonatal infection (Chen et al., 2005). Bezirtzoglou (1997) reported that *Clostridium tertium* is more frequently isolated from infants breast fed with iron supplement, whereas *Clostridium difficile* and *Clostridium paraputrificum* were not isolated. Although *Clostridium difficile* has been associated with NEC, colonization by this organism is not exclusive to formula-fed neonates (Penders et al., 2005).

In a review of related studies, Haenel (1970) reported differences between breast- and formula-fed neonates in the first 6 days of life, including the predominance of "*Lactobacillus bifidus*" (proposed name) in breast-fed infants. The microbial flora of 21 infants from 6 to 22 months of age was present in the stomach through to the distal colon. The *Bacteroides* group organisms, which were isolated only from the large intestines, were found at ca. 10^9 CFU/g, whereas the numbers of coliforms ranged from 10^6 (stomach) to 10^8 (large intestines) CFU/g, with the *Klebsiella-Enterobacter* group being found at 10^5 to 10^6 CFU/g. The study also included the postmortem flora of a 3-month-old infant who had died of toxic dyspepsia and endocardial fibrosis.

In breast-fed, full-term infants, gram-positive bifidobacteria are commonly referred to as becoming dominant along with lactobacilli and streptococci. In contrast, in formula-fed infants, the predominant flora is a mixture of *Enterobacteriaceae* (*Escherichia coli* and *Klebsiella* spp.), *Staphylococcus*, *Clostridium*, *Bifidobacterium*, *Enterococcus*, and *Bacteroides* species (Rubaltelli et al., 1998; Harmsen et al., 2000). The microbial diversity undergoes further changes when the diet changes at weaning, with the introduction of solid food. Subsequently, the fecal flora of the two groups becomes indistinguishable and resembles the adult flora. Wang et al. (2004) studied in detail, using T-RFLP of the fecal flora, the microbial diversity of two infants during breast-feeding and weaning periods. The infants were initially colonized mostly by *Enterobacteriaceae*, *Veillonella*, *Enterococcus*, *Streptococcus*, *Staphylococcus*, and *Bacteroides* species. The members of *Enterobacteriaceae* and *Bacteroides* were predominant during breast-feeding in both infants. After weaning, the levels of *Enterobacteriaceae* organisms decreased, whereas the levels of clostridia increased.

Although a number of studies have reported a tendency of greater numbers of bifidobacteria in breast-fed as opposed to formula-fed infants, it is not reported in all studies. Hopkins et al. (2005) used real-time PCR to compared the microbial diversity of 40 infants from birth to 24 months and the effect of feeding. The numbers of *Bacteroides* and *Desulfovibrio* spp. increased considerably in the 7- to 24-month age groups, whereas numbers of *Enterococcus faecalis* decreased. There were increased levels of bifidobacteria in breast-fed babies and higher levels of desulfovibrios in bottle-fed children. Penders et al. (2005) used real-time PCR to quantify *Bifidobacterium* spp., *Escherichia coli*, and *Clostridium difficile* in fecal samples of 50 breast-fed and 50 formula-fed infants. In contrast to some earlier studies, all infants were colonized by *Bifidobacterium* species at ca. 10^{10} CFU/g. *Clostridium difficile* was detected in both cohorts, i.e., 14% of the breast-fed and 30% of the formula-fed infants. The *Clostridium difficile* counts were significantly lower for breast-fed infants (10^3 CFU/g) than for the formula-fed group (10^7 CFU/g). The prevalence of *Escherichia coli* in the breast-fed and formula-fed group was 80% and 94%, respectively, at ca. 10^9 CFU/g. These authors concluded that the prevalence and numbers of *Clostridium difficile* and *Escherichia coli* were significantly lower in breast-fed than in formula-fed infants, whereas the prevalence and counts of *Bifidobacterium* spp. were similar in both groups.

Heavey and Rowland (1999) reviewed a number of studies comparing the microbial flora of breast- and formula-fed infants and reported that the early reports of bifidobacterium dominance in breast-fed infants appear to have decreased. They attributed this to changes in obstetric practices, the environment, bacterial detection techniques, and improvements in formula feeds. Human milk contains nucleotides and gangliosides which, if added to formula milk, increases the colonization by bifidobacteria (Balmer et al., 1994; Rueda et al., 1998).

Despite the differing reports concerning the predominance of bifidobacteria in breast-fed neonates, it is accepted that the intestinal microbial flora of formula-fed infants is initially more diverse than that of breast-fed infants. Since *E. sakazakii* infections are most notable in low-birth-weight neonates (Food and Agriculture Organization-World Health Organization, 2006), it is of particular interest to consider the gut flora of such infants (see chapter 4). Preterm neonates with low birth weights differ from full-term neonates in a number of aspects. They are compromised by deficiencies in humoral and cellular immune responses (Table 2) and are exposed to increased microbial hazards due to prolonged stays in the hospital and handling and feeding requirements. They may be fed orally (enteral) or intravenously (parenteral) and may receive antibiotic therapy for prolonged periods. The nutrition source may be formula or breast milk, though the latter may require fortification. So far, only a few studies have monitored the bacterial succession in preterm infants.

Blakey et al. (1982) studied the microbial colonization of the throat, stomach, and large intestines over the first 3 weeks of life for 28 preterm neonates. The throat was frequently colonized by *Staphylococcus epidermidis*, especially 9 to 12 days after birth. *Escherichia coli* was a slower colonizer, and *Klebsiella* species were common in neonates >16 days of age. *Enterobacter* species were isolated from up to a quarter of neonates. The stomach was colonized predominantly by *Bacteroides* species in up to half of the neonates, and various *Enterobacteriaceae*, including *Enterobacter* species, were present in up to

Table 2 Summary of premature neonate immune deficiencies[a]

Defense mechanism	Premature neonate immune deficiency	Susceptibility
Epidermal and epithelial barriers	Immature, thin skin	Environment conducive to bacterial colonization and growth
Intact endothelial and epithelial tissues	Trauma caused by invasive medical procedures	Damage to barriers may facilitate infection
Gastrointestinal mucosa	Decreased mucus, gastric acid, proteolytic enzymes, and reduced gut mobility	Facilitates adherence and survival in the gut; decreased breakdown of toxins
Complement	Lower levels with decreasing gestational age	Decreased bactericidal activity, decreased chemoattractant ability
Cytokines	Decreased production of IL-1, IL-8, IFN-γ, TNF-α, Th2-biased immune response	Increased susceptibility to gram-negative infections
Defensins	Lower levels with decreasing gestational age	Decreased bactericidal activity
Neutrophils	Diminished bone marrow stores, increased fluidity of membranes, immature oxidative burst, decreased azurophilic granules	Decreased bactericidal activity
Monocytes	Decreased in number and function, decreased adherence	Decreased antigen presentation and cytokine release results in insufficient stimulation of humoral immune response
T cells, B cells, and antibodies	Decreased lymphocytes in gut, less maternally transferred IgG, reduced antibody production and T-cell proliferation	Increased susceptibility to infection, reduced protective immune response

[a]Modified from Kaufman and Fairchild (2004).

one-third of neonates. The large intestines were colonized in most neonates in the first 4 days of life. *Bacteroides* species were the most common organisms isolated and declined in incidence after day 12. *Escherichia coli* and *Klebsiella* species were common after day 4. Infants receiving parenteral feeding had delayed bacterial colonization compared to that of enteral-fed infants, and this was not related to antibiotic therapy.

Gewolb (1999) studied stool samples from 29 extremely low-birth-weight infants (<1,000 g) that were collected on days 10, 20, and 30 after birth. By day 30, the predominant species were *Enterococcus faecalis*, *Escherichia coli*, *Staphylococcus epidermidis*, *E. cloacae*, *K. pneumoniae*, and *Staphylococcus haemolyticus*. Members of the *Lactobacillus* and *Bifidobacterium* genera were identified in only one infant. The total counts and numbers of species identified increased in breast milk-fed infants over time. The total number of species/stool sample increased from 2.5 on day 10 to 4.27 on day 30. Hence, these neonates were colonized by a very small number of bacterial species. During this study, two formula-fed neonates developed NEC, one of whom died following intestinal perforation. The neonate who died was colonized by *Escherichia coli*, *Enterococcus faecalis*, and *Staphylococcus haemolyticus* on the day-20 sample obtained shortly before the baby developed NEC and died. Whether any of these organisms was causative agent of the NEC is unknown.

A more recent study by Schwiertz et al. (2003) included 29 preterm infants, including samples from antibiotic-treated infants and one with NEC. The microbial diversities (determined using PCR-DGGE profiles and cultivation from feces) of these infants were compared to those of 15 breast-fed, full-term infants. The DGGE profiles of preterm infants were simple in the first days of life, with only a few DGGE bands. After 4 days, the banding patterns became more constant in most infants, and 10 days after birth the banding pattern had stabilized, with only minor changes. The number of major bands in individual babies ranged from 5 to 20 and became more similar in most infants after 2 weeks. The most frequently identified organisms were *Escherichia coli*, *Enterococcus* spp., and *K. pneumoniae*. Bifidobacteria were detected in breast-fed, full-term infants but not in the preterm infants. There was a greater diversity of profiles in the preterm infants than in those of breast-fed, full-term infants. The authors concluded that the initial colonization of the neonate intestinal tract is highly dependent on the neonate's environment. Rosberg-Cody et al. (2004) isolated bifidobacteria from 6 out of 24 neonates aged between 3 days and 2 months in a neonatal unit. Four of these were full-term infants, and the other two were premature neonates 3 and 6 weeks after birth.

OVERVIEW OF NEONATAL INFECTIONS

Neonatal Infections

Emori and Gaynes (1993) reviewed the most frequently isolated pathogens in nosocomial infections and reported them to the U.S. National Nosocomial Infections Surveillance System. From 1986 to 1989, *Escherichia coli* was the most common isolate (16%) reported to the U.S. National Nosocomial Infections Surveillance System, followed by enterococci (12%), *Pseudomonas aeruginosa* (11%), *Staphylococcus aureus* (10%), and coagulase-negative staphylococci (CoNS) (9%). There was a significant change in the pathogens associated with nosocomial infections in the 1980s, and these, for example, *P. aeruginosa* and *Enterobacter* spp., were often difficult to treat with antibiotics. The number of CoNS reports increased dramatically from 9% of all pathogens in 1980 to 31% during 1990 to 1992. This was most notable for blood isolates. The authors commented that although the data probably

Table 3 Organisms associated with neonatal bloodstream infections

Organism(s)	% of total causative organisms as determined by: Lee et al. (2004)	Kaufman and Fairchild (2004) Early onset[a]	Kaufman and Fairchild (2004) Late onset[b]
CoNS	29	10.7	47.9
Staphylococcus aureus	22	1.2	7.8
Group B streptococci	2	10.7	2.3
Enterococcus and *Streptococcus* spp.	2	7.4	3.3
Listeria monocytogenes		2.4	
Escherichia coli		44	4.9
Enterobacter spp.			2.5
Enterobacter cloacae	17		
Enterobacter sakazakii	2		
Enterobacter aerogenes	1		
Klebsiella spp.	11	1.2	4.0
Serratia marcescens	1		2.2
Acinetobacter baumannii	5		
Fungi	7	2.4	12.2

[a]Less than 72 h since birth.
[b]More than 3 days since birth.

reflect the true incidence of infections, nevertheless there has been a tendency to report CoNS in cultures as true pathogens rather than sample contaminants, as previously practiced.

More recent reviews by Kaufman and Fairchild (2004) and Lee et al. (2004) are summarized in Table 3. Lee et al. (2004) reviewed the incidence of bacteremias in a neonatal intensive care unit (NICU). Fifty-eight out of 623 (9.31%) patients had bacteremia, with 12 deaths. The main bacterial pathogens isolated were CoNS (29%), *Staphylococcus aureus* (22%), and *E. cloacae* (17%). The main causative agents of bacteremia were oxacillin-resistant *Staphylococcus epidermidis*, oxacillin-resistant *Staphylococcus aureus*, and multidrug-resistant *Enterobacteriaceae*.

Bloodstream infections caused by gram-negative pathogens are the leading cause of morbidity and mortality in infants in the NICU. The source of the organisms can be the environment, hands of caregivers, and endogenously from the intestinal tract. Larson et al. (2005) studied 77 incidences (26% of the total) of bloodstream infections due to gram-negative bacteria among 2,935 neonatal admissions. Among the 77 episodes caused by gram-negative bacilli, 47 (61%) were catheter-related bloodstream infections. The most common pathogens were *K. pneumoniae* (38.7%), *Escherichia coli* (21%), *E. cloacae* (11%), and *Serratia marcescens* (11%). The authors concluded that the source of pathogens was most likely the endogenous intestinal flora rather than the hands of caregivers.

Bacterial Pathogenesis and NEC

NEC is a disease of the immature intestinal tract. The accepted case definition was proposed by Bell et al. (1978). It is characterized by systemic symptoms ranging from apnea, bradycardia, and temperature instability to diffuse intravascular coagulation and septic shock, intestinal symptoms (e.g., abdominal distention and bloody stools), and radiological findings such as pneumatosis intestinalis and gas in the portal vein. There are three clinical stages of NEC, given in Table 4 (Boccia et al., 2001). The pathological changes include mucosal edema, hemorrhage, ischemia, necrosis, and bacterial overgrowth.

The mechanisms responsible for the ischemic intestinal necrosis are not fully understood. However, the roles of the inflammatory mediators platelet-activating factor (PAF) and TNF-α have been studied. PAF is a potent phospholipid inflammatory mediator that can be produced by a variety of cell types, including endothelial cells, neutrophils, platelets, and macrophages. PAF is not stored in cells but is synthesized in response to certain stimuli, including endotoxins and hypoxia. The effects of PAF include neutrophil margination and activation, macrophage activation, capillary leakage, and hypotension. In animal studies, PAF either alone or in combination with low-dose endotoxin results in intestinal injury resembling NEC. It has also been shown that hypoxia-induced intestinal necrosis in the rat is mediated by PAF. TNF-α is a cytokine released by endotoxin-stimulated macrophages. It has many of the same pathophysiologic effects as PAF and may stimulate PAF production. TNF-α administration to rats produced intestinal injury closely resembling NEC, which appears to be mediated by PAF (Shah and Walker, 2000).

LPS is also known as an endotoxin because of its pyrogenic nature. Free LPS stimulates mediators from host cells and may lead to septic shock. LPS-responsive cell types include monocytes/macrophages, polyleukocytes, and endothclial and epithelial cells. Endotoxins may compromise basal colonic water and electrolyte transport as well as cause increased intestinal permeability. It can also cause increased release of cytokines, TNF, IL-1, PAF, and oxygen-derived free radicals from the intestinal tract. LPS recognition triggers the production of cytokines, adhesive proteins, and enzymes that produce low-molecular-weight inflammatory mediators.

The SCFA butyrate is another bacterial factor that may play an important role in the intestinal epithelium. Butyrate is an energy source for colonocytes and also stimulates epithelial cell proliferation and cytokine release. Sodium

Table 4 Clinical stages of NEC[a]

Stage	Systemic signs	Intestinal sign(s)	Radiological signs
I, Suspected NEC			
IA	Temperature instability, apnea, bradycardia, lethargy	Elevated pregavage residuals, mild abdominal distention, emesis, heme-positive stools	Normal or intestinal dilation, mild ileus
IB	Same as IA	Bright red blood from rectum	Same as IA
II, Definite NEC			
IIA, Mildly ill	Same as IA	Same as IA, plus absent bowel sounds, without abdominal tenderness	Intestinal dilation, ileus, pneumatosis intestinalis
IIB, Moderately ill	Same as IIA, plus mild metabolic acidosis, mild thrombocytopenia	Same as IA, plus absent bowel sounds, definite abdominal tenderness, without abdominal cellulitis or right-lower-quadrant mass	Same as IIA, plus portal vein gas, without ascites
III, Advanced NEC			
IIIA, Severely ill, bowel intact	Same as IIB, plus hypotension, bradycardia, severe apnea, combined respiratory and metabolic acidosis, DIC, neutropenia	Same as IIB, plus signs of generalized peritonitis, marked tenderness, and distention of abdomen	Same as IIB, plus definite ascites
IIIB, Severely ill, bowel perforated	Same as IIIA	Same as IIA	Same as IIIA, plus pneumoperitoneum

[a]Modified from Walsh and Kliegman (1986) and Boccia et al. (2001).

butyrate may increase significantly the release of IL-8 from intestinal epithelial cells, and this effect is particularly marked when the cells also are stimulated with IL-1β or LPS.

A number of factors have been implicated in the pathogenesis of NEC. These include formula feeding, intestinal ischemia, bacterial adherence, invasion, and proliferation, along with host factors such as gut immaturity. It mainly affects low-birth-weight preterm neonates (62 to 94%). Cases are mainly sporadic, though outbreaks do occur. The incidence of NEC is 0.3 to 2.4% per 1,000 live births and 1 to 5% of neonates in NICUs. Lucas and Cole (1990) reported a greater incidence of NEC with formula-fed infants than with breast-fed infants. NEC is a leading cause of morbidity and mortality in newborns (9 to 28% of cases). Understandably, there is a considerable amount of research in this area that can only be summarized in this chapter.

The pathogenesis of NEC is unclear. Prematurity is the most important risk factor for NEC. However, case-control studies have not identified any consistent risk factors for preterm neonates. In contrast, term and near-term infants with NEC often have associated conditions that may lead to impaired gastrointestinal oxygen delivery, and therefore NEC in term neonates may be linked to a specific ischemic insult to the intestinal tract. Hence, the pathogenesis of NEC may be different between term and preterm neonates. Claud and Walker (2001) proposed that NEC in preterm neonates was due to the combined effects of prematurity, enteral feeding, and bacterial colonization.

Bacterial colonization of the intestinal tract is probably a prerequisite for NEC development, as the disease has not been reported for a neonate less than 7 days old. Immature intestinal host defenses may permit aberrant bacterial colonization or inadequate neutralization of bacterial toxins or allow translocation of bacteria across the intestinal wall. Term and specifically preterm neonates have lower rates of gastric acid secretion than do adults. Such hypochlorydia is associated with enteric bacterial colonization of the stomach in the majority of enterally fed preterm infants (Blakey et al., 1982). As reviewed earlier, neonates have specific intestinal immunologic deficiencies that may predispose them to aberrant bacterial overgrowth and injury. In brief, the intestinal host defense is impaired at many levels: decreased acid secretion, decreased proteolytic enzyme activity, poor intestinal motility, reduced IgA secretion, reduced intestinal T-cell numbers, and increased mucosal permeability. Translocation of intestinal bacteria occurs with relative ease in neonates (Duffy, 2000). Additionally, enterally fed preterm infants have been shown to have a substantial incidence of spontaneous endotoxinemia, presumably of intestinal origin (Shah and Walker, 2000). Townsend et al. (2007) reported increased intestinal translocation of

E. sakazakii in rat pups following administration of LPS (endotoxin). This paper also surveyed the levels of endotoxin in enteral feeds by using the *Limulus* amoebocyte lysate method. Endotoxins can result in widespread tissue injury and systemic toxicity through activation of the inflammatory cascade. The amount of endotoxin in the enteral feeds varied 500-fold in 75 samples, from 40 to 5.5×10^4 endotoxin units/g.

No single infectious agent has been consistently associated with the development of NEC. Instead, the organisms cultured from the peritoneal fluid of neonates with NEC have probably translocated from injured intestines. However, several species of bacteria and viruses have been associated with particular epidemics of NEC. This suggests that certain organisms have a greater propensity to cause intestinal injury. Bacteria producing toxins associated with NEC include the clostridia (enterotoxins), and endotoxins from gram-negative bacteria have little direct cytotoxicity but may cause widespread tissue injury and systemic toxicity through activation of the inflammatory cascade. In summary, there is no evidence suggesting that NEC is the result of infection with a single organism, although certain organisms may be responsible for selected outbreaks of NEC.

Hoy et al. (2000) studied the duodenal microbial flora in VLBW neonates and NEC. Four hundred twenty-two duodenal aspirates were collected from 122 VLBW (<1,500 g) neonates. Fifty percent of samples contained *Enterobacteriaceae*, predominantly *Escherichia coli*, *Klebsiella*, and *Enterobacter* species, with counts of up to 10^8 CFU/g. During the study, duodenal samples were collected before 13 incidences of NEC. Seven of these yielded *Enterobacteriaceae*, of which five strains were also isolated from infants without NEC. There was no association between duodenal colonization with particular groups of *Enterobacteriaceae* and the development of NEC.

Boccia et al. (2001) reviewed 17 epidemics of NEC in the literature between 1973 and 1999. The average mortality rate was 6.25%, and the mean age at disease onset was 9.5 days. Most of the infants had low birth weights (mean, 1.395 kg). The main risk factors associated with NEC were low birth weight, low gestational age, low Apgar score, perinatal complications, hyaline membrane disease, and umbilical catherization. The bacteria involved often included *Enterobacteriaceae*, particularly *Escherichia coli*, *K. pneumoniae*, and *E. cloacae*. Several viruses were isolated from some of the epidemics, i.e., coronavirus, rotavirus, and enteroviruses, such as echovirus type 22. They reported that the causative role of clostridia was controversial, although the organism may be unreported due to the lack of anaerobic culturing of samples. Interestingly, de la Cochetiere et al. (2004) were the first to report the detection of *Clostridium perfringens* in the first 2 weeks of life for three neonates who later developed NEC. The organism was absent from nine control neonates.

Neonatal Infection through Parenteral and Enteral Feeding

Contaminated enteral nutrition can cause nosocomial bloodstream infections and may contribute to NEC. The topic of neonatal infections through contaminated formula feeding is covered in detail elsewhere in this book (chapter 4) and will not be reviewed here. Instead, this section will focus on the general topic of infections through contaminated enteral feeding.

Levy et al. (1989) surveyed 309 formula bottles over a 1-year period and found that 83 (27%) were contaminated. Unfortunately, the source of contamination remained unidentified; however, plasmid profiles of *E. cloacae* isolated from feeding solutions remained identical for several months, which may indicate a common source. Plasmid profiling, however, is not as discriminatory as PFGE, which would be applied to similar studies these days. The authors report that *E. cloacae* nosocomial sepsis cultures over a 7-year period (1979 to 1985) had plasmid profiles that linked them to contaminated enteral nutrition solutions. Two (out of seven) *K. pneumoniae* bacteremias over a 6-month period could also be linked to contaminated enteral feeds.

Enteral feeding tubes can act as loci for pathogenic bacteria. Mehall et al. (2002b) reported that 71 of 125 tubes from 50 neonates had bacterial counts of >1,000 CFU/ml. During the study, seven neonates developed NEC. All were fed formula contaminated with greater than 100,000 CFU of gram-negative bacteria/ml (see chapter 9 regarding microbiological criteria). Mehall and coworkers (2002a) also have shown that enteral feeding tubes can act as reservoirs for nosocomial antibiotic-resistant pathogens. Methicillin-resistant *Staphylococcus aureus* (MRSA) strains were cultured from enteral feeding tubes used in the same NICU. Twenty-three *Staphylococcus aureus* isolates were obtained, of which all 23 were MRSA. During this study there were four MRSA infections in neonates who were not being tube fed. However, the same MRSA strains also were found in the feeding tubes of other babies.

Enterobacteriaceae Causing Outbreaks in NICUs

E. sakazakii can cause septicemia and meningitis in preterm and full-term infants (Forsythe, 2005). The bacterium also has been associated with NEC in neonates (Van Acker et al., 2001) as well as infections in elderly immunocompromised individuals (Hawkins et al., 1991; Lai, 2001). Powdered infant formula and powdered milk have been epidemiologically implicated as sources of the pathogen. The organism will not be covered in any further detail; instead, the reader is directed to chapter 4.

While neonatal infections due to *E. sakazakii* have been the motivation for this book, the incidence of infections should be appreciated. Stool et al. (2004) reported the rates of *E. sakazakii* infections among VLBW infants by using culture data that came from the National Institute of Child Health and Human Development Neonatal Research Network. Only one *E. sakazakii* case (a breast-fed neonate) was identified from 10,660 neonates. These data suggest that outside of epidemic situations, *E. sakazakii* is very rare in VLBW infants. Between 1 September 1998 and 21 December 2001, 11,936 infants were born and/or cared for at 19 National Institute of Child Health and Human Development Neonatal Research Network centers. Analysis showed that 6,825 infants (out of 10,600) had one or more blood or cerebrospinal fluid (CSF) cultures performed after day 3 of life. Twenty percent of blood cultures (6,323 out of 31,777) and 6% of CSF cultures (314 out of 5,433) were positive for bacterial or fungal agents. One hundred thirty-five infants had a blood or CSF culture positive for *Enterobacter* species; 2% of those were evaluated with culture (135 out of 6,825), or 1% of all infants (135 out of 10,660). The most common *Enterobacter* species were *E. cloacae* (101 out of 135; 75%) and *E. aerogenes* (20 out of 135; 15%). Only one case of *E. sakazakii* septicemia (a breast-fed neonate) and no case of meningitis was identified among this cohort of 10,660 VLBW infants. Table 3 in chapter 4 summarizes the reported cases of *E. sakazakii* infections in neonates and infants. Although to the authors' knowledge no surveys of *E. sakazakii* incidence in the healthy neonatal intestinal tract have been published, a number of outbreak investigations have detected the organism in asymptomatic neonates.

E. cloacae is in Category B of the Food and Agriculture Organization-World Health Organization 2004 and 2006 descriptions of bacteria associated with neonatal infections, but epidemiological evidence is not available. It is commonly isolated from powdered infant formula, as well as being part of the normal human intestinal flora (Muytjens et al., 1988; Gewolb, 1999; Forsythe, 2005). It is associated with outbreaks in NICUs (Gaston, 1987), surgical wards (Burchard et al., 1986), and burn units (Markowitz et al., 1983). This bacterium emerged in the 1980s as an important cause of nosocomial infections and as a therapeutic problem due to multiple antibiotic resistance (Modi et al., 1987; Pitout et al., 1997; Sanders and Sanders, 1997). Plasmid-mediated extended-spectrum β-lactamases (ESBLs), first reported in the 1980s, have caused several outbreaks due to *Escherichia coli* and *K. pneumoniae*. *Enterobacter* species have been associated with several outbreaks due to either chromosomal β-lactamases or, less frequently, ESBLs (Cantón et al., 2002). In England, Wales, and Northern Ireland there has been an increase in bacteremia cases due to *Enterobacter*, as well as *Klebsiella*, *Serratia*, and *Citrobacter* species (Anonymous, 2004). From 2001 to 2003, *E. cloacae* cases rose from 1,207 to 1,762, and *E. sakazakii* cases rose from 53 to 67.

E. cloacae is genetically diverse, with at least five DNA relatedness groups, and it can be divided further into genetic clusters based on the housekeeping genes *hsp60*, *rpoB*, and *hemB* (Hoffman and Roggenkamp, 2003). Given the close relatedness with *E. sakazakii* (Farmer et al., 1980), it is not surprising that similar DNA fingerprinting methods have been applied to the analysis of strains from both *E. sakazakii* and *E. cloacae* NICU outbreaks. These methods vary in their discriminatory power. Methods include plasmid profiling, randomly amplified polymorphic DNA (RAPD) (arbitrarily primed PCR), enterobacterial repetitive intergenic consensus (ERIC)-PCR, PFGE, and ribotyping (Garaizar et al., 1991; Clementino et al., 2001; Fernández-Baca et al., 2001). Repetitive extragenic palindromic PCR (as opposed to ERIC-PCR) with PFGE confirmation has been proposed by Stumpf et al. (2005) for the analysis of *E. cloacae* outbreak strains.

Grattard et al. (1994) were the first to apply RAPD in an *E. cloacae* NICU outbreak involving six ventilated neonates. DNA typing showed that two clones could be isolated from neonates in the NICU and children from another ward. Shi et al. (1996) used plasmid profiling and RAPD to determine that, during a 4-month outbreak, strains were transmitted among infants via the hands of personnel. Peters et al. (2000) used ERIC-PCR to show that the infections were due to endogenous (intestinal) bacteria in one neonate and horizontal transmission between seven other neonates.

Hervas et al. (2001) undertook a retrospective study of neonatal sepsis and meningitis cases over a 22-year period and ribotyped *Enterobacter* strains isolated from 1995 to 1997. *K. pneumoniae* and *Staphylococcus epidermidis* were the most frequent late-onset infections from 1977 to 1991, whereas *Enterobacter* spp. were the most common isolates from 1992 to 1998. There were 45 *Enterobacter* sepsis cases between 1977 and 1998, 3 of which involved meningitis and 7 of which involved urinary tract infections. Twenty-six of the neonates were born preterm. *E. cloacae* caused the majority of the cases (93.3%), with the remainder being due to *E. aerogenes* (20% were polymicrobial). Five *E. cloacae* clones caused sepsis cases between 1995 and 1997, and they were eliminated following increased hygiene management. Fernández-Baca et al. (2001) used molecular epidemiology to study *E. cloacae* strains isolated from a NICU over a 3-year period. PFGE was the most discriminatory typing method compared to ribotyping, ERIC-PCR, and plasmid profiling. Strains were also characterized by antibiotic resistance patterns and the production of chromosomal β-lactamases, ESBLs, and inducible β-lactamases. Improved hygiene measures resulted in the sudden disappearance of *E. cloacae* and indicates a patient-to-patient transmission route of infection. Similarly, Talon et al. (2004) reported the disappearance of an infectious *E. cloacae* strain following improved hygienic practices. PFGE analysis showed that 10 out of 25 neonates were colonized by *E. cloacae* and that some strains were the same clone as that causing three cases of bacteremia. Van den Berg et al. (1999)

traced the source of an *E. cloacae* outbreak in a NICU to inadequately disinfected thermometers. Van Nierop et al. (1998) found the same pulsotype on the hands of a staff member as that in neonatal blood cultures, an amino acid perfusate, and environmental sources. Other pulsotypes of *E. cloacae* also were found. Drastic improvements in hygiene were required to control this outbreak.

Transmission between patients and wards was also reported for a NICU outbreak caused by *Serratia marcescens* (Miranda et al., 1996), in which only four biotypes were identified from 33 patients. One biotype accounted for 84% of the strains. The reservoir of *Serratia marcescens* was not identified. Neonatal outbreaks can also be caused by *K. pneumoniae*, *K. oxytoca,* and, less frequently, by *K. planticola* (Westbrook et al., 2000). Like *Enterobacter* species, these organisms can be isolated from powdered infant formula (Muytjens et al., 1988; Iversen and Forsythe, 2004) and form part of the endogenous intestinal tract flora of neonates (Blakey et al., 1982), and they may cause infections due to intestinal injury and spread via poor hygiene control.

Staphylococcus aureus and CoNS as Causative Agents of Neonatal Infections

In 2006, the Food and Agriculture Organization-World Health Organization added CoNS to Category C organisms (causality less plausible or not yet demonstrated). Whereas *Staphylococcus aureus* is well known as an opportunistic pathogen of adults that frequently colonizes the skin, CoNS cause the majority of nosocomial infections in preterm neonates. Although CoNS are common commensal organisms with little pathogenicity in immunocompetent hosts, premature neonates are particularly susceptible to invasive infection. CoNS adhere to skin, mucosal surfaces, or in-dwelling artificial devices, such as intravascular catheters and central nervous system shunts, which are commonly used in preterm infants. Adherence and biofilm formation of CoNS are facilitated by the production of capsular material composed of poly-*N*-succinyl glucosamine. Of the 13 species of CoNS that are able to colonize human skin, the species reported to cause disease in infants include *Staphylococcus epidermidis*, *Staphylococcus haemolyticus*, *Staphylococcus hominis*, *Staphylococcus warneri*, *Staphylococcus saprophyticus*, *Staphylococcus cohnii,* and *Staphylococcus capitis*. The major species involved in neonatal infection is *Staphylococcus epidermidis*, which causes 60 to 93% of CoNS bacteremias. Occasionally, CoNS are acquired from the mother at birth, but the majority of CoNS colonization is acquired nosocomially, predominantly from the hands of health care workers.

This chapter has reviewed our understanding of the neonatal immune status and intestinal bacterial flora. The neonate, especially the VLBW neonate, is more susceptible to bacterial infections due to the immature immune system. Shortly after birth, the neonate intestine begins to be colonized by bacteria from feeding, the caregiver, and the environment. This flora changes over time and is partially dependent upon the source of the feed. The intestinal flora can include members of the *Enterobacteriaceae*, including *Enterobacter* species. The application of molecular techniques to neonatal gut ecology will enable us to monitor this colonization process. Consequently, the extent to which intestinal organisms such as the opportunistic pathogen *E. sakazakii* can invade due to risk factors such as increased host susceptibility by immune immaturity, antibiotic therapy, or other intestinal trauma will be better understood.

REFERENCES

Adkins, B., C. Leclerc, and S. Marshall-Clarke. 2004. Neonatal adaptive immunity comes of age. *Nat. Rev. Immunol.* **4**:553–564.

Ahman, H., H. Kayhty, P. Tamminen, A. Vuorela, F. Malinoski, and J. Eskola. 1996. Pentavalent pneumococcal oligosaccharide conjugate vaccine PncCRM is well-tolerated and able to induce an antibody response in infant. *Pediatr. Infect. Dis. J.* **15**:134–139.

Akira, S., and K. Takeda. 2004. Toll-like receptor signalling. *Nat. Rev. Immunol.* **4**:499–511.

Anonymous. 2004. *Klebsiella*, *Enterobacter*, *Serratia*, and *Citrobacter* spp. bacteraemia, England, Wales, and Northern Ireland: 2003. *CDR Wkly.* **14**:1–6.

Arias, E., M. F. MacDorman, D. M. Strobino, and B. Guyer. 2003. Annual summary of vital statistics—2002. *Pediatrics* **112**:1215–1230.

Bach, J. F. 2002. The effect of infections on susceptibility to autoimmune and allergic diseases. *N. Engl. J. Med.* **347**:911–920.

Balmer, S. E., L. S. Hanvey, and B. A. Wharton. 1994. Diet and faecal flora in the newborn: nucleotides. *Arch. Dis. Child. Fetal Neonatal Ed.* **70**:F137–F140.

Bartosch, S., A. Fine, G. T. Macfarlane, and M. E. T. McMurdo. 2004. Characterization of bacterial communities in feces from healthy elderly volunteers and hospital elderly patients by using real-time PCR and effects on antibiotic treatment on the fecal microbiota. *Appl. Environ. Microbiol.* **70**:3572–3581.

Bell, M. J., J. L. Ternberg, and R. D. Feigin. 1978. Neonatal necrotizing enterocolitis therapeutic decisions based upon clinical staging. *Ann. Surg.* **187**:1–7.

Bezirtzoglou, E. 1997. The intestinal microflora during the first weeks of life. *Anaerobe* **3**:173–177.

Blakey, J. L., L. Lubitz, G. L. Barnes, R. F. Bishop, N. T. Campbell, and G. L. Gilliam. 1982. Development of gut colonization in pre-term neonates. *J. Med. Microbiol.* **15**:519–529.

Blanco-Quiros, A., E. Arranz, et al. 2000. Cord blood interleukin-10 levels are increased in preterm newborns. *Eur. J. Pediatr.* **159**:420–423.

Block, C., O. Peleg, N. Minster, B. Bar-Oz, A. Simhon, I. Arad, and M. Shapiro. 2002. Cluster of neonatal infections in Jerusalem due to unusual biochemical variant of *Enterobacter sakazakii*. *Eur. J. Clin. Microbiol. Infect. Dis.* **21**:613–616.

Brown, A. P., S. J. Burakoff, and A. Sher. 1980. Specificity of alloreactive T lymphocytes that adhere to lung stage schistosomula of *Schistosoma mansoni. J. Immunol.* **124:**2516–2518.

Boccia, D., I. Stolfi, S. Lana, and M. L. Moro. 2001. Nosocomial necrotizing enterocolitis outbreaks: epidemiology and control measures. *Eur. J. Pediatr.* **160:**385–391.

Buhrer, C., J. Graulich, D. Stibenz, J. W. Dudenhausen, and M. Obladen. 1994. L-Selectin is down-regulated in umbilical cord blood granulocytes and monocytes of newborn infants with acute bacterial infection. *Pediatr. Res.* **36:**799–804.

Burchard, K. W., D. T. Barrall, M. Reed, and G. J. Stotman. 1986. *Enterobacter cloacae* bacteremia in surgical patients. *Surgery* **100:**857–862.

Burns, A. J., and I. R. Rowland. 2000. Anti-carcinogenicity of probiotics and prebiotics. *Curr. Issues Intest. Microbiol.* **1:**13–24.

Cantón, R., A. Oliver, T. M. Coque, M. del Carmen Varela, J. Pérez-Díaz, and F. Baquero. 2002. Epidemiology of extended-spectrum β-lactamase-producing *Enterobacter* isolates in a Spanish hospital during a 12-year period. *J. Clin. Microbiol.* **40:**1237–1243.

Carneiro, L. A. M., M. L. P. Queiroz, and V. L. C. Merquior. 2004. Antimicrobial-resistance and enterotoxin-encoding genes among staphylococci isolated from expressed human breast milk. *J. Med. Microbiol.* **53:**761–768.

Cerbulo-Vazquez, A., R. Ramos, and L. Santos-Argumedo. 2003. Activated umbilical cord blood cells from pre-term and term neonates express CD69 and synthesize IL-2 but are unable to produce IFN-γ. *Arch. Med. Res.* **34:**100–105.

Chelvarajan, R. L., S. M. Collins, I. E. Doubinskaia, S. Goes, J. Van Willigen, D. Flanagan, W. J. De Villiers, J. S. Bryson, and S. Bondada. 2004. Defective macrophage function in neonates and its impact on unresponsiveness of neonates to polysaccharide antigens. *J. Leukoc. Biol.* **75:**982–994.

Chen, T.-L., P.-F. Thien, S.-C. Liaw, C.-P. Fung, and L. K. Siu. 2005. First report of *Salmonella enterica* serotype Panama meningitis associated with consumption of contaminated breast milk by a neonate. *J. Clin. Microbiol.* **43:**5400–5402.

Christensen, R. D., and G. Rothstein. 1980. Exhaustion of mature marrow neutrophils in neonates with sepsis. *J. Pediatr.* **96:**316–318.

Christensen, R. D., J. L. MacFarlane, N. L. Taylor, H. R. Hill, and G. Rothstein. 1982. Blood and marrow neutrophils during experimental group B streptococcal infection: quantification of the stem cell, proliferative, storage and circulating pools. *Pediatr. Res.* **16:**549–553.

Christensen, R. D., H. R. Hill, H. B. Anstall, and G. Rothstein. 1984. Exchange transfusion as an alternative to granulocyte concentrate administration in neonates with bacterial sepsis and profound neutropenia. *J. Clin. Apher.* **2:**177–183.

Chu, S. W., and W. A. Walker. 1989. Development of the mucosal barrier: bacterial toxin interaction with the immature enterocyte. *Immunol. Investig.* **18:**405–416.

Clarridge, J. E., III. 2004. Impact of 16S rRNA gene sequence analysis for identification of bacteria on clinical microbiology and infectious diseases. *Clin. Microbiol. Rev.* **17:**840–862.

Claud, E. C., and W. A. Walker. 2001. Hypothesis: inappropriate colonization of the premature intestine can cause neonatal necrotizing enterocolitis. *FASEB J.* **15:**1398–1403.

Clementino, M. M., I. de Filippis, C. R. Nascimento, R. Branquinho, C. L. Rocha, and O. B. Martins. 2001. PCR analysis of tRNA intergeneric spacer, 16S-23S internal transcribed spacer, and randomly amplified polymorphic DNA reveal inter- and intraspecific relationships of *Enterobacter cloacae* strains. *J. Clin. Microbiol.* **39**:3865–3870.

Das, R. M., P. F. Holt, and M. C. Horne. 1977. The formation of asbestos bodies. *Med. Lav.* **68**:431–436.

de la Cochetiere, M. F., H. Piloquet, C. des Robert, D. Darmaun, J. P. Galmiche, and J. C. Roze. 2004. Early intestinal bacterial colonization and necrotizing enterocolitis in premature infants: the putative role of *Clostridium. Pediatr. Res.* **56**:366–370.

Drasar, B. S., M. Shiner, and G. M. McLeod. 1969. Studies on the intestinal flora. 1. The bacterial flora of the gastrointestinal tract in healthy and achlorhydric persons. *Gastroenterology* **56**:71–79.

Duffy, L. C. 2000. Interactions mediating bacterial translocation in the immature intestine. *J. Nutr.* **130**:432S–436S.

Edwards, C. A., and A. M. Parrett. 2002. Intestinal flora during the first months of life: new perspectives. *Br. J. Nutr.* **88**:S11–S18.

Egert, M., A. A. de Graaf, H. Smidt, W. M. de Vos, and K. Venema. 2006. Beyond diversity: functional microbiomics of the human colon. *Trends Microbiol.* **14**:86–91.

Ehlers, S., and K. A. Smith. 1991. Differentiation of T cell lymphokine gene expression: the in vitro acquisition of T cell memory. *J. Exp. Med.* **173**:25–36.

Eisenfeld, L., P. J. Krause, V. Herson, J. Savidakis, P. Bannon, E. Maderazo, C. Woronick, C. Giuliano, and L. Banco. 1990. Longitudinal study of neutrophil adherence and motility. *J. Pediatr.* **117**:926–929.

Emori, T. G., and R. P. Gaynes. 1993. An overview of nosocomial infections, including the role of the microbiology laboratory. *Clin. Microbiol. Rev.* **6**:428–42.

Evans, N. J., and N. Rutter. 1986. Development of the epidermis in the newborn. *Biol. Neonate* **49**:74–80.

Faix, R. G., and J. T. Adams. 1994. Neonatal necrotizing enterocolitis: current concepts and controversies. *Adv. Pediatr. Infect. Dis.* **9**:1–36.

Farmer, J. J., II, M. A. Asbury, F. W. Hickman, D. J. Brenner, and The *Enterobacteriaceae* Study Group. 1980. *Enterobacter sakazakii*: a new species of "*Enterobacteriaceae*" isolated from clinical specimens. *Int. J. Syst. Bacteriol.* **30**:569–584.

Favier, C. F., E. E. Vaughan, W. M. De Vos, and A. D. L. Akkermans. 2002. Molecular monitoring of succession of bacterial communities in human neonates. *Appl. Environ. Microbiol.* **68**:219–226.

Fernández-Baca, V., F. Ballesteros, J. A. Hervás, P. Villalón, M. A. Domínguez, V. J. Benedí, and S. Albertí. 2001. Molecular epidemiological typing of *Enterobacter cloacae* isolates from a neonatal intensive care unit: three-year prospective study. *J. Hosp. Infect.* **49**:173–182.

Fishman, P. H., and E. E. Atikkan. 1980. Mechanism of action of cholera toxin: effect of receptor density and multivalent binding on activation of adenylate cyclase. *J. Membr. Biol.* **54**:51–60.

Flesch, I. E., and S. H. Kaufmann. 1994. Role of macrophages and alpha beta T lymphocytes in early interleukin 10 production during *Listeria monocytogenes* infection. *Int. Immunol.* **6**:463–468.

Food and Agriculture Organization-World Health Organization (FAO-WHO). 2004. *Enterobacter sakazakii* and other microorganisms in powdered infant formula: meeting report. *Microbiological risk assessment series 6.* World Health Organization-Food and Agriculture Organization of the United Nations, Geneva and Rome. WHO Press, Geneva, Switzerland. http://www.who.int/foodsafety/publications/micro/mra6/en/index.html.

Food and Agriculture Organization-World Health Organization (FAO-WHO). 2006. *Enterobacter sakazakii* and *Salmonella* in powdered infant formula: meeting report. *Microbiological risk assessment series 10.* World Health Organization-Food and Agriculture Organization of the United Nations, Geneva and Rome. WHO Press, Geneva, Switzerland. http://www.who.int/foodsafety/publications/micro/mra10/en/index.html.

Forsythe, S. J. 2005. *Enterobacter sakazakii* and other bacteria in powdered infant milk formula. *Matern. Child Nutr.* **1**:44–50.

Gall, L. S. 1970. Normal fecal flora of man. *Am. J. Clin. Nutr.* **23**:1457–1465.

Garaizar, J., M. E. Kaufmann, and T. L. Pitt. 1991. Comparison of ribotyping with conventional methods for the type identification of *Enterobacter cloacae. J. Clin. Microbiol.* **29**:1303–1307.

Garty, B. Z., A. Ludomirsky, Y. L. Danon, J. B. Peter, and S. D. Douglas. 1994. Placental transfer of immunoglobulin G subclasses. *Clin. Diagn. Lab. Immunol.* **1**:667–669.

Gasparoni, A., L. Ciardelli, A. Avanzini, A. M. Castellazzi, R. Carini, G. Rondini, and G. Chirico. 2003. Age-related changes in intracellular TH1/TH2 cytokine production, immunoproliferative T lymphocyte response and natural killer cell activity in newborns, children and adults. *Biol. Neonate* **84**:297–303.

Gaston, M. A. 1987. Isolation and selection of a bacteriophage-typing set for *Enterobacter cloacae. J. Med. Microbiol.* **24**:285–290.

Gewolb, I. H. 1999. Stool microflora in extremely low birthweight infants. *Arch. Dis. Child. Fetal Neonatal Ed.* **80**:F167–F173.

Grattard, F., B. Pozzetto, P. Berthelot, I. Rayet, A. Ros, B. Lauras, and O. G. Gaudin. 1994. Arbitarily primed PCR, ribotyping, and plasmid pattern analysis applied to investigation of a nosocomial outbreak due to *Enterobacter cloacae* in a neonatal intensive care unit. *J. Clin. Microbiol.* **32**:596–602.

Griffioen, A. W., E. A. Toebes, B. J. Zegers, and G. T. Rijkers. 1992. Role of CR2 in the human adult and neonatal in vitro antibody response to type 4 pneumococcal polysaccharide. *Cell Immunol.* **143**:11–22.

Haenel, H. 1970. Human normal and abnormal gastrointestinal flora. *Am. J. Clin. Nutr.* **23**:1433–1439.

Hanson, L. A., S. Ahlstedt, B. Andersson, B. Carlsson, S. P. Fallstrom, L. Mellander, O. Porras, T. Soderstrom, and C. S. Eden. 1985. Protective factors in milk and the development of the immune system. *Pediatrics* **75**:172–176.

Hariharan, D., W. Ho, J. Cutilli, D. E. Campbell, and S. D. Douglas. 2000. C-C chemokine profile of cord blood mononuclear cells: selective defect in RANTES production. *Blood* **95**:715–718.

Harmsen, H. J. M., A. C. M. Wideboer-Veloo, G. C. Raangs, A. A. Wagendorp, N. Klijn, J. G. Bindels, and G. W. Welling. 2000. Analysis of intestinal flora development in breast-fed and formula-fed infants by using molecular identification and detection methods. *J. Pediatr. Gastr. Nutr.* **30**:61–67.

Harris, M. C., M. Shalit, and F. S. Southwick. 1993. Diminished actin polymerization by neutrophils from newborn infants. *Pediatr. Res.* **33**:27–31.

Hasegawa, T., K. Sakurai, Y. Kambayashi, A. R. Saniabadi, H. Nagamoto, K. Tsukada, A. Takahashi, H. Kuwano, and M. Nakano. 2003. Effects of OPC-6535 on lipopolysaccharide-induced acute liver injury in the rat: involvement of superoxide and tumor necrosis factor-alpha from hepatic macrophages. *Surgery* **134**:818–826.

Hawkins, R. E., C. R. Lissner, and J. P. Sanford. 1991. *Enterobacter sakazakii* bacteremia in an adult. *South. Med. J.* **84**:793–795.

Hayward, A. R. 1983. The human fetus and newborn: development of the immune response. *Birth Defects Orig. Artic. Ser.* **19**:289–294.

Heavey, P. M., and I. R. Rowland. 1999. The gut microflora of the developing infant: microbiology and metabolism. *Microb. Ecol. Health Dis.* **11**:75–83.

Heilig, H. G., E. G. Zoetendal, E. E. Vaughan, P. Marteau, A. D. Akkermans, and W. M. de Vos. 2002. Molecular diversity of *Lactobacillus* spp. and other lactic acid bacteria in the human intestine as determined by specific amplification of 16S ribosomal DNA. *Appl. Environ. Microbiol.* **68**:114–123.

Hervas, J. A., F. Ballesteros, A. Alomar, J. Gil, V. J. Benedi, and S. Alberti. 2001. Increase of *Enterobacter* in neonatal sepsis: a twenty-two-year study. *Pediatr. Infect. Dis. J.* **20**:134–140.

Hill, V. T., S. M. Townsend, R. S. Arias, J. M. Jenabi, I. Gomez-Gonzalez, H. Shimada, and J. L. Badger. 2004. TraJ-dependent *Escherichia coli* K1 interactions with professional phagocytes are important for early systemic dissemination of infection in the neonatal rat. *Infect. Immun.* **72**:478–488.

Hirsch, C. S., R. Hussain, Z. Toossi, G. Dawood, F. Shahid, and J. J. Ellner. 1996. Cross-modulation by transforming growth factor beta in human tuberculosis: suppression of antigen-driven blastogenesis and interferon gamma production. *Proc. Natl. Acad. Sci. USA* **93**:3193–3198.

Hoffmann, H., and A. Roggenkamp. 2003. Population genetics of the nomenspecies *Enterobacter cloacae. Appl. Environ. Microbiol.* **69**:5306–5318.

Holdeman, L. V., E. P. Cato, and W. E. C. Moore (ed.). 1977. *Anaerobe Laboratory Manual*, 4th ed. Anaerobe Laboratory, Virginia Polytechnic Institute and State University, Blacksburg, VA.

Hooper, L. V., M. H. Wong, A. Thelin, L. Hansson, P. G. Falk, and J. I. Gordon. 2001. Molecular analysis of commensal host-microbial relationships in the intestine. *Science* **291**:881–884.

Hopkins, M. J., G. T. Macfarlane, E. Furrie, A. Fite, and S. Macfarlane. 2005. Characterisation of intestinal bacteria in infant stools using real-time PCR and northern hybridization analyses. *FEMS Microbiol. Ecol.* **54**:77–85.

Hoy, C. M., C. M. Wood, P. M. Hawkey, and J. W. L. Puntis. 2000. Duodenal microflora in very-low-birth-weight neonates and relation to necrotizing enterocolitis. *J. Clin. Microbiol.* **68**:4539–4547.

Iversen, C., and S. Forsythe. 2004. Isolation of *Enterobacter sakazakii* and other *Enterobacteriaceae* from powdered infant formula milk and related products. *Food Microbiol.* **21**:771–777.

Joyner, J. L., N. H. Augustine, K. A. Taylor, T. R. La Pine, and H. R. Hill. 2000. Effects of group B streptococci on cord and adult mononuclear cell interleukin-12 and interferon-gamma mRNA accumulation and protein secretion. *J. Infect Dis.* **182**:974–977.

Kassim, O. O., D. H. Raphael, A. K. Ako-Nai, O. Taiwo, S. E. Torimiro, and O. O. Afolabi. 1989. Class-specific antibodies to *Bordetella pertussis, Haemophilus influenzae* type b, *Streptococcus pneumoniae* and *Neisseria meningitidis* in human breast-milk and maternal-infant sera. *Ann. Trop. Paediatr.* **9**:226–232.

Kaufman, D., and K. D. Fairchild. 2004. Clinical microbiology of bacterial and fungal sepsis in very-low-birth-weight infants. *Clin. Microbiol. Rev.* **17**:638–680.

Kaufmann, S. H., H. R. Rodewald, E. Hug, and G. De Libero. 1988. Cloned *Listeria monocytogenes* specific non-MHC-restricted Lyt-2$^+$ T cells with cytolytic and protective activity. *J. Immunol.* **140**:3173–3179.

Kien, C. L., E. A. Liechty, D. Z. Myerberg, and M. D. Mullett. 1987. Dietary carbohydrate assimilation in the premature infant: evidence for a nutritionally significant bacterial ecosystem in the colon. *Am. J. Clin. Nutr.* **46**:456–460.

Kirjavainen, P. V., and G. R. Gibson. 1999. Healthy gut microflora and allergy: factors influencing development of the microbiota. *Ann. Med.* **31**:288–292.

Kotiranta-Ainamo, A., J. Rautonen, and N. Rautonen. 2004. Imbalanced cytokine secretion in newborns. *Biol. Neonate* **85**:55–60.

Lai, K. K. 2001. *Enterobacter sakazakii* infections among neonates, infants, children, and adults. Case reports and a review of the literature. *Medicine* (Baltimore) **80**:113–122.

La Pine, T. R., J. L. Joyner, N. H. Augustine, S. D. Kwak, and H. R. Hill. 2003. Defective production of IL-18 and IL-12 by cord blood mononuclear cells influences the T helper-1 interferon gamma response to group B streptococci. *Pediatr. Res.* **54**:276–281.

Larson, E. L., J. P. Cimiotti, J. Haas, M. Nesin, A. Allen, P. Della-Latta, and L. Saiman. 2005. Gram-negative bacilli associated with catheter-associated and non-catheter-associated bloodstream infections and hand carriage by healthcare workers in neonatal intensive care units. *Pediatr. Crit. Care Med.* **6**:457–461.

Lee, N., S. Chen, R. Tang, and B. Hwang. 2004. Neonatal bacteremia in a neonatal intensive care unit: analysis of causative organisms and antimicrobial susceptibility. *J. Chin. Med. Assoc.* **67**:15–20.

Levy, J., Y. Van Leathem, G. Verhaegen, C. Perpete, J. P. Butzler, and R. P. Wenzel. 1989. Contaminated enteral nutrition solutions as a cause of nosocomial bloodstream infection: a study using plasmid fingerprinting. *J. Parenter. Enteral. Nutr.* **13**:228–234.

Lucas, A., and T. J. Cole. 1990. Breastmilk and neonatal necrotizing enterocolitis. *Lancet* **336**:1519–1523.

MacKendrick, W., and M. Caplan. 1993. Necrotizing enterocolitis: new thoughts about pathogenesis and potential treatments. *Pediatr. Clin. N. Am.* **40**:1047–1059.

Mallow, E. B., A. Harris, N. Salzman, J. P. Russell, R. J. DeBerardinis, E. Ruchelli, and C. L. Bevins. 1996. Human enteric defensins. Gene structure and developmental expression. *J. Biol. Chem.* **271**:4038–4045.

Markowitz, S. M., S. M. Smith, and D. S. Williams. 1983. Retrospective analysis of plasmid patterns in a study of burn unit outbreaks of infection due to *Enterobacter cloacae. J. Infect. Dis.* **148**:18–23.

Marodi, L. 2006. Innate cellular immune responses in newborns. *Clin. Immunol.* **118**:137–144.

Marodi, L., K. Goda, A. Palicz, and G. Szabo. 2001. Cytokine receptor signalling in neonatal macrophages: defective STAT-1 phosphorylation in response to stimulation with IFN-γ. *Clin. Exp. Immunol.* **126**:456–460.

Marodi, L., R. Kaposzta, and E. Nemes. 2000. Survival of group B streptococcus type III in mononuclear phagocytes: differential regulation of bacterial killing in cord macrophages by human recombinant gamma interferon and granulocyte-macrophage colony-stimulating factor. *Infect. Immun.* **68**:2167–2170.

Mehall, J. R., C. A. Kite, C. H. Gilliam, R. J. Jackson, and S. D. Smith. 2002a. Enteral feeding tubes are a reservoir for nosocomial antibiotic-resistant pathogens. *J. Pediatr. Surg.* **37**:1011–1012.

Mehall, J. R., C. A. Kite, D. A. Saltzman, T. Wallett, R. J. Jackson, and S. D. Smith. 2002b. Prospective study of the incidence and complications of bacterial contamination of enteral feeding in neonates. *J. Pediatr. Surg.* **37**:1177–1182.

Miranda, G., C. Kelly, F. Solorano, B. Leans, R. Coria, and J. E. Patterson. 1996. Use of pulsed-field gel electrophoresis typing to study an outbreak of infection due to *Serratia marcescens* in a neonatal intensive care unit. *J. Clin. Microbiol.* **34**:3138–3141.

Miyano, A., T. Miyamichi, M. Nakayama, H. Kitajima, and A. Shimizu. 1996. Effect of chorioamnionitis on the levels of serum proteins in the cord blood of premature infants. *Arch. Pathol. Lab. Med.* **120**:245–248.

Modi, N., V. Damjanovic, and R. W. I. Cooke. 1987. Outbreak of cephalosporin resistant *Enterobacter cloacae* infection in a neonatal intensive care unit. *Arch. Dis. Child.* **62**:148–151.

Muytjens, H. L., H. Roelofs-Willemse, and G. H. Jaspar. 1988. Quality of powdered substitutes for breast milk with regard to members of the family *Enterobacteriaceae*. *J. Clin. Microbiol.* **26**:743–746.

Muyzer, G. 1999. DGGE/TGGE a method for identifying genes from natural ecosystems. *Curr. Opin. Microbiol.* **2**:317–322.

Muyzer, G., and K. Smalla. 1998. Application of denaturing gradient gel electrophoresis (DGGE) and temperature gradient gel electrophoresis (TGGE) in microbial ecology. *Antonie Leeuwenhoek* **73**:127–141.

Muyzer, G., E. C. De Waal, and G. A. Uitterlinden. 1993. Profiling of complex populations by denaturing gradient gel electrophoresis analysis of polymerase chain reaction-amplified genes coding for 16S rRNA. *Appl. Environ. Microbiol.* **59**:695–700.

Neish, A. S., A. T. Gewirtz, H. Zeng, A. N. Young, M. E. Hobert, V. Karmali, A. S. Rao, and J. L. Madara. 2000. Prokaryotic regulation of epithelial responses by inhibition of IkB-alpha ubiquitination. *Science* **289**:1560–1563.

Nomura, H., E. Sato, S. Koyama, M. Haniuda, K. Kubo, S. Nagai, and T. Izumi. 2001. Histamine stimulates alveolar macrophages to release neutrophil and monocyte chemotactic activity. *J. Lab. Clin. Med.* **138**:226–235.

Oddy, W. H. 2002. The impact of breastmilk on infant and child health. *Breastfeeding Rev.* **10**:5–18.

Pahwa, S. G., R. Pahwa, E. Grimes, and E. Smithwick. 1977. Cellular and humoral components of monocyte and neutrophil chemotaxis in cord blood. *Pediatr. Res.* **11**:677–680.

Penders, J., C. Vink, C. Dressen, N. London, C. Thijs, and E. E. Stobbergh. 2005. Quantification of *Bifidobacterium* spp., *Escherichia coli* and *Clostridium difficile* in faecal samples of breast-fed and formula-fed infants by real-time PCR. *FEMS Microbiol. Lett.* **243**:141–147.

Peters, S., J. Bryan, and M. Cole. 2000. Enterobacterial repetitive intergenic consensus polymerase chain reaction typing of isolates of *Enterobacter claocae* from an outbreak of infection in a neonatal intensive care unit. *Am. J. Infect. Control* **28**:123–129.

Pitout, J. D. D., E. S. Moland, K. S. Thomson, C. C. Sanders, and S. R. Fitzsimmons. 1997. β-lactamases and detection of β- resistance in *Enterobacter* spp. *Antimicrob. Agents Chemother.* **41**:35–39.

Poltorak, A., P. Ricciardi Ricciardi-Castagnoli, S. Citterio, and B. Beutler. 2000. Physical contact between lipopolysaccharide and toll-like receptor 4 revealed by genetic complementation. *Proc. Natl. Acad. Sci. USA* **97**:2163–2167.

Porcu, P., J. Gaddy, and H. E. Broxmeyer. 1998. Alloantigen-induced unresponsiveness in cord blood T lymphocytes is associated with defective activation of Ras. *Proc. Natl. Acad. Sci. USA* **95**:4538–4543.

Reina, J., F. Parras, J. Gil, F. Salva, and P. Alomar. 1989. Human infections caused by *Enterobacter sakazakii*. Microbiologic considerations. *Enferm. Infecc. Microbiol. Clin.* **7**:147–150.

Reuter, G. 2001. The *Lactobacillus* and *Bifidobacterium* microflora of the human intestine: composition and succession. *Curr. Issues Intest. Microbiol.* **2**:43–53.

Ribeiro-do-Couto, L. M., L. C. Boeije, J. S. Kroon, B. Hooibrink, B. S. Breur-Vriesendorp, L. A. Aarden, and C. J. Boog. 2001. High IL-13 production by human neonatal T cells: neonate immune system regulator? *Eur. J. Immunol.* **31**:3394–3402.

Rognum, T. O., S. Thrane, L. Stoltenberg, A. Vege, and P. Brandtzaeg. 1992. Development of intestinal mucosal immunity in fetal life and the first postnatal months. *Pediatr. Res.* **32**:145–149.

Rosberg-Cody, E., R. P. Ross, S. Hussey, C. A. Ryan, B. P. Murphy, G. F. Fizgerald, R. Devery, and C. Stanton. 2004. Mining the microbiota of the neonatal gastrointestinal tract for conjugated linoleic acid-producing bifidobacteria. *Appl. Environ. Microbiol.* **70**:4635–4641.

Rubaltelli, F. F., R. Biadaioli, P. Pecile, and P. Nicoletti. 1998. Intestinal flora in breast- and bottle-fed infants. *J. Perinatol. Med.* **26**:186–91.

Ruderman, J. W., J. B. Peter, R. C. Gall, M. E. Stewart, J. J. Pomerance, and E. R. Stiehm. 1990. Prevention of hypogammaglobulinemia of prematurity with intravenous immune globulin. *J. Perinatol.* **10**:150–155.

Rueda, R., J. L. Sabatel, J. Maldonado, J. A. Molina, and A. Gil. 1998. Addition of gangliosides to an adapted milk formula modifies levels of fecal *Escherichia coli* in preterm newborn infants. *J. Pediatr.* **133**:90–94.

Salyers, A. 1979. Energy sources of major intestinal fermentative anaerobes. *Am. J. Clin. Nutr.* **32**:158–163.

Sanders, W. E., Jr., and C. Sanders. 1997. *Enterobacter* spp.: pathogens poised to flourish at the turn of the century. *Clin. Microbiol. Rev.* **10**:220–241.

Satokari, R. M., E. E. Vaughan, A. D. Akkermans, M. Saarela, and W. M. de Vos. 2001. Bifidobacterial diversity in human feces detected by genus-specific PCR and denaturing gradient gel electrophoresis. *Appl. Environ. Microbiol.* **67**:504–513.

Savage, D. C. 2001. Microbial biota of the human intestine: a tribute to some pioneering scientists. *Curr. Issues Intest. Microbiol.* **2**:1–15.

Schelonka, R. L. and A. J. Infante. 1998. Neonatal immunology. *Semin. Perinatol.* **22**:2–14.

Schelonka, R. L., B. A. Yoder, S. E. desJardins, R. B. Hall, and J. Butler. 1994. Peripheral leukocyte count and leukocyte indexes in healthy newborn term infants. *J. Pediatr.* **125**:603–606.

Schmitz, J., M. Assenmacher, and A. Radbruch. 1993. Regulation of T helper cell cytokine expression: functional dichotomy of antigen-presenting cells. *Eur. J. Immunol.* **23**:191–199.

Schuit, K. E., and D. A. Powell. 1980. Phagocytic dysfunction in monocytes of normal newborn infants. *Pediatrics* **65**:501–504.

Schwiertz, A., B. Gruhl, M. Lobnitz, P. Michel, M. Radke, and M. Blaut. 2003. Development of the intestinal bacterial composition in hospitalized preterm infants in comparison with breast-fed, full-term infants. *Pediatr. Res.* **54**:393–399.

Shah, U., and W. A. Walker. 2000. Adverse host responses to bacterial toxins in human infants. *J. Nutr.* **130**:420S–425S.

Sher, A., R. T. Gazzinelli, I. P. Oswald, M. Clerici, M. Kullberg, E. J. Pearce, J. A. Berzofsky, T. R. Mosmann, S. L. James, and H. C. Morse III. 1992. Role of T-cell derived cytokines in the downregulation of immune responses in parasitic and retroviral infection. *Immunol. Rev.* **127**:183–204.

Shi, Z.-H., Y.-F. Liu, Y.-J. Lau, Y.-H. Lin, and B.-S. Hu. 1996. Epidemiological typing of isolates from an outbreak of infection with multidrug-resistant *Enterobacter cloacae* by repetitive extragenic palindromic unit b1-primed PCR and pulsed-field gel electrophoresis. *J. Clin. Microbiol.* **34**:2784–2790.

Shiokawa, S., F. Mortari, J. O. Lima, C. Nunez, F. E. Bertrand III, P. M. Kirkham, S. Zhu, A. P. Dasanayake, and H. W. Schroeder, Jr. 1999. IgM heavy chain complementarity-determining region 3 diversity is constrained by genetic and somatic mechanisms until two months after birth. *J. Immunol.* **162**:6060–6070.

Smith, C. J., and M. P. Bryant. 1979. Introduction to metabolic activities of intestinal bacteria. *Am. J. Clin. Nutr.* **32**:149–157.

Spencer, A. J., M. P. Osborne, S. J. Haddon, J. Collins, W. G. Starkey, D. C. Candy, and J. Stephen. 1990. X-ray microanalysis of rotavirus-infected mouse intestine: a new concept of diarrhoeal secretion. *J. Pediatr. Gastroenterol. Nutr.* **10**:516–529.

Stoll, B. J., and N. Hansen. 2003. Infections in VLBW infants: studies from the NICHD Neonatal Research Network. *Semin. Perinatol.* **27**:293–301.

Stool, B. J., N. Hansen, A. A. Fanaroff, and J. A. Lemons. 2004. *Enterobacter sakazakii* is a rare cause of neonatal septicemia or meningitis in VLBW infants. *J. Pediatr.* **144**:821–823.

Stumpf, A. N., A. Roggenkamp, and H. Hoffmann. 2005. Specificity of enterobacterial repetitive intergenic consensus and repetitive extragenic palindromic polymerase chain reaction for the detection of clonality within the *Enterobacter cloacae* complex. *Diagn. Microbiol. Infect. Dis.* **53**:9–16.

Tallant, T., A. Deb, N. Kar, J. Lupica, M. J. de Veer, and J. A. DiDonato. 2004. Flagellin acting via TLR5 is the major activator of key signaling pathways leading to NF-κB and proinflammatory gene program activation in intestinal epithelial cells. *BMC Microbiol.* **4**:33.

Talon, D., P. Menget, M. Thouverez, G. Thiriez, H. G. Haore, C. Fromentin, A. Muller, and X. Bertrand. 2004. Emergence of *Enterobacter cloacae* as a common pathogen in neonatal units: pulsed-field gel electrophoresis analysis. *J. Hosp. Infect.* **57:**119–125.

Taylor, S., and Y. J. Bryson. 1985. Impaired production of gamma-interferon by newborn cells in vitro is due to a functionally immature macrophage. *J. Immunol.* **134:**1493–1497.

Thomas, E. D., R. E. Ramberg, G. E. Sale, R. S. Sparkes, and D. W. Golde. 1976. Direct evidence for a bone marrow origin of the alveolar macrophage in man. *Science* **192:** 1016–1018.

Townsend, S., J. Caubilla-Barron, C. Loc-Carrillo, and S. Forsythe. 2007. The presence of endotoxin in powdered infant formula milk and the influence of endotoxin and *Enterobacter sakazakii* on bacterial translocation in the infant rat. *Food Microbiol.* **24:**67–74.

Townsend, S. M., H. A. Pollack, I. Gonzalez-Gomez, H. Shimada, and J. L. Badger. 2003. *Citrobacter koseri* brain abscess in the neonatal rat: survival and replication within human and rat macrophages. *Infect. Immun.* **71:**5871–5880.

Trinchieri, G. 1997. Cytokines acting on or secreted by macrophages during intracellular infection (IL-10, IL-12, IFN-γ). *Curr. Opin. Immunol.* **9:**17–23.

Udall, J. N., Jr. 1990. Gastrointestinal host defense and necrotizing enterocolitis. *J. Pediatr.* **117:**S33–S43.

Van Acker, J., F. De Smet, G. Muyldermans, A. Bougatef, A. Naessens, and S. Lauwers. 2001. Outbreak of necrotizing enterocolitis associated with *Enterobacter sakazakii* in powdered milk formula. *J. Clin. Microbiol.* **39:**293–297.

Van den Berg, R. W. A., H. L. Claahsen, M. Niessen, H. L. Muytjens, K. Liem, and A. Voss. 1999. *Enterobacter cloacae* outbreak in the NICU related to disinfected thermometers. *J. Hosp. Infect.* **45:**29–34.

Van Nierop, W. H., A. G. Duse, R. G. Stewart, Y. R. Bilgeri, and H. J. Koornhof. 1998. Molecular epidemiology of an outbreak of *Enterobacter cloacae* in the neonatal intensive care unit of a provincial hospital in Gauteng, South Africa. *J. Clin. Microbiol.* **36:**3085–3087.

Vaughan, E. E., F. Schut, H. G. H. J. Heilig, E. G. Zoetendal, W. M. de Vos, and A. D. L. Akkermans. 2000. A molecular view of the intestinal ecosystem. *Curr. Issues Intest. Microbiol.* **1:**1–12.

Verreck, F. A., T. de Boer, D. M. Langenberg, M. A. Hoeve, M. Kramer, E. Vaisberg, R. Kastelein, A. Kolk, R. de Waal-Malefyt, and T. H. Ottenhoff. 2004. Human IL-23-producing type 1 macrophages promote but IL-10-producing type 2 macrophages subvert immunity to (myco)bacteria. *Proc. Natl. Acad. Sci. USA* **101:**4560–4565.

Walker, V., G. A. Mills, M. A. Hall, and J. A. Lowes. 1989. Carbohydrate fermentation by gut microflora in preterm neonates. *Arch. Dis. Childhood* **64:**1367–1373.

Walsh, M. C., and R. M. Kliegman. 1986. Necrotizing enterocolitis: treatment based on staging criteria. *Pediatr. Clin. N. Am.* **33:**179–201.

Walter, J., C. Hertel, G. W. Tannock, C. M. Lis, K. Munro, and W. P. Hammes. 2001. Detection of *Lactobacillus*, *Pediococcus*, *Leuconostoc*, and *Weissella* species in human feces by using group-specific PCR primers and denaturing gradient gel electrophoresis. *Appl. Environ. Microbiol.* **67:**2578–2585.

Wang, M., S. Ahrné, M. Antonsson, and G. Molin. 2004. T-RFLP combined with principal component analysis and 16S rRNA gene sequencing: an effective strategy for comparison of fecal microbiota in infants of different ages. *J. Microbiol. Methods* **59**:53–69.

Westbrook, G. L., C. M. O'Hara, S. B. Roman, and J. M. Miller. 2000. Incidence and identification of *Klebsiella planticola* in clinical isolates with emphasis on newborns. *J. Clin. Microbiol.* **38**:1495–1497.

Wilson, C. B., J. Westall, L. Johnston, D. B. Lewis, S. K. Dower, and A. R. Alpert. 1986. Decreased production of interferon-gamma by human neonatal cells. Intrinsic and regulatory deficiencies. *J. Clin. Investig.* **77**:860–867.

Wintzingerode, F., U. B. Göbel, and E. Stackebrandt. 1997. Determination of microbial diversity in environmental samples: pitfalls of PCR-based rRNA analysis. *FEMS Microbiol. Rev.* **21**:213–229.

Wolach, B., M. Ben Dor, O. Chomsky, R. Gavrieli, and M. Shinitzky. 1992. Improved chemotactic ability of neonatal polymorphonuclear cells induced by mild membrane rigidification. *J. Leukoc. Biol.* **51**:324–328.

Wold, A. E., and I. Adlerberth. 2000. Breast feeding and the intestinal microflora of the infant implications for protection against infectious diseases. *Adv. Exp. Med. Biol.* **478**:77–93.

Yan, S. R., G. Qing, D. M. Byers, A. W. Stadnyk, W. Al-Hertani, and R. Bortolussi. 2004. Role of MyD88 in diminished tumor necrosis factor alpha production by newborn mononuclear cells in response to lipopolysaccharide. *Infect. Immun.* **72**:1223–1229.

Enterobacter sakazakii
Edited by Jeffrey M. Farber and Stephen J. Forsythe
© 2008 ASM Press, Washington, D.C.

Enterobacter sakazakii Disease and Epidemiology

4

Anna B. Bowen and Christopher R. Braden

EPIDEMIOLOGY

Although *Enterobacter sakazakii* is best known as a cause of life-threatening infection for neonates (Himelright et al., 2002), it can lead to a variety of infections in many age groups. The epidemiology of *E. sakazakii* is poorly understood, because infection is rare and is not a reportable condition in most countries. However, incidence appears to differ by demographic group and underlying health status; infants and immunocompromised persons appear to be at higher risk of invasive infections than other persons. While at least 47 invasive (bloodstream, central nervous system [CNS], or urinary tract) infections and 31 noninvasive infections of infants have been reported (Urmenyi and Franklin, 1961; Joker et al., 1965; Monroe and Tift, 1979; Adamson and Rogers, 1981; Jimenez and Gimenez, 1982; Muytjens et al., 1983; Arseni et al., 1987; Biering et al., 1989; Reina et al., 1989; Simmons et al., 1989; Clark et al., 1990; Noriega et al., 1990; Gallagher and Ball, 1991; Nazarowec-White and Farber, 1997; Bar-Oz et al., 2001; Van Acker et al., 2001; Himelright et al., 2002; Block et al., 2002; Barreira et al., 2003; Stoll et al., 2004; Centers for Disease Control and Prevention [CDC], unpublished data), only 8 cases for toddlers and older children (Arseni et al., 1987; Reina et al., 1989; Murray et al., 1990; Lai, 2001) and 11 cases for adults (Jimenez and Gimenez, 1982; Pribyl et al., 1985; Reina et al., 1989; Hawkins et al., 1991; Lai, 2001; Dennison and Morris, 2002; Ongradi, 2002) are documented in the literature. Israel, Brazil, New Zealand, 11 European countries, 1 Canadian province, and 16 U.S. states reported cases between 1958

Anna B. Bowen and Christopher R. Braden, Foodborne and Diarrheal Diseases Branch, Division of Bacterial and Mycotic Diseases, National Center for Infectious Diseases, Centers for Disease Control and Prevention, Atlanta, GA 30333.

101

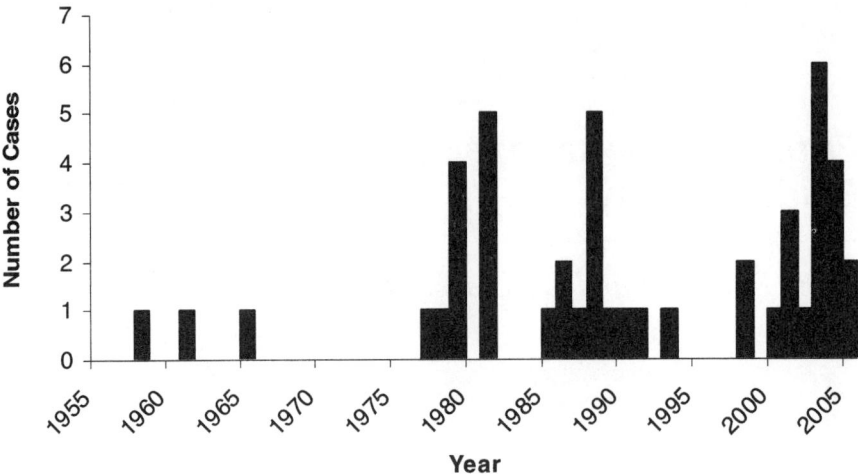

Figure 1 Number of invasive infant *E. sakazakii* cases by year. The year of onset was used when available; otherwise, the report year was used.

and 2005. Incidence of invasive infant infection may be increasing, according to reported cases (Fig. 1). However, the apparent geographic and temporal trends in *E. sakazakii* infection may reflect differences in surveillance and reporting by countries and changes in surveillance over time rather than real trends.

The incidence of all forms of *E. sakazakii* invasive infection appears to be higher among infants than older age groups. Additionally, other than a single case in a child with a dermoid cyst, *E. sakazakii* meningitis has been reported exclusively among infants. A survey conducted by the Foodborne Diseases Active Surveillance Network (FoodNet), a network of 10 surveillance sites in the United States, identified four cases of *E. sakazakii* meningitis or bacteremia among the catchment population during 2002, an incidence rate of one case per 100,000 infants per year (CDC, unpublished). Similarly, a survey of 19 U.S. neonatology centers revealed one episode of *E. sakazakii* bacteremia among 10,660 infants with very low birth weight (VLBW; <1,500 g), an estimated annual incidence of 9.4 cases per 100,000 VLBW infants (Stoll et al., 2004).

Characteristics of infants with invasive disease are shown in Table 1. Approximately half of the cases were male. Eight (40%) of 20 with delivery route information were born via Cesarean section. Nearly 70% were infected nosocomially, and the majority were born prematurely. Four (9%) had underlying medical conditions, including trisomy 21 with imperforate anus,

Table 1 Characteristics of infants with invasive infection[a]

Characteristic	No. positive/Total no.	%
Male	21/39	54
Cesarean delivery	8/20	40
Nosocomial onset	29/42	69
Premature birth	21/38	55

[a]Adapted from Bowen and Braden (2006).

omphalocele, jejunal atresias, and meningomyelocele (Muytjens et al., 1983; Biering et al., 1989; Noriega et al., 1990).

The incidence of infection among toddlers and older children is unknown. Three cases of invasive *E. sakazakii* infection among children >12 months of age have been reported in the literature: two 3-year-old children with malignancies developed sepsis, and a 20-month-old child with a subdural dermoid cyst developed a posterior fossa abscess (Arseni et al., 1987; Tekkok et al., 1996; Lai, 2001). Additionally, peritonitis following appendiceal rupture and osteomyelitis subsequent to a traumatic foot wound occurred in otherwise healthy children (Reina et al., 1989; Murray et al., 1990). Three cases of diarrhea or fecal carriage among children >12 months of age have been reported in the literature (Arseni et al., 1987).

Among adults the incidence is unknown, but infection is likely uncommon. In the United Kingdom, 53, 53, and 67 cases were observed in 2001, 2002, and 2003, respectively (Anonymous, 2004). Of the 11 detailed adult cases reported in the literature, 4 were associated with *E. sakazakii* bacteremia; all occurred in persons aged >70 years (Jimenez and Gimenez, 1982; Hawkins et al., 1991; Lai, 2001). Two occurred nosocomially following complicated abdominal surgeries (Lai, 2001); a third case had no clear source (Hawkins et al., 1991); and the fourth patient had a history of urinary retention and developed urosepsis (Jimenez and Gimenez, 1982). Among the seven remaining adult cases reported without bacteremia, two had pneumonia, with *E. sakazakii* isolated from sputum (Lai, 2001), two had *E. sakazakii* isolated from soft tissue infections (Reina et al., 1989; Dennison and Morris, 2002), and one had diabetes-associated osteomyelitis of the foot, with a bone biopsy that yielded *E. sakazakii*, *Staphylococcus epidermidis*, and enterococci (Pribyl et al., 1985). These five patients had underlying medical conditions, such as malignancy, peripheral vascular disease, diabetes mellitus, or recent abdominal surgery. Two of the seven cases without bacteremia occurred in persons without reported comorbidities; these included an infected soft-tissue injury and vaginitis (Reina et al., 1989; Ongradi, 2002).

CLINICAL PRESENTATION AND COURSE

E. sakazakii causes a variety of infections. Clinical syndromes include meningitis, bacteremia, urinary tract infections, and wound infections. *E. sakazakii* also has been isolated in association with pneumonia, conjunctivitis, vaginitis, appendicitis, and, among premature infants, necrotizing enterocolitis (NEC).

Among the syndromes, CNS infection is the best described. One 20-month-old child with a dermal sinus and subdural dermoid cyst developed a brain abscess (Tekkok et al., 1996). All 33 other reported CNS infections manifested as meningitis with or without cerebritis in infants <6 weeks of age (median age, 6 days) (Table 2). Approximately half of affected infants were premature and had low birth weight. Spread is likely hematogenous, and *E. sakazakii* is frequently cultured from blood and cerebrospinal fluid (CSF) concurrently. Affected infants develop typical signs of sepsis, including irritability or lethargy, temperature instability, and feeding intolerance, generally during the first week of life. A recent analysis found that seizures rapidly developed in approximately 33% of reported cases; occasionally these became intractable (CDC, unpublished). Additional complications included ventriculitis and brain abscess, which may be difficult to distinguish clinically and radiographically from brain cysts (Kleiman et al., 1981; Kline, 1988; Willis and Robinson, 1988; Gallagher and Ball, 1991; Burdette and Santos, 2000). Meningitis often rapidly cascades into cerebral hemorrhage, infarct, necrosis, liquefaction, and, eventually, cavitation and cyst formation (Fig. 2). Brain cysts or abscesses occurred in 21% of infants, and ventricular shunts for hydrocephalus were placed in 42% of infants with meningitis (Bowen and Braden, 2006). Among infants with *E. sakazakii* meningitis, 42% died; of survivors, 74% exhibited adverse neurological outcomes, such as developmental delays, according to subsequent examinations (Bowen and Braden, 2006).

E. sakazakii bacteremia has occurred in all age groups and presents with symptoms typical of gram-negative bloodstream infections, specifically temperature instability or fever, altered mental status, and hypotension. Reported cases of bacteremia without meningitis have occurred in 11 infants with a median age of greater than 1 month; however, the majority of these infants

Table 2 Characteristics of infants with bacteremia or meningitis[a]

Characteristic	Bacteremia (*n* = 12)	Meningitis (*n* = 33)	P
Age (median, days)	35	6	<0.001
Gestational age (median, wks)	27.8	37	0.02
Birth weight (median, g)	850	2,454	<0.01

[a]Adapted from Bowen and Braden (2006).

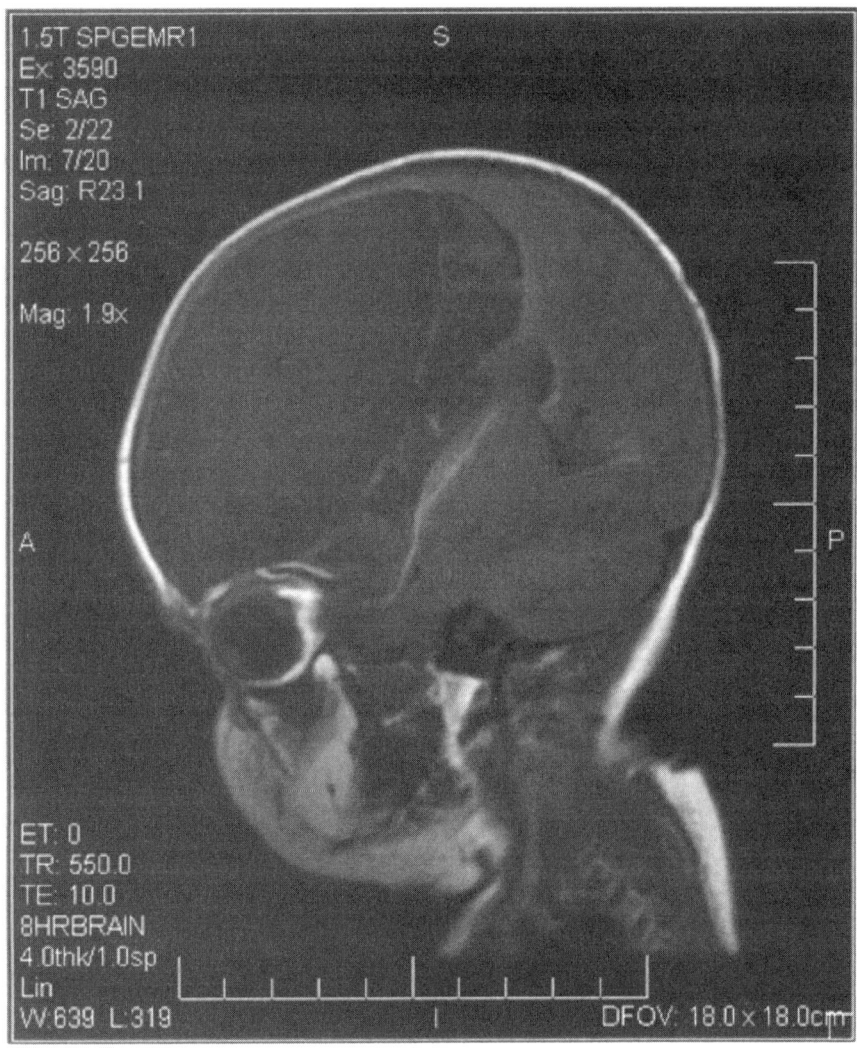

Figure 2 Encephalomalacia following *E. sakazakii* meningitis. Courtesy of Michelle Hulse, Children's Hospital of Saint Paul, Saint Paul, MN.

were born prematurely and at low birth weight (Table 2). Among infants who developed bacteremia without meningitis, 1 (9%) of 11 experienced seizures and a second infant (9%) died, although the latter infant simultaneously experienced NEC (Bowen and Braden, 2006). Among four reported adult cases of bacteremia, two had surgical procedures with complications shortly before infection, and one of these patients died despite treatment with appropriate antimicrobial agents (Lai, 2001). Another adult developed urosepsis secondary to urinary retention, and a fourth developed sepsis without an obvious source; both of these patients were treated with appropriate antimicrobial agents and survived (Jimenez and Gimenez, 1982; Hawkins et al., 1991).

Wound infections have been reported only among adolescents and adults. Patients developed osteomyelitis secondary to a diabetic foot ulcer (Pribyl et al., 1985), infections of a vascular graft and a surgical wound in a patient with diabetes mellitus and peripheral vascular disease (Dennison and Morris, 2002), infections of traumatic groin and finger injuries in healthy young adults (Reina et al., 1989), and osteomyelitis following a traumatic foot injury in an adolescent (Murray et al., 1990).

Isolation of *E. sakazakii* from stool or gastric aspirates has been associated with a range of gastrointestinal syndromes. Reports include fever with diarrhea in a child (Arseni et al., 1987) and a spectrum ranging from bloody diarrhea to fatal NEC in infants (Muytjens et al., 1983; Simmons et al., 1989; Van Acker et al., 2001). Simultaneous gastrointestinal symptoms and *E. sakazakii* bacteremia in infants have been described as well (Muytjens et al., 1983; Simmons et al., 1989; Van Acker et al., 2001). Asymptomatic carriage has also been reported (Biering et al., 1989; Bar-Oz et al., 2001; Himelright et al., 2002; Loc-Carillo et al., 2006).

RESERVOIRS AND TRANSMISSION

Many possible sources and modes of transmission for *E. sakazakii* infections have been debated. However, the only vehicle that has been epidemiologically and microbiologically associated with infection is powdered infant formula or the equipment used to prepare it (Muytjens et al., 1983; Muytjens et al., 1988; Biering et al., 1989; Simmons et al., 1989; Noriega et al., 1990; Van Acker et al., 2001; Iversen and Forsythe, 2004). Several investigations have identified powdered infant formula as a source of *E. sakazakii* infections (see the review of outbreaks of *E. sakazakii* infections below). Although *E. sakazakii* is often difficult to detect in powdered infant formula due to sporadic or low-level contamination (Muytjens et al., 1988), it has been identified as an intrinsic rare contaminant of powdered infant formula from multiple manufacturers (Muytjens et al., 1988; Nazarowec-White and Farber, 1997; Iversen and Forsythe, 2004; Leuschner and Bew, 2004) and has been found in unopened cans of powdered infant formula associated with outbreaks of disease (Van Acker et al., 2001; Himelright et al., 2002). Further, investigators have found contamination of blenders and other equipment used to mix infant formula (Muytjens et al., 1983; Simmons et al., 1989; Noriega et al., 1990; Block et al., 2002). Although formula used during these outbreaks was found not to be intrinsically contaminated with *E. sakazakii*, powdered infant formula used previously may have contaminated the equipment; *E. sakazakii* may have amplified in the moist environments, and subsequent preparations of formula may have become contaminated. Thus, the most likely source of *E. sakazakii* contamination of the formula preparation equipment in these instances remains the powdered formula rather than extrinsic contamination from the environment. However, not all investigations have identified powdered infant formula as the source of infections in infants, and some infants who became infected were not fed powdered formula products (Barreira et al., 2003; Stoll et al., 2004). Additionally, documented *E. sakazakii* infections of adults indicate that sources of infections other than powdered infant formula exist. Other potential sources include maternal transmission, nosocomial transmission, home environments, food and food-manufacturing environments, and insect or animal reservoirs.

Since many reported cases occurred among neonates, vertical transmission has been suggested as a source of infection. However, vaginal and intestinal carriage have not been reported from mothers of affected neonates. A single report of vaginal infection or colonization exists: *E. sakazakii* was isolated from the vaginal discharge of a nonpregnant 26-year-old woman with vaginitis (Ongradi, 2002). Infections among infants during the first 3 days of life are rare; further, infections have been noted in infants beyond the neonatal period (Simmons et al., 1989), including in a previously healthy 8-month-old infant who developed *E. sakazakii* bacteremia in the outpatient setting (CDC, unpublished). Finally, at least six infected infants were delivered by Cesarean section (Urmenyi and Franklin, 1961; Muytjens et al., 1983; Bar-Oz et al., 2001; Himelright et al., 2002; CDC, unpublished). Vertical transmission, therefore, is unlikely to be a source of infection (Monroe and Tift, 1979; Adamson and Rogers, 1981; Muytjens and Kollee, 1990).

Nosocomial sources of infection also have been proposed, but no hospital sources other than equipment used to prepare formula have been implicated. During outbreaks of illness, isolettes, isolette humidifiers (Arseni et al., 1987), nursery and neonatal intensive care unit (NICU) surfaces, formula preparation areas, medical fluids (Biering et al. 1989), nursery water used to prepare formula (Van Acker et al., 2001; Block et al., 2002), and hand cultures and rectal swabs from nursery staff (Biering et al., 1989; Van Acker et al., 2001; Block et al., 2002) all failed to yield *E. sakazakii*. Further, during a cohort study conducted by CDC in 2002, ventilator use, humidified isolette use, and oral medications were not significantly associated with *E. sakazakii* illness or colonization (Himelright et al., 2002). Transmission of *E. sakazakii* from one patient to another, either directly or indirectly, was also a concern to outbreak investigators. However, of the three outbreak investigations which systematically analyzed the source of *E. sakazakii* infections and colonization among infant cohorts, all found that each infected or colonized infant was exposed to infant formula prior to illness or colonization (Simmons et al., 1989; Van Acker et al., 2001; Himelright et al., 2002). Although this finding does not exclude the possibility that infant cross-contamination or health care provider transmission occurred concurrent to implicated formula exposure, it is unlikely that this was a primary mode of *E. sakazakii* transmission. Moreover, person-to-person transmission of *E. sakazakii* has not been documented.

E. sakazakii has also been identified in a wide range of non-hospital environments. *E. sakazakii* has been detected in dried infant foods, milk powders, cheese products, herbs and spices, soy protein, ground rice and maize, minced beef, sausage, and vegetables (Leclercq et al., 2002; Iversen and Forsythe, 2003; Iversen and Forsythe, 2004). Kandhai et al. (2004) found 30 (23%) of 131 households and food-manufacturing plants tested to be contaminated. E. sakazakii has also been recovered from the midgut of the stable fly, *Stomoxys calcitrans*, which has a worldwide distribution and is typically associated with farm animals (Hamilton et al., 2003). Contamination of powdered infant formula during the manufacturing process is well established as a source of infection in infants (see chapter 6). The relationship between other identified sources of *E. sakazakii* and human illness is less well understood.

INCUBATION AND CARRIAGE

Because the route of exposure is generally unclear at the time of infection, the incubation period is unknown. However, in outbreaks associated with powdered infant formula, illnesses began as soon as 3 to 4 days after initial exposure to the implicated formula product (Simmons et al., 1989; Van Acker et al., 2001). Rectal carriage has been documented as persisting for 8 to 18 weeks, even following treatment with cefotaxime (Arseni et al., 1987; Bar-Oz et al., 2001).

DIAGNOSIS

Standard blood, CSF, urine, and tissue culture specimens can be used to diagnose *E. sakazakii* infection, and isolates grow well on standard nonselective media. However, identification of *E. sakazakii* can be complicated by the occasional occurrence of white colonies and inaccurate results using the API 20E system (see chapter 2). Serial radiographic studies may be warranted for infants with meningitis, because *E. sakazakii* appears to have a propensity for brain necrosis and abscess formation (Gallagher and Ball, 1991; Burdette and Santos, 2000; Bar-Oz et al., 2001).

TREATMENT

There are no comparative studies guiding treatment options for *E. sakazakii* infections. However, inference from antimicrobial susceptibility studies may be useful. The original description of *E. sakazakii* as a species in 1980 included antibiotic disk diffusion testing of over 100 isolates and MIC testing for a small subset of 10 isolates (Farmer et al., 1980). Antimicrobial resistance was uncommon, with only one strain showing resistance to multiple antibiotics. In fact, the authors suggested that greater disk diffusion zone sizes for ampicillin and cephalothin might be used to differentiate *E. sakazakii* from *Enterobacter cloacae*. However, testing of more recent clinical isolates has shown increasing antimicrobial resistance. Clinical and environmental isolates from three outbreaks of *E. sakazakii* were resistant to cefazolin or cephalothin but were susceptible to ampicillin, more advanced penicillins and cephalosporins, carbapenems, fluoroquinolones, aminoglycosides, tetracycline, trimethoprim-sulfamethoxazole, and chloramphenicol (Clark et al., 1990; Block et al., 2002). Block et al. (2002) found that all isolates associated with their outbreak produced β-lactamase, most likely the Bush group 1 β-lactamase. Similar findings were reported for nine clinical isolates from four hospitals in Canada (Nazarowec-White and Farber, 1999). In a report of five cases of *E. sakazakii* infection involving one child and four adults from one U.S. hospital during a 2-year period, Lai (2001) described uniform resistance to ampicillin, cefazolin, and extended-spectrum penicillins, and some were not susceptible to expanded-spectrum cephalosporins or quinolones. These isolates were uniformly susceptible to the aminoglycosides and trimethoprim-sulfamethoxazole. These data suggest that antimicrobial resistance in *E. sakazakii* is increasing, paralleling the similar increase in resistance of other *Enterobacter* species (Lai, 2001).

Present standards for empirical therapy for acute meningitis or sepsis (Tunkel and Scheld, 2005), including combinations of ampicillin and gentamicin (neonates), ampicillin and a third-generation cephalosporin (infants), or a carbapenem plus vancomycin (noninfants), would provide adequate therapy for *E. sakazakii* infections according to the limited susceptibility data available. In the setting of a severe infection, combination therapy including an aminoglycoside or trimethoprim-sulfamethoxazole should be considered. Antimicrobial susceptibility testing of *E. sakazakii* isolates should be conducted to help guide therapy. Little information is available to determine the optimal duration of therapy. Severe infections such as cerebritis, ventriculitis, or brain abscesses may require prolonged therapy, and ventricular shunting may be necessary if increased intracranial pressures or hydrocephalus develops.

PREVENTION

Exclusive breastfeeding during the first 6 months of life has been recommended by the World Health Organization and the American Academy of Pediatrics (Kramer and Kikumo, 2002). Promotion of exclusive breastfeeding during this period could protect infants from *E. sakazakii* disease by limiting their exposure to the only source of infection presently known to cause illness in this group, namely, powdered infant formula. Prenatal breastfeeding education, perinatal support from family and health care providers, increased workplace and societal provisions for breastfeeding, proscription of formula gifts during the prenatal and neonatal periods, and restriction of infant formula advertising claims comparing formula to breast milk could help increase breastfeeding rates (Howard et al., 1997; DiGirolamo et al., 2001; Sheehan et al., 2001; Ryan et al., 2002; Merewood et al., 2003; Taveras et al., 2004; Anonymous, 2005).

Educating infant caregivers and health care providers about the risks of powdered infant formula, practices for safer preparation and feeding, and alternative feeding choices, such as breastfeeding and sterile formulas, is also important. Education would likely reduce the risk to infants through decreased use of powdered infant formula products and safer preparation and delivery practices when powdered infant formula is used. Infant formula manufacturers could provide warnings about the nonsterility of powdered infant formula as well as information about safer handling practices and alternative feeding options to consumers on powdered infant formula labels and during infant formula advertising. Expanding nutrition curricula for health care providers would also be helpful (Howard et al., 1997; Sheehan et al., 2001; Ryan et al., 2002; Taveras et al., 2004).

However, due to formula pricing and availability, use of powdered formula products may be necessary in some circumstances. Metabolic formulas and some breast milk fortifiers are available only in powdered form. In developing countries, infant formula may be available exclusively in powdered form. When powdered formula must be used, following guidelines created by the American Dietetic Association (2003) for the preparation of infant formula in health care settings also will help minimize risks associated with powdered infant formula (Fig. 3). Similar practices would minimize risks when preparing powdered infant formula in private homes. Adapting manufacturing processes to improve powdered formula safety would further help prevent illness (see chapter 6).

1. Formula products should be chosen based on nutritional needs; powdered formulas should be used only when alternative sterile, liquid formulas are not available.

2. Trained personnel should prepare powdered formula under aseptic technique in a designated formula preparation area.

3. Powdered formula should be measured by weight, not volume, and mixed with sterile water according to manufacturer's instructions.

4. Product should be chilled to 4°C immediately after preparation, and discarded if not used within 24 h.

5. The administration or "hang" time for continuous enteral feeding in neonatal intensive care units should not exceed 4 h; tubing and syringes must be changed every 4 h for feeds which contain expressed human milk.

6. Written guidelines should be available to hospital personnel for reporting and follow-up of formula products recalled by the manufacturer.

Figure 3 Recommendations for preparation and administration of powdered infant formula in health care settings (adapted from the American Dietetic Association [2003]).

In the hospital setting, additional practices may reduce the risk of infection. Strict adherence to hand hygiene, cleaning, and disinfection protocols might reduce the risk for nosocomial transmission. Substituting sterile, liquid nutrition for prepared nonsterile, powdered meal replacements when feeding immunosuppressed infants, children, and adults may also protect patients' health. Use of acid-suppressing medications has also been associated with increased risk for *Clostridium difficile*-associated disease and salmonellosis (Dial et al., 2005; Bowen et al., 2005), and limiting their use may help prevent disease due to other enteric organisms such as *E. sakazakii*.

Timely reporting of *E. sakazakii* cases to local public health officials, with subsequent rapid epidemiologic and environmental investigation, may help identify additional illness risk factors and clarify prevention methods. When a nosocomial case is identified, laboratory screening (e.g., obtaining rectal or pharyngeal specimens) of the relevant hospital cohort may identify additional cases, while epidemiologic investigation may elucidate the vehicle of infection and lead to its removal from circulation.

REVIEW OF OUTBREAKS OF *E. SAKAZAKII* INFECTIONS

Although many individual cases of *E. sakazakii* infection have been reported in the literature and public health and infection control officials have investigated others, the source and mode of infection are often difficult to establish for individual cases. There are often many potential sources of infection, and it can be challenging to focus an investigation. Environmental investigations in these circumstances are less likely to successfully identify the source of infections. During outbreaks, however, investigators can conduct epidemiologic investigations, examine common exposures, and statistically implicate sources based on probability assessments. Thus, outbreaks of infection permit development and testing of exposure hypotheses. Once exposures are identified this way, food and environmental testing can be conducted with an intensive and systematic approach; targeted testing is much more likely to verify the source and mode of infections. Later, molecular subtyping of *E. sakazakii* clinical and environmental isolates can link environmental sources and infections based on common clonal origins. Thus, investigations of outbreaks have provided the most information about the epidemiology of *E. sakazakii* infections. In this section, we review published investigations of outbreaks of *E. sakazakii* disease (Table 3).

United Kingdom, 1961

The earliest reported cluster of *E. sakazakii* infections was published in 1961, when *E. sakazakii* was known as a yellow-pigmented variant of *E. cloacae* (Urmenyi and Franklin, 1961). Urmenyi and Franklin described two infants from the same nursery who died of sepsis and meningitis within days of each other. One infant was delivered vaginally at term and did well during her first 10 days of life. She was readmitted to the hospital with sepsis on day 11 of life and died within 48 hours. The second infant was premature, born at 32 weeks gestation by Cesarean delivery, and weighed 2,017 g. On day 5 of life, she developed fulminant sepsis and meningitis and died within several hours. The autopsies revealed hemorrhagic necrosis of the brain in both cases. Cultures of the meninges, CSF, and blood yielded yellow-pigmented gram-negative rods now recognized as *E. sakazakii*. Because the infants might have shared the same incubator, the incubator and heated humidifier apparatus were swabbed for culture. The offending organism was not identified from these environments, and the source of infections was not determined.

Table 3 Published outbreaks due to *E. sakazakii*

Publication yr	Country	No. with meningitis/ cerebritis	No. with bacteremia	No. with colonization	No. with NEC	No. that died	No. that survived with neurological deficit	Infant formula implicated as source	Reference
1961	United Kingdom	2	NS[a]	NS	NS	2	NS	N	Urmenyi and Franklin (1961)
1983	The Netherlands	5	NS	NS	NS	4	1	Y[b]	Muytjens et al. (1983)
1987	Greece	NS	NS	11[c]	NS	4	NS	N	Arseni et al. (1987)
1989	Iceland	3	NS	1	NS	1	2	Y	Biering et al. (1989)
1989	United States	NS	2	2	NS	NS	NS	Y	Simmons et al. (1989)
2001	Belgium	NS	1[d]	NS	12	2	NS	Y	Van Acker et al. (2001)
2002	Israel	1	2	3	NS	0	1	Y[d,e]	Block et al. (2002)
2002	United States	1	NS	8[f]	NS	1	NS	Y[g]	Himelright et al. (2002)
2006	France	2	0	7	0	2	0	Y[g]	Coignard et al. (2006)

[a]NS, not specified.

[b]Plasmid profiles of *E. sakazakii* recovered from clinical specimens and formula were distinct.

[c]Several other gram-negative organisms also were identified in surveillance cultures of throat, trachea, and anus.

[d]Two distinct *E. sakazakii* strains were isolated from the blood of one infant.

[e]*E. sakazakii* was isolated from a sample of prepared formula and formula mixer but not from formula powder.

[f]Two infants had illnesses concurrent with positive *E. sakazakii* surveillance cultures but no sterile site culture yielding *E. sakazakii*.

[g]Infecting strain of *E. sakazakii* was isolated from unopened cans of implicated powdered infant formula.

The Netherlands, 1983

In 1983, Muytjens et al. reported an investigation of eight cases of neonatal meningitis due to *E. sakazakii* in The Netherlands. Among these eight infants was a cluster of five from the same hospital; the other three infected neonates were from three different hospitals in distant parts of the country. These eight neonates all suffered meningitis and cerebritis due to *E. sakazakii*; six died and two survived but developed hydrocephalus and mental retardation. Environmental sampling of the pediatric ward at the hospital housing the five clustered cases yielded *E. sakazakii* from prepared formula, a dish brush, and a stirring spoon, though *E. sakazakii* was not recovered from the dry formula or the water used in formula preparation. Plasmid profile comparisons among the isolates from the five neonates revealed three with identical profiles and two with varying profiles. Three formula isolates were identical to each other but were different from the isolates from the five patients, and the dish brush isolate had a unique plasmid profile. The plasmid profiles of the remaining three patients from other parts of the country were each unique. The authors concluded that the environmental strains were not the cause of infections in the neonates based on their plasmid profile analyses.

This is the first report in which infant formula was found to be contaminated with *E. sakazakii* in association with an outbreak. Though the authors did not conclude that the formula was the source of infections, they did not fully address the possibility. The infants were infected by multiple strains, and at least two clinical strains were associated temporally, and in the same facility, with the identification of *E. sakazakii* strains from formula or instruments used to prepare it. It is possible that they did not identify the strains responsible for infections among the various strains contaminating the formula. More recent studies have demonstrated multiple different strains of *E. sakazakii* cultured from powdered infant formula from the same manufacturer (Nazarowec-White and Farber, 1999). The investigators did not recover the organism from dry infant formula or the water used to prepare it, but they did culture the organism from prepared formula. In this case, the prepared formula likely acted as an enriching medium for the growth of the organism originally present in the powdered formula. Culturing *E. sakazakii* directly from dry formula powder is difficult due to typically low levels of contamination; more modern culture techniques are more sensitive due to the use of larger quantities of formula, enrichment techniques, and selective media (Muytjens et al., 1988; Nazarowec-White and Farber, 1997; Iversen and Forsythe, 2004).

Greece, 1987

In 1987, Arseni et al. reported a cluster of neonates colonized with *E. sakazakii* in one of two NICUs in a hospital in Greece. Routine surveillance cultures of the rectum and throat were obtained just after admission, at 3 to 4 days, and every week thereafter. Blood cultures were also performed, but their timing and frequency were not indicated. Repeat blood cultures and other specimens were obtained according to clinical indication. During a 5-week period, they identified 11 neonates with surveillance cultures positive for *E. sakazakii*; 5 of the 11 had clinical signs of severe sepsis, and 4 died. *E. sakazakii* was not isolated from blood or other sterile sites for any of the neonates, but one ill neonate had *Pseudomonas aeruginosa* bacteremia and another had *Klebsiella pneumoniae* bacteremia; both survived. Other gram-negative bacteria, including *Klebsiella* spp., *Serratia* spp., and *Enterobacter* spp., also were isolated from surveillance cultures, complicating the clinical picture. In addition, half of the *E. sakazakii* isolates recovered from surveillance cultures were moderately resistant to amikacin, and all were resistant to cefoxitin and penicillins. These drugs were generally used for empiric treatment of infants with suspected sepsis. None of the neonates not colonized with *E. sakazakii* during this time period died. Environmental investigations to identify a source of *E. sakazakii* included 100 swab cultures of environmental surfaces and medical fluids and 77 cultures of staff fingerprint impressions on blood agar; none yielded *E. sakazakii*. Analytic case-control or cohort studies of exposures were not performed, and infant formulas were not tested.

This investigation was the first to describe *E. sakazakii* colonization of the throat and rectum of neonates as an indicator of an outbreak. Two had evidence of sepsis, one had meningitis with no organisms cultured from blood or CSF, and four died after treatment with antibiotics to which the colonizing *E. sakazakii* was resistant, suggesting that *E. sakazakii* may have caused their illness. No source of *E. sakazakii* colonization was identified, despite extensive culturing of the unit environment and staff. In this circumstance, an epidemiologic study comparing exposures among colonized and noncolonized neonates may have been helpful in identifying a source but was not done.

Iceland, 1989

A report published in 1989 from Iceland described three neonates with *E. sakazakii* meningitis and cerebritis who fell ill at the same hospital during a 9-month period beginning in 1986 (Biering et al., 1989). Two infants were born after term gestation, and the third was born at 36 weeks of gestation; all were of normal birth weight. One child had trisomy 21 and associated anomalies. All fell ill at day 5 of life; one died and the two others suffered permanent neurological deficits, despite appropriate antibiotic therapy. In addition to these three cases, *E. sakazakii* was isolated from the urine of another neonate in the nursery. Although repeat cultures of the urine were negative, *E. sakazakii* was isolated from swabs of the groin and anus. This infant did not show signs of illness and was considered colonized. Owing to previous reports of isolating *E. sakazakii* from infant formula, and because each of these neonates received infant formula prior to illness onset, the investigators cultured one bottle of previously prepared formula stored for an unknown period of time, freshly prepared formula directly plated onto blood agar plates, and freshly prepared formula incubated for 4 h at 36 °C. Sampling included unopened cans of the available powdered infant formula from five different manufactured lots. *E. sakazakii* was recovered from the previously prepared formula and from all lots of formula freshly prepared and incubated; *E. sakazakii* was not isolated from formula freshly prepared but not incubated. Cultures of the ward environment, kitchen where formula was prepared, and formula preparation utensils did not yield *E. sakazakii*. The clinical isolates and all but one of 23 formula isolates demonstrated indistinguishable biotypes, antibiograms, plasmid profiles, multilocus enzyme electrophoresis patterns, and chromosomal restriction endonuclease patterns (Clark et al., 1990).

It was notable that two out of the three affected infants were born at term and were otherwise healthy. Also, the powdered formula was likely contaminated at a low level, since recovery of the organism required preplating incubation. Perhaps these illnesses were due to a combination of inappropriate handling and storage of formula (as the authors suspected) and providing the formula to infants within the first few days of life, when they were more susceptible to enteric colonization and CNS infection. The fourth infant with colonization indicates that additional infants may have been exposed to contaminated formula but either did not become ill or did not have cultures done to detect it. Twenty-two of 23 formula isolates were highly related according to multiple analyses (biotypes, antibiograms, plasmid profiles, multilocus enzyme electrophoresis, and chromosomal endonuclease analysis); one formula isolate was clearly unrelated to the others. This finding underscores the observation that powdered formula may be contaminated with multiple strains of *E. sakazakii*.

United States, 1989

Simmons et al. reported an outbreak of *E. sakazakii* involving four infants in an intensive care unit in the United States (1989). All four infants were premature and had low birth weight; two had bacteremia, and one had a urinary tract infection with *E. sakazakii*. The fourth infant had bloody diarrhea; all had *E. sakazakii* cultured from the stool. The illnesses occurred over a 6-week period; patient ages at the time of illness ranged from 13 to 57 days. The authors did not indicate whether any of these infants died. Because all infants had *E. sakazakii* isolated from their stool, the investigators reviewed feeding procedures. All four were fed a powdered protein hydrolysate formula mixed in a blender; culture of the blender yielded a heavy growth of *E. sakazakii*, *Pseudomonas maltophilia,* and *P. fluorescens*. One infant also had *P. maltophilia* grown from stool and another from the CSF (thought to be a contaminant). A cohort study was conducted that included all infants in the NICU during the dates of illness onset for the four cases. A retrospective review of cultures obtained for the evaluation of clinical illness revealed that four of five infants who received the hydrolysate formula had positive cultures for *E. sakazakii;* none of the 40 other infants in the cohort who did not receive this formula had positive cultures ($P = 0.0006$). Cultures of the powder from the opened can of implicated formula yielded *E. sakazakii* and *E. cloacae. E. sakazakii* isolates from three of the four infants and the powdered formula had the same plasmid profile, were the same multilocus enzyme electrophoresis type, and demonstrated indistinguishable chromosomal DNA restriction endonuclease and ribotype patterns. The fourth patient isolate was not analyzed (Clark et al., 1990).

This was the first reported outbreak of *E. sakazakii* infections in which powdered infant formula was identified as the source of infections on the basis of statistical association in a cohort study, culture of the implicated formula, and molecular characterization of the recovered *E. sakazakii* isolates from the formula and clinical samples. It also highlighted the potential role played by a heavily contaminated formula blender. Because the blender was apparently used only for the preparation of the implicated formula and the formula powder was shown to be contaminated with *E. sakazakii*, the formula was very likely the source of blender contamination and, subsequently, infections in the infants. The blender may have acted to potentiate the contamination of prepared formula fed to the infants. The blender was rinsed with tap water between uses and occasionally washed with hand cleansers. Once this practice was corrected with proper sanitation of the blender between uses, no further *E. sakazakii* clinical isolates were identified from the NICU patients.

Belgium, 2001

In 1998, a cluster of NEC was observed among 12 infants in a NICU in Belgium over a 2-month period (Van Acker et al., 2001). All infants with NEC were premature and weighed <2,000 g at birth. Four infants required operative treatment, and two (twins) died as a result of NEC. The investigators identified a cohort of 50 infants admitted to the NICU during the 2-month period concurrent with the onset of the NEC outbreak. All 12 infants with NEC were fed formula orally before illness onset; 10 of 12 received the same powdered semielemental formula, whereas only 4 of the 38 infants without NEC received that formula ($P < 0.0001$). Six of 14 infants who had received the semielemental formula had cultures yielding *E. sakazakii*, whereas none of the 36 who did not receive that formula had cultures yielding *E. sakazakii* ($P = 0.0002$). Surveillance cultures, including anal swabs, stomach aspirates, and blood cultures, were obtained from each patient with NEC if possible and if ordered by the physician. Eleven surveillance cultures from six patients yielded *E. sakazakii*. One infant with NEC had *E. sakazakii* isolated from the blood; this culture demonstrated two morphologically distinct *E. sakazakii* strains. The NICU stopped using the implicated formula early in the investigation, but because *E. sakazakii* was not initially identified in the powdered formula, it was used to feed one additional patient approximately 10 days later. This infant developed symptoms of NEC after 3 days of feedings, and *E. sakazakii* was isolated from a stomach aspirate and an anal swab. No further NEC cases or *E. sakazakii* clinical isolates were observed once the implicated formula was finally removed from use.

Extra bottles of formula prepared in the formula preparation area during the time of the outbreak were cultured. *E. sakazakii* was identified from several preparations of the implicated semielemental formula; none of the other brands of formula or water used to mix formula or rinse formula preparation equipment yielded *E. sakazakii*. In addition, *E. sakazakii* was recovered from one of two separate lots of unopened cans of the implicated powdered formula. A total of 14 isolates of *E. sakazakii* were obtained from the formula.

Nine isolates from five patients and all 14 formula isolates were analyzed by arbitrarily primed PCR (AP-PCR). Three distinct AP-PCR types were observed among the patient isolates, including two types from the blood culture from one patient. One AP-PCR type was identified among all the formula isolates and matched the type seen in three patients. Four patients had *E. sakazakii* isolates with AP-PCR types not identified in formula isolates.

This outbreak was the first to demonstrate an association between *E. sakazakii* infection and colonization and the development of NEC. Although two neonates with *E. sakazakii* infections in the case series reported by Muytjens et al. (1983) also developed NEC, the investigators of the present outbreak were able to demonstrate a statistically significant association between NEC and *E. sakazakii* infection or colonization by conducting a cohort study. Six of 12 infants with NEC had cultures which yielded *E. sakazakii*, whereas none of the 39 infants without NEC had cultures which yielded *E. sakazakii* ($P < 0.0001$). While *E. sakazakii* infection or colonization alone is not likely to be sufficient to cause NEC, it may be a factor among the triad of intestinal ischemia, excess protein substrate in the intestinal lumen, and intestinal microbial colonization that seem to be prerequisite for the pathogenesis of NEC (Kosloske, 1984).

The clinical isolates from patients with NEC in this outbreak revealed three different AP-PCR profiles, only one of which was identified among isolates from the implicated powdered formula. Although it is theoretically possible that the two other clinical strains originated from sources unrelated to the implicated formula, it seems unlikely. As indicated earlier, other studies have demonstrated multiple strains of *E. sakazakii* as contaminants of powdered infant formula from the same manufacturer (Nazarowec-White and Farber, 1999). As the authors point out, the *E. sakazakii* strain identified in the present investigation may be the predominant, but not the only, strain present in the contaminated powdered formula.

Israel, 2001

Block and colleagues reported two neonates infected and three colonized with *E. sakazakii* at a Jerusalem hospital (2002). Within 1 month, one neonate was diagnosed with sepsis and bacteremia and the other was diagnosed with bacteremia, meningitis, and cerebritis. The additional three cases of *E. sakazakii* colonization were identified during a subsequent investigation. Colonization of the infants was documented to persist for 8 to 18 weeks; two were still stool culture positive on discharge from the hospital. *E. sakazakii* isolates showed an unusual biochemical variant: all were nitrite negative. An investigation of the source of *E. sakazakii* included cultures of prepared formulas, powdered formulas in present use, water used to prepare the formulas, and the blender used to mix formulas. A sample of prepared formula and the blender yielded the same strain of *E. sakazakii* identified in ill and colonized neonates, as determined by SpeI endonuclease restriction and pulsed-field gel electrophoresis (PFGE) of isolate DNA. *E. sakazakii* was not recovered from formula powder or water used to prepare formula. Of note, the blender was first cultured after thorough cleansing and disinfection with boiling water and 70% alcohol, and still *E. sakazakii* was recovered. The blender, although not in use, remained culture positive for at least 5 months, even though it was decontaminated before each sampling with 70% alcohol followed by a boiling water rinse (C. Block, personal communication). It was noted to have a small crack in its base. The authors suggested that the original source of *E. sakazakii* may have been powdered formula that contaminated the formula blender, where *E. sakazakii* established an environmental niche. However, an environmental source other than powdered formula could not be excluded.

United States, 2002

The death of a neonate due to *E. sakazakii* meningitis prompted an investigation at a Tennessee hospital in 2001 (Himelright et al., 2002). Screening cultures were performed among a cohort of 49 infants hospitalized at the time of the index illness. Two additional patients with respiratory illnesses had tracheal aspirates that yielded *E. sakazakii*, and six neonates had asymptomatic *E. sakazakii* colonization. A review of medical records to assess possible risk factors identified only one statistically associated exposure; all 9 ill or colonized infants had been fed a specific powdered formula product, whereas 21 of 40 other infants had been similarly fed ($P < 0.01$). Factors not associated with illness or colonization included ventilator use, humidified incubator use, oral medications, and other types of feedings (other powdered formulas, liquid formulas, and breast milk). Microbiological studies of the implicated formula identified *E. sakazakii* in opened and unopened cans from one of two manufactured lots of the implicated formula; these isolates were indistinguishable by PFGE pattern from the isolates obtained from the CSF of the index case. *E. sakazakii* was not identified from environmental cultures of the surfaces where the formula was prepared or from commercially sterile water used to prepare the formula. After discontinuation of the implicated formula, no other cases of illness or colonization were detected at the affected hospital.

This investigation demonstrated a complete investigation, including the screening of infants concurrently hospitalized in the unit to identify cases, a cohort study of potential risk exposures, and a microbiological study of formula and associated environments. The results of the cohort study that implicated a specific formula focused the microbiological study. Finally, molecular subtyping of the powdered formula and clinical isolates confirmed the formula as the source of *E. sakazakii* infection. The manufacturer voluntarily recalled the lot of implicated formula and disseminated a letter to health care providers about the risks of powdered infant formulas.

France, 2004

During one week in December 2004, French public health officials were notified of two deaths due to *E. sakazakii* meningitis and two cases of *E. sakazakii* colonization detected among infants hospitalized in two institutions between October 25 and December 7 (Coignard et al., 2006). That day, they launched a national investigation to identify additional cases, common risk factors, and control measures. In total, investigators detected two invasive cases and seven noninvasive cases associated with five hospitals. Two neonates with meningitis died; infants with conjunctivitis (one), enteritis (one), and gastrointestinal carriage (five) recovered. Although some of the hospitals were not following formula-handling standards (holding prepared bottles of formula up to 24 h, using noncommercial refrigerators for storage of prepared formula), no environmental samples yielded *E. sakazakii*. However, all seven infants were fed the same brand of powdered infant formula. Four lots of the powdered infant formula consumed by the ill infants were identified and recalled on the first day of the investigation. Public health officials isolated *E. sakazakii* from 36 (14%) of 252 samples obtained from four lots of unopened powdered formula tins associated with the outbreak. PFGE patterns were indistinguishable among the nine available clinical specimens and 21 of the powdered formula isolates. Subsequently, the formula manufacturer issued a recall of the implicated brand worldwide and temporarily ceased production in The Netherlands and the United States. Public health officials also distributed notices to hospitals regarding safe infant formula preparation and handling practices.

New Zealand, 2004

In a NICU, a 6-day-old, formula-fed female infant born at 33 weeks of gestation developed *E. sakazakii* bacteremia. Although the CSF was sterile, a cerebral ultrasound revealed a brain abscess. The infant died on the 18th day of life. Screening cultures were performed on surviving patients in the NICU; *E. sakazakii* was detected in the stool of 4 (17%) of 23 additional patients. During an environmental investigation, two brands of powdered infant formula used in the NICU yielded *E. sakazakii*; other environmental samples were negative. Clinical isolates were indistinguishable from each other and from isolates from one formula brand based on PFGE testing. Isolates from the second formula brand yielded a distinct PFGE pattern. Although no infant formula was recalled because of the outbreak, on 21 July 2005 New Zealand became the first country to make *E. sakazakii* infection a notifiable disease (Donald Campbell, New Zealand Food Safety Authority, personal communication, 2006).

The above review of outbreaks focuses on instances where multiple patients with infections or colonization were affected and investigated. Although there have been multiple reports of individual *E. sakazakii* cases, outbreaks provide the best opportunities to identify potential risk factors and sources of infections through cohort studies and focused microbiological sampling. These studies are often difficult to conduct, however. In the United States, *E. sakazakii* infection is not a nationally notifiable condition, so public health authorities may not learn of cases unless astute care providers take the initiative to call on public health authorities or investigate the incident themselves. Often, by the time the significance of an *E. sakazakii* infection is realized, too much time has passed to identify and study a cohort of potentially exposed patients in order to conduct appropriate studies of risk factors for infection. Also, many infants in the NICUs where infections are most likely to be recognized and investigated have prolonged and complicated medical courses, making the identification of specific risk factors difficult. Formula feeding is common and may be customized at the bedside for the most ill and vulnerable populations; identifying the exact type and lot of formulas which infants have been fed is often difficult, if not impossible, given the generally nonspecific documentation of feedings. These investigations also are very time intensive, technically challenging, and costly endeavors, outstripping the capacity of many local care providers and public health authorities. The outbreak investigations reviewed above represent the successful investigations among many other infections and outbreaks that have not been reported or successfully investigated. Making *E. sakazakii* a regionally or nationally notifiable disease may improve identification and timeliness of outbreak investigations and permit greater understanding of disease incidence, risk factors, and control measures.

ACKNOWLEDGMENTS

Sincere thanks are extended to Michelle Hulse, Matthew Kuehnert, Matthew Arduino, and Clifford McDonald for providing clinical and microbiological information.

REFERENCES

Adamson, D., and J. Rogers. 1981. *Enterobacter sakazakii* meningitis with sepsis. *Clin. Microbiol. Newsl.* 3:19–20.

American Dietetic Association. 2003. *Infant feedings: guidelines for preparation of formula and breast milk in health care facilities.* Pediatric Nutrition Practice Group, American Dietetic Association, Chicago, IL.

Anonymous. 2005. Policy statement. Breastfeeding and the use of human milk. *Pediatrics* 115:496–506.

Arseni, A., E. Malamou-Ladas, C. Koutsia, M. Xanthou, and E. Trikka. 1987. Outbreak of colonization of neonates with *Enterobacter sakazakii. J. Hosp. Infect.* 9:143–150.

Bar-Oz, B., O. Peleg, C. Block, and I. Arad. 2001. *Enterobacter sakazakii* infection in the newborn. *Acta Paediatr.* 90:356–358.

Barreira, E., D. de Souza, P. Gois, and J. Fernandez. 2003. Meningite por *Enterobacter sakazakii* em recen-nascido: relato de caso. *Pediatria* 25:65–70.

Biering, G., S. Karlsson, N. Clark, K. Jonsdottir, P. Ludvigsson, and O. Steingrimsson. 1989. Three cases of neonatal meningitis caused by *Enterobacter sakazakii* in powdered milk. *J. Clin. Microbiol.* 27:2054–2056.

Block, C., O. Peleg, N. Minster, B. Bar-Oz, A. Simhon, I. Arad, and M. Shapiro. 2002. Cluster of neonatal infections in Jerusalem due to unusual biochemical variant of *Enterobacter sakazakii. Eur. J. Clin. Microb. Infect. Dis.* 21:613–616.

Bowen, A., A. Newman, C. Estivariz, N. Gilbertson, J. Archer, A. Srinivasan, M. Lynch, and J. Painter. 2005. The role of acid-suppressing medication in a sustained outbreak of salmonellosis in a long-term care facility. *Abstr. Infect. Dis. Soc. Am. 43rd Annu. Meet.*, San Francisco, CA.

Bowen, A., and C. Braden. 2006. Clinical characteristics and outcomes of infants with invasive *Enterobacter sakazakii* disease. *Emerg. Infect. Dis.* **12**:1185–1189.

Burdette, J., and C. Santos. 2000. *Enterobacter sakazakii* brain abscess in the neonate: the importance of neuroradiologic imaging. *Pediatr. Radiol.* **30**:33–34.

Clark, N., B. Hill, C. O'Hara, O. Steingrimsson, and R. Cooksey. 1990. Epidemiologic typing of *Enterobacter sakazakii* in two neonatal nosocomial outbreaks. *Diagn. Microbiol. Infect. Dis.* **13**:467–472.

Coignard, B., V. Vaillant, J.-P. Vincent, A. Leflèche, P. Mariani-Kurkdjian, C. Bernet, F. L'Hériteau, H. Sénéchal, P. Grimont, and E. D. Bingen. 2006. Infections sévères à *Enterobacter sakazakii* chez des nouveau-nés ayant consommé une préparation en poudre pour nourrissons, France, octobre-décembre 2004. *Bull. Épidémiol. Hebdomadaire* **2–3**:10–13.

Dennison, S., and J. Morris. 2002. Multiresistant *Enterobacter sakazakii* wound infection in an adult. *Infect. Med.* **19**:533–535.

Dial, S., J. Delaney, A. Barkun, and S. Suissa. 2005. Use of acid-suppressive medications and the risk of community-acquired *Clostridium difficile*-associated disease. *JAMA* **294**:2989–2995.

DiGirolamo, A., L. Grummer-Strawn, and S. Fein. 2001. Maternity care practices: implications for breastfeeding. *Birth* **28**:94–100.

Farmer, J., M. Asbury, F. Hickman, and D. Brenner. 1980. *Enterobacter sakazakii*: a new species of "*Enterobacteriaceae*" isolated from clinical specimens. *Int. J. Syst. Bacteriol.* **30**:569–584.

Gallagher, P., and W. Ball. 1991. Cerebral infarctions due to CNS infection with *Enterobacter sakazakii*. *Pediatr. Radiol.* **21**:135–136.

Hamilton, J., M. Lehane, and H. Braig. 2003. Isolation of *Enterobacter sakazakii* from midgut of *Stomoxys calcitrans*. *Emerg. Infect. Dis.* **9**:1355–1356.

Hawkins, R., C. Lissner, and J. Sanford. 1991. *Enterobacter sakazakii* bacteremia in an adult. *South. Med. J.* **84**:793–795.

Himelright, I., E. Harris, V. Lorch, and M. Anderson. 2002. *Enterobacter sakazakii* infections associated with the use of powdered infant formula–Tennessee, 2001. *JAMA* **287**:2204–2205.

Howard, C., S. Schaffer, and R. Lawrence. 1997. Attitudes, practices, and recommendations by obstetricians about infant feeding. *Birth* **24**:240–246.

Iversen, C., and S. Forsythe. 2004. Isolation of *Enterobacter sakazakii* and other *Enterobacteriaceae* from powdered infant formula milk and related products. *Food Microbiol.* **21**:771–777.

Iversen, C., and S. Forsythe. 2003. Risk profile of *Enterobacter sakazakii*, an emergent pathogen associated with infant milk formula. *Trends Food Sci. Technol.* **14**:443–454.

Jimenez, E., and C. Gimenez. 1982. Septic shock due to *Enterobacter sakazakii*. *Clin. Microbiol. Newsl.* **4**:30.

Joker, R., T. Norholm, and K. Siboni. 1965. A case of neonatal meningitis caused by a yellow *Enterobacter*. *Danish Med. Bull.* **12**:128–130.

Kandhai, M., M. Reij, L. Gorris, O. Guillaume-Gentil, and M. van Schothorst. 2004. Occurrence of *Enterobacter sakazakii* in food production environments and households. *Lancet* **363**:39–40.

Kleiman, M., S. Allen, P. Neal, and J. Reynolds. 1981. Meningoencephalitis and compartmentalization of the cerebral ventricles caused by *Enterobacter sakazakii*. *J. Clin. Microbiol.* 14:352–354.

Kline, M. 1988. Pathogenesis of brain abscess caused by *Citrobacter diversus* or *Enterobacter sakazakii*. *Pediatr. Infect. Dis. J.* 7:891–892.

Kosloske, A. 1984. Pathogenesis and prevention of necrotizing enterocolitis: a hypothesis based on personal observation and review of the literature. *Pediatrics* 74:1086–1092.

Kramer, M., and R. Kikumo. 2002. The optimal duration of exclusive breastfeeding: a systematic review. World Health Organization, Geneva, Switzerland.

Lai, K. 2001. *Enterobacter sakazakii* infections among neonates, children, and adults. *Medicine* 80:113–122.

Leclercq, A., C. Wanegue, and P. Baylac. 2002. Comparison of fecal coliform agar and violet red bile lactose agar for fecal coliform enumeration in foods. *Appl. Environ. Microbiol.* 68:1631–1638.

Leuschner, R., and J. Bew. 2004. A medium for the presumptive detection of *Enterobacter sakazakii* in infant formula: interlaboratory study. *J. AOAC* 83:604–613.

Merewood, A., B. Philipp, N. Chawla, and S. Cimo. 2003. The baby-friendly hospital initiative increases breastfeeding rates in a US neonatal intensive care unit. *J. Hum. Lact.* 19:166–171.

Monroe, P., and W. Tift. 1979. Bacteremia associated with *Enterobacter sakazakii* (yellow pigmented *Enterobacter cloacae*). *J. Clin. Microbiol.* 10:850–851.

Murray, M., D. Welch, and T. Kuhls. 1990. *Serratia* osteochondritits after puncture wounds of the foot. *Pediatr. Infect. Dis. J.* 9:523–524.

Muytjens, H., and L. Kollee. 1990. *Enterobacter sakazakii* meningitis in neonates: causative role of formula? *Pediatr. Infect. Dis. J.* 9:372–373.

Muytjens, H., H. Roelofs-Willemse, and G. Jaspar. 1988. Quality of powdered substitutes for breast milk with regard to members of the family *Enterobacteriaceae*. *J. Clin. Microbiol.* 26:743–746.

Muytjens, H., H. Zanen, H. Sonderkamp, L. Kollee, I. Wachsmuth, and J. Farmer. 1983. Analysis of eight cases of neonatal meningitis and sepsis due to *Enterobacter sakazakii*. *J. Clin. Microbiol.* 18:115–120.

Nazarowec-White, M., and J. Farber. 1997. *Enterobacter sakazakii*: a review. *Int. J. Food. Microbiol.* 34:103–113.

Nazarowec-White, M., and J. Farber. 1999. Phenotypic and genotypic typing of food and clinical isolates of *Enterobacter sakazakii*. *J. Med. Microbiol.* 48:559–567.

Noriega, F., K. Kotloff, M. Martin, and R. Schwalbe. 1990. Nosocomial bacteremia caused by *Enterobacter sakazakii* and *Leuconostoc mesenteroides* resulting from extrinsic contamination of infant formula. *Pediatr. Infect. Dis. J.* 9:447–449.

Ongradi, J. 2002. Vaginal infection by *Enterobacter sakazakii*. *Sex. Transm. Infect.* 78:467–468.

Pribyl, C., R. Salzer, J. Beskin, R. Haddad, B. Pollock, R. Beville, B. Holmes, and W. Mogabgab. 1985. Aztreonam in the treatment of serious orthopedic infections. *Am. J. Med.* 78:51–56.

Reina, J., F. Parras, J. Gil, F. Salva, and P. Alomar. 1989. Human infections caused by *Enterobacter sakazakii.* Microbiologic considerations. *Enferm. Infecc. Microbiol. Clin.* **7:** 147–150.

Ryan, A., Z. Wenjun, and A. Acosta. 2002. Breastfeeding continues to increase into the new millennium. *Pediatrics* **110:**1103–1109.

Sheehan, D., P. Krueger, S. Watt, W. Sword, and B. Bridle. 2001. The Ontario mother and baby study: breastfeeding outcomes. *J. Hum. Lact.* **17:**211–219.

Simmons, B., M. Gelfand, M. Haas, L. Metts, and J. Ferguson. 1989. *Enterobacter sakazakii* infections in neonates associated with intrinsic contamination of a powdered milk formula. *Infect. Control. Hosp. Epidemiol.* **10:**398–401.

Stoll, B., N. Hansen, A. Fanaroff, and J. Lemons. 2004. *Enterobacter sakazakii* is a rare cause of neonatal septicemia or meningitis in VLBW infants. *J. Pediatr.* **144:**821–823.

Taveras, E., R. Li, L. Grummer-Strawn, M. Richardson, R. Marshall, V. Rego, I. Miroschnik, and T. Lieu. 2004. Opinions and practices of clinicians associated with continuation of exclusive breastfeeding. *Pediatrics* **113:**e283–e290.

Tekkok, I., S. Baessa, M. Higgins, and E. Ventureyra. 1996. Abscedation of posterior fossa dermoid cysts. *Child. Nerv. Syst.***12:**318–322.

Tunkel, A. R., and W. M. Scheld. 2005. Acute meningitis, p. 1083-1084. *In* G. L. Mandell, J. E. Bennett, and R. Dolin (ed.), *Principles and Practice of Infectious Diseases*, 6th ed. Churchill Livingstone, New York, NY.

Urmenyi, A., and A. Franklin. 1961. Neonatal death from pigmented coliform infection. *Lancet* **i:**313–315.

Van Acker, J., F. De Smet, G. Muyldermans, A. Bougatef, A. Naessens, and S. Lauwers. 2001. Outbreak of necrotizing enterocolitis associated with *Enterobacter sakazakii* in powdered milk formula. *J. Clin. Microbiol.* **39:**293–297.

Willis, J., and J. Robinson. 1988. *Enterobacter sakazakii* meningitis in neonates. *Pediatr. Infect. Dis. J.* **7:**196–199.

Enterobacter sakazakii
Edited by Jeffrey M. Farber and Stephen J. Forsythe
© 2008 ASM Press, Washington, D.C.

Pathogenicity of *Enterobacter sakazakii*

5

Franco Pagotto, Jeffrey M. Farber, and Raquel Lenati

INTRODUCTION

Enterobacter sakazakii is an emerging food-borne pathogen that has increasingly raised interest among the scientific community, health care providers, and the food industry since the early 1980s, when it was accepted as a new species. Nevertheless, there is still a lack of information concerning its natural habitat, mechanisms of virulence and pathogenicity, and its dose response for human infections. In 2002, the International Commission for Microbiological Specifications for Foods classified *E. sakazakii* as a severe hazard for restricted populations, causing life-threatening or substantial chronic sequelae or illness of long duration, with the high-risk populations being newborns and immunocompromised infants. In 2004, the Food and Agriculture Organization of the United Nations (FAO) and the World Health Organization (WHO) jointly held an expert meeting on *E. sakazakii* and other microorganisms of concern in powdered infant formula (PIF), aiming to gather information for the revision of the Recommended International Code of Hygienic Practice for Foods for Infants and Children (FAO, 1994). The FAO-WHO expert meeting of 2004 further agreed on a list of recommendations to the scientific community and infant formula manufacturers. They recommended a focus on a better understanding of *E. sakazakii* and potentially other microorganisms that could be found in infant formula. Specifically, one recommendation focused on the need for a better understanding of the ecology, taxonomy, virulence, and other characteristics of this emerging pathogen. This was also an emerging recommendation of the FAO-WHO during their second meeting, held in February 2006. This

Franco Pagotto, Jeffrey M. Farber, and Raquel Lenati, Bureau of Microbial Hazards, Health Products and Food Branch, Health Canada, Ottawa, Ontario, Canada.

chapter attempts to summarize research endeavors aimed at understanding the pathogenesis of this important food-borne pathogen.

Enterobacter species are known as opportunistic pathogens that generally cause infection in debilitated patients with an underlying illness, patients on immunosuppressive medication, or those taking antimicrobial agents, all factors that may facilitate the colonization of *Enterobacter* spp. in humans (Borderon et al., 1996). A member of the genus *Enterobacter*, *E. sakazakii* has been associated with infant gastroenteritis, bacteremia, and meningitis, and it has been linked in many cases to the consumption of reconstituted powdered infant formula (Biering et al., 1989; Block et al., 2002; Clark et al., 1990; Muytjens and Kollee, 1982; van Acker et al., 2001; Weir, 2002; Noriega et al., 1990; Simmons et al., 1989). It is possible that the immature neonatal immune system increases the risk of acquiring an *E. sakazakii* infection (Hammerman and Kaplan, 2006). However, it is not known exactly what host and environmental factors need to be present in order to cause infection in neonates.

E. sakazakii infections in newborns were, at one time, suspected to occur via passage of the organism through the mother's birth canal, similar to newborn infections caused by other pathogens transmitted from mother to child (Tift, 1977; Monroe and Tift, 1979). This hypothesis lost favor when *E. sakazakii* infections in neonates born by Cesarean section were reported (Bar-Oz et al., 2001; Muytjens and Kolle, 1990; Muytjens et al., 1983) and when neonates were diagnosed with *E. sakazakii* infection within weeks after birth (Kleiman et al., 1981; Willis and Robinson, 1988; Noriega et al., 1990). To date, the host and environmental factors needed to be present in order to cause infection in neonates remain unclear.

The first two known cases of *E. sakazakii* meningitis date from 1958 and were first reported in 1961. Subsequently, cases of meningitis, septicemia, and necrotizing enterocolitis due to this organism have been reported worldwide. While the overall number of cases reported worldwide has been small, the consequences can be very severe, with as many as one of two infants dying during an outbreak. Although most reported cases have involved infants, there have also been infections in children (Block et al., 2002; Lai et al., 2001) and adults (Dennison and Morris, 2002; Hawkins et al., 1991; Jimenez and Gimenez, 1982; Lai et al., 2001). Infections with *E. sakazakii* can occur as single cases or as part of an outbreak. In a number of cases, these outbreaks were found to be caused by contaminated powdered infant formula, most often in neonatal intensive care units. At present, research is being started in a number of laboratories around the world, but the impact of these studies will not be felt for a while. Therefore, initial steps must be taken to help regulatory agencies develop policies and control strategies based on sound science.

The pathogenesis of neonatal *E. sakazakii* meningitis has not been fully defined. A possible process is the translocation of the bacterium through the chordus plexus and subsequent cellular invasion by means of pathogenic secretory factors (e.g., elastases, glycopeptides, endotoxins, collagenases, and proteases) used to increase blood-brain barrier permeability, thus gaining access to the nutrient-rich cerebral matter (Iversen and Forsythe, 2003; Hurrell et al., 2006). Others have reported a similarity between the tropism of *E. sakazakii* and *Citrobacter koseri* (formerly *C. diversus*) for invasion of and infection in the central nervous system, although *C. koseri* shows no involvement of chordus plexus (Willis and Robinson, 1988; Kline, 1988; Burdette and Santos, 2000; Townsend et al., 2003).

VIRULENCE FACTORS: HEAT-LABILE, HEAT-STABLE ENTEROTOXINS

While *Enterobacter cloacae* pathogenicity and production of exotoxins, aerobactin, and hemagglutinin have been previously documented (Keller et al., 1998), the virulence factors of *E. sakazakii* remain unclear. Our laboratory was one of the first to attempt to describe a putative mechanism of pathogenesis for *E. sakazakii*, demonstrating the organism's potential for producing cytotoxins and/or enterotoxins, using Vero, CHO, and Y-1 cell lines, as well as a suckling mice assay (Pagotto et al., 2003).

In the study aimed at defining the presence of potential virulence factors, eight food strains of *E. sakazakii* isolated in our laboratory from various dried infant formulas available on the Canadian retail market were evaluated for enterotoxin production (Pagotto et al., 2003). In addition, nine clinical isolates implicated in human illness that were obtained from Canadian hospital culture collections were also tested for enterotoxin production. Of the eight food, nine clinical, and one type strain evaluated, four (22%) were positive for enterotoxin production. Of these four positive strains, three were clinical strains (MONT, LA, and SK92) isolated in three different hospitals in three different geographic locations in Canada (Montreal, Quebec; London, Ontario; and Toronto, Ontario, respectively). Only one (MNW5) of the eight *E. sakazakii* food isolates produced enterotoxin. Two other food strains isolated from different lots of the same dried infant formula were negative in the enterotoxin assay. The type strain, ATCC 29544, was not toxigenic.

VARIATION IN VIRULENCE EXPRESSION

In our initial studies (Pagotto et al., 2003), 10^8 cells of *E. sakazakii* administered orally to mice did not cause death in the majority of cases, and therefore some host factor(s) required for infection of that particular animal model may not have been present. Clinical strains used in this study may have lost some of their virulence factors typically carried on plasmids in other bacteria or at least lost the ability to express those virulence factors. In contrast, other animal models, such as those discussed later in this chapter, may provide more useful information on the regulation of genes implicated in the pathogenicity of the organism.

Among the features that allow *Escherichia coli* K1 to survive hostile environments are the K1 capsular polysaccharide, O-lipopolysaccharides (O-LPS), and outer membrane protein A (OmpA). It has been shown that the deletion of *Escherichia coli ompA* results in significant reduction of bacteremia in neonatal rat models (Wang and Kim, 2002). Interestingly, it has recently been demonstrated that *ompA* of *E. sakazakii* had a high degree of homology to the *ompA* genes of other gram-negative *Enterobacteriaceae* (Mohan Nair and Venkitanarayanan, 2006).

Little work has focused on gene expression of *E. sakazakii*, mainly due to the fact that virulence factors have not been clearly shown to be necessary for infection. To our knowledge, no gene knockouts, transposon mutants, etc., have been described (G. J. Phillips and E. Grace, Isolation of pigment mutants of *Enterobacter sakazakii*, abstr. D-098. Abstr. 104th Gen. Meet. Am. Soc. Microbiol., 2004). Our laboratory has recently begun using suppressive subtractive hybridization techniques to attempt to investigate differences that may occur among clinical (i.e., strains associated with *E. sakazakii* infant infections), environmental, and food strains. A strain, ATCC BAA-894, was sequenced by the Genome Sequencing Center at the Washington University School of Medicine in St. Louis. Bioinformatics (see http://genome.wustl .edu/home.cgi) analyses may help in the identification of putative virulence factors that may then be studied using a variety of methods.

TISSUE CULTURE STUDIES

In a first attempt to investigate the pathogenesis of *E. sakazaki* (previously conducted in this laboratory), Vero, CHO, and Y-1 cell lines were infected with *E. sakazakii* supernatant free of bacterial cells, and suckling mice were inoculated orally and intraperitoneally (i.p.) with up to 10^8 cells of *E. sakazakii*. Interestingly, not all *E. sakazakii* isolates tested demonstrated cytopathic effects against the cell lines, and some of the strains that tested positive in the enterotoxin assay performed with suckling mice did not cause any cytopathic effects in the cell lines tested (Pagotto et al., 2003). In addition, bacterial culture filtrates from *E. sakazakii* strain LA were found to be toxic to three cell lines. *E. sakazakii* LA filtrates had a more pronounced effect on Y-1 and Vero cells than CHO cells. For all cell lines, boiling of culture filtrates for 20 min did not have any effects, except on Vero cells, where there was an observed decrease in cytopathic effect. Interestingly, other isolates of *E. sakazakii*, such as MONT or ATCC 29544, did not demonstrate any cytopathic effects against any of the cell lines, even though strain MONT was positive for enterotoxin production using the suckling mouse assay (see below).

An in vitro assay described by Hurrell (2006) suggests that *E. sakazakii* is toxic to N2a neuroblastoma cells using the 3-(4,5-dimethylthiazol-2-yl)-2,5-diphenyl tetrazolium bromide cell proliferation assay. They theorized that protease, phosphatase, and lipase activities may contribute to host cell death during an *E. sakazakii* infection. Recently, in vitro assays have shown the ability of *E. sakazakii* to adhere to epithelial cells (Collado et al., 2005; Mange et al., 2006). Not surprisingly, as observed for most bacterial pathogens, this bacterium seems to be able to adhere to cells, an important first step in pathogenesis. Collado et al. (2005) demonstrated the adherence of the *E. sakazakii* type strain (ATCC 29544) to immobilized human intestinal mucus and its displacement by bifidobacteria.

Mange et al. (2006) described the adhesive properties of 50 *E. sakazakii* strains, at different phases of growth, to human epithelial (Hep-2 and Caco-2) and brain microvascular (HBMEC) cell lines. They observed two distinctive adherence patterns, diffuse and localized clusters of adhesion to the three cell lines. *E. sakazakii* adherence to the cell lines increased with a higher multiplicity of infection and was maximal at the late bacterial exponential growth phase. Interestingly, adhesion to the epithelial and endothelial cells was suggested to be mostly nonfimbria based.

Hurrell et al. (2006) initiated a comparative investigation into a range of potential virulence factors, such as attachment and invasion in Caco-2 cell lines, macrophage uptake (using human macrophage-like cell line U937), and serum resistance studies, as well as the potential role of motility and capsule formation. They compared the virulence of 12 non-*E. sakazakii Enterobacteriaceae* species to 27 strains of *E. sakazakii*. The *E. sakazakii* strains were from the four 16S cluster groups and included clinical, environmental, and food isolates. Levels of bacterial attachment and invasion of mammalian intestinal cells (Caco-2 and HT29), macrophage survival, and serum resistance for *E. sakazakii* were comparable to those of other *Enterobacteriaceae*, such as *E. cloacae* and *C. freundii*, but were less than those of *Salmonella enterica* serovar Typhimurium. Strains from 16S cluster 1 showed higher Caco-2 invasion rates than the other clusters. Similarly, the survival of *E. sakazakii* in macrophages varied, with strains from cluster 1 persisting or replicating, whereas those from cluster 3 were killed. In addition, strains from cluster 4 were serum sensitive. Hence, *E. sakazakii* 16S clusters 1 and 2 showed greater virulence potential than clusters 3 and 4. Visualization in both macrophages and Caco-2 culture cells showed the uptake of *E. sakazakii* into phagosomes.

ANIMAL MODELS

The mechanism(s) that *E. sakazakii* uses to reach the brain and cause meningitis has not been well investigated. However, Sondheimer et al. (1985) showed that 2-week-old infants have a more acidic gastric pH (ranging from 2.9 to 5.2) than 1-week-old infants (ranging from pH 4.6 to 5.8), while Dinsmore et al. (1997) and Mehall et al. (2001) demonstrated in the neonatal rabbit model that less acidic conditions in the gut facilitated the translocation of *E. cloacae* from the intestine to the mesenteric lymph nodes, spleen, and cecum. Since most of the *E. sakazakii* infections involve newborns, it is possible that the lower level of acidity in the neonatal gastrointestinal tract plays a role in the colonization and translocation of this organism. Once *E. sakazakii* makes its way past the gastrointestinal tract to the bloodstream, it likely attaches to endothelial cells and then somehow crosses through the blood-brain barrier to infect the meninges and brain.

Due to a lack of available animal models for the study of *E. sakazakii* pathogenesis, our laboratory was funded to screen several nonprimate species in order to identify the most suitable model that could be used to extrapolate results to human infections. From the results obtained in this study, young pigs, rabbits, and guinea pigs were not seen as good candidates to mimic the infectious process seen in human neonates. However, young chicks and gerbils were seen as potentially useful models, with neonatal gerbils being the most promising. Interestingly, neonatal rats are presently used as models for neonatal meningitis studies, and another group successfully conducted a study of *E. sakazakii* infections (Townsend et al., 2007). However, under the conditions of our study (Fig. 1), neonatal rats did not seem to be suitable for investigating the pathogenesis of *E. sakazakii* (data not shown).

Infant Sprague-Dawley rats have been used for more than three decades as animal models for the study of neonatal bacterial meningitis. These animals are recognized models for meningitis caused by *Escherichia coli* K1 and *Citrobacter* species (Xie et al., 2004; Xie et al., 2006; Hill et al., 2004; Glode et al., 2004; Townsend et al., 2003) and are often inoculated on day 2 of age via the oral, intranasal (i.n.), i.p., or cardiac route. It appears that, similar to observations of human infant clinical cases, *Citrobacter* meningitis in these animal models is age dependent, whereas compared to that of 2-day-old rats, 5-day-old rats are resistant to infection (Kline, 1988). Neonatal rats inoculated both i.n. (10^5 CFU) and i.p. (10^5 CFU) with *C. koseri* have been reported to develop bacteremia, meningitis, and death within 24 to 72 h, with some of the animals surviving up to day 8 postinoculation (p.i.) (Kline, 1988); at this point, the brain abscesses were more advanced than they were early on in infection. Both *Citrobacter* and *Escherichia coli* K1 meningitis appears to be hematogenous, since there is a correlation between the levels of bacteremia and the development of meningitis (Xie et al., 2004). The infection of the meninges caused by these two pathogens seems to occur more often (in both human infants and the neonatal rat model) when the levels of bacteria in the blood are $>10^3$ CFU/ml (Kline, 1988; Xie et al., 2004). This type of data presently is not available for *E. sakazakii* infections. Due to the similarities that exist among *E. sakazakii* and the latter organisms that also cause neonatal meningitis, some of the mechanisms of pathogenesis established for *Citrobacter* and *Escherichia coli* K1 may be similar in *E. sakazakii*.

In our work (Pagotto et al., 2003), the minimum lethal doses for LA, MONT, and ATCC 29544 were the same in challenge studies using either oral or i.p. inoculation routes, with lower doses required for the i.p. trials. It is possible that the differences noted in these experiments indicate differences in receptor-toxin binding or a complex regulation system. Challenge studies using i.p. injections showed that all of the tested strains of *E. sakazakii* were lethal to suckling mice at 10^8 CFU/mouse. All deaths occurred within 3 days postdosing, typically within 24 to 48 h. Strains SK92 (enterotoxin positive) and MNW6 (enterotoxin negative) had the lowest minimum lethal doses by the i.p. route but were nonlethal at high oral doses. These strains, therefore, may be missing some accessory virulence factors that would allow them to survive passage through the stomach and/or translocate across the intestinal wall. Additionally, it is quite possible that the virulence factors are under the control of a regulatory protein, such as a two-component system, that senses when *E. sakazakii* is in the presence of the host (associated with attachment to the intestinal wall versus growth in tryptic soy broth, for example) and then turns on the virulence genes. In comparison, only two strains of *E. sakazakii*, one clinical isolate (SK81) and one food isolate (MNW2), caused death by the peroral route, i.e., the clinical and food isolates were lethal at 10^7 CFU to one out of four mice after 48 and 72 h, respectively. Thus, it appears that under the conditions tested, these two strains may be more virulent than the others. Interestingly, however, under the conditions used, neither of these strains produced enterotoxin. Further, we do not rule out the possibility that *E. sakazakii* possesses more than one cytotoxin or that a horizontal gene transfer event transferring cytotoxin (and related genes) occurred only in certain isolates.

In our present studies, animals were challenged with the same dose (10^9 CFU). Although none of the young gerbils that were inoculated died or presented visible symptoms of infection, many of them contained *E. sakazakii*

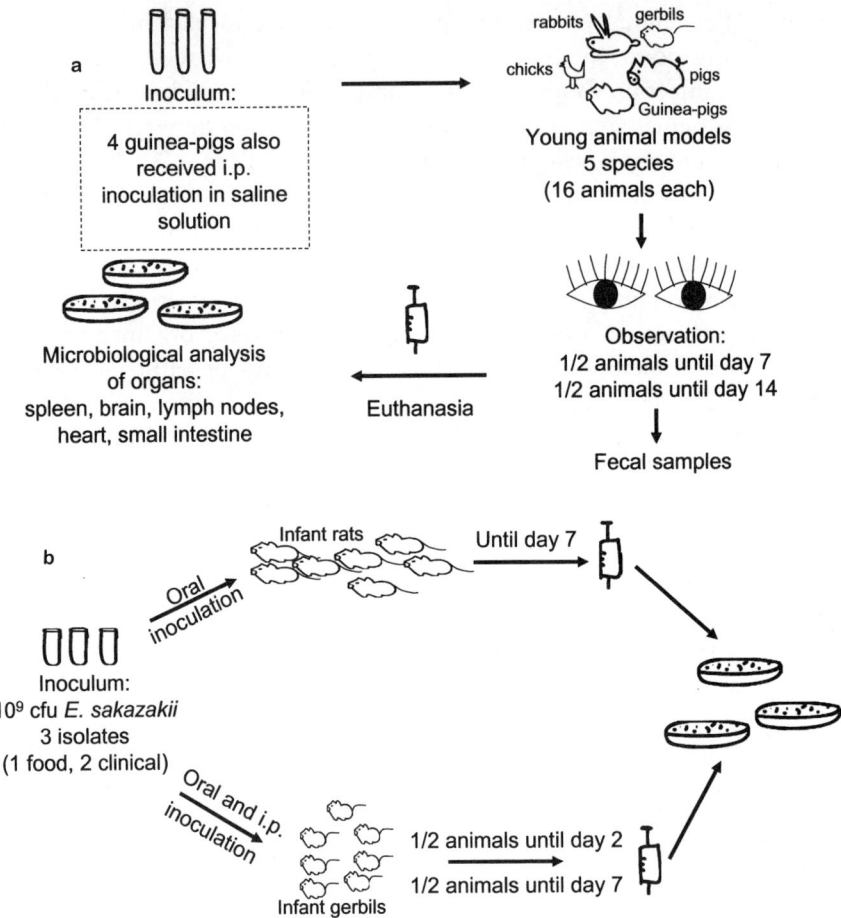

Figure 1 Schematic diagram illustrating young (a) and neonatal (b) experimental designs for assessing the virulence and pathogenicity of *E. sakazakii*.

cells in their organs (Table 1). However, infant gerbils (6 out of 26) died within 48 h p.i., and *E. sakazakii* was isolated from all of the infant gerbils inoculated. Because *E. sakazakii* was isolated from all infant gerbils for which high doses were used, further investigations are warranted to see if lower doses also can cause infection in infant gerbils. Furthermore, additional studies with this model would be interesting regarding the different indicators and end points that could be considered during investigations of *E. sakazakii* pathogenesis.

The food (2871) and clinical (2855) isolates used in the above study were previously used to assess *E. sakazakii* pathogenesis in the suckling mouse assay

Table 1 Recovery of *E. sakazakii* from the organs of young gerbils on days 7 and 14 p.i.

Time of euthanasia (day p.i.)	Gerbil no. (*E. sakazakii* strain)	Amt of *E. sakazakii* (CFU/g) recovered from:					
		Brain	Heart	Spleen	Liver	Kidney	Intestine
7	01 (2855)	25×10^1	NG	NG	NG	NG	+
	02 (2855)	NG[a]	NG	1.6×10^2	NG	NG	+
	03 (2871)	+[b]	NG	NG	+	NG	+
	04 (2871)	+	+	+	+	NG	+
	05 (3290)	$>3.0 \times 10^3$	5.0×10^2	3.3×10^2	1.6×10^3	3.0×10^2	NG
	06 (3290)	$>3.0 \times 10^3$	1.0×10^1	9.0×10^1	1.1×10^3	8.5×10^1	NG
14	07 (2855)	NG	NG	NG	NG	NG	NG
	08 (2855)	NG	NG	NG	NG	NG	NG
	09 (2871)	5.0×10^1	NG	NG	NG	NG	NG
	10 (2871)	NG	NG	NG	NG	NG	+
	11 (3290)	NG	NG	NG	NG	NG	NG
	12 (3290)	NG	NG	NG	NG	NG	NG

[a]NG, no growth.
[b]+, positive upon selective enrichment.

(Pagotto et al., 2003). While isolate 2871 was the only one to cause the death of neonatal gerbils in the second study after oral inoculation with 10^9 cells (Table 2), in our first study, three of four and one of four neonatal mice died after oral inoculation with this same isolate at concentrations of 10^8 and 10^7 CFU, respectively. Moreover, strain 2855 was among the five isolates that caused the death of suckling mice when inoculated via the i.p. route at a lower dose (10^7 CFU). However, isolates 2871 and 2855 did not test positive in the suckling mice enterotoxin assay, which is based on the ratio of intestine-to-carcass weight. In our present animal study, strain 2871 was isolated more often than strain 2855 from young gerbils on day 7 p.i., but only after selective enrichment. Interestingly, clinical isolate 3290, not included in our first study, was found more often in the organs of the young chicks and in higher numbers from the organs of the young gerbils than strains 2871 and 2855. While there were no major differences in the levels of the three strains in the intestines and brains of neonatal gerbils, strain 2855 was observed more often and in higher numbers in the spleen, liver, heart, and kidneys.

Gerbils have been a well-established model for *Helicobacter pylori* infection (Kodama et al., 2005) and have also been used as a model for *Streptococcus pneumoniae* meningitis. However, this is the first time that gerbils have been used as a model for gram-negative neonatal meningitis. The results obtained in our work indicate that the neonatal gerbil model is a suitable model for studies of *E. sakazakii* infection. Furthermore, the young gerbils, perhaps

Table 2 Presence of *E. sakazakii* in the organs of neonatal gerbils after oral and i.p. inoculation with 10^9 cells of strain 2871

Time of death/euthanasia (p.i.)	Mode of inoculation	Neonatal gerbil no.	Amt of *E. sakazakii* (CFU/g) recovered from:					
			Brain	Heart	Spleen	Liver	Kidney	Intestine
24 h (death)	i.p.	01	TNTC[a]	TNTC	TNTC	TNTC	TNTC	TNTC
	i.p.	02	TNTC	TNTC	TNTC	TNTC	TNTC	TNTC
	i.p.	03	TNTC	TNTC	TNTC	TNTC	TNTC	TNTC
48 h (death)	Oral	04	2×10^1	5×10^2	1×10^3	1×10^3	1×10^3	TNTC
	Oral	05	3×10^2	2×10^3	NC[b]	TNTC	NC	TNTC
	Oral	06	5×10^2	2×10^3	TNTC	TNTC	TNTC	TNTC
48 h (euthanasia)	Oral	07	NG[c]	NG	1.0×10^0	NG	5.0×10^0	2×10^3
	Oral	08	NG	NG	3.4×10^2	2.0×10^0	3×10^1	4×10^1
	Oral	09	NG	NG	1.4×10^1	1.0×10^0	3×10^1	2×10^3
	Oral	10	3.0×10^1	5.0×10^0	NG	NG	5.0×10^0	1×10^3
7 days (euthanasia)	Oral	11	NG	NG	6×10^1	3.0×10^0	3×10^1	2×10^3
	Oral	12	NG	NG	5×10^1	NG	3×10^2	TNTC

[a]TNTC, too numerous to count.
[b]NC, not collected.
[c]NG, no growth.

with some further investigations, may also give useful information, since *E. sakazakii* caused infection and was recovered from the organs of both young and neonatal gerbils. Interestingly, *E. sakazakii* could not be recovered from some of the intestines (7 out of 12) of the young gerbils, even when their fecal specimens were positive for the bacteria up to day 7 p.i. We have hypothesized that this may be attributable to the background microbiota present in the intestine, which may compete with *E. sakazakii* for growth on the laboratory media. Moreover, since *E. sakazakii* was recovered from the intestines of 5 out of 12 young gerbils, it is possible that variability exists among the young animals' intestinal microbiota. This variability should be less important in the neonatal gerbils, since they are fed only by the mother and have been less exposed to the environment. Further studies using the neonatal gerbil model will be important in order to investigate whether differences among *E. sakazakii* isolates could be consistently identified in challenge studies as well as to improve the protocol used herein.

The causes of the differences in numbers of *E. sakazakii* cells isolated from the organs of the different animal models in this study remain unclear. Factors such as previous exposure of young gerbils to *Enterobacter* spp. from the environment may have played a role in the clearing of *E. sakazakii* cells. However, the neonatal gerbils were inoculated at day 2 of age and most likely did not live long enough to develop any immunocompetent cells. In order to better clarify the reason for the recovery of different levels of *E. sakazakii* from the organs of neonatal gerbils, aspects such as the amount of mother's milk being fed to the neonates or the timing of inoculation in relation to the feeding time could be further assessed.

Townsend et al. (2007), using an animal assay model similar to that previously used for *C. koseri* (Townsend et al., 2003), studied the permeability of the intestinal tract and blood-brain barrier by orally challenging neonatal rats with 10^8 cells of *E. sakazakii* (ATCC 12868) with or without a dosing with LPS. Consequently, they found cerebrospinal fluid to be positive for bacteria from the intestinal tract. This occurred in 25% of the animals challenged with *E. sakazakii* only and for 40% of the animals in which *E. sakazakii* inoculation was preceded by an LPS dose. *E. sakazakii* was isolated from the spleen. Interestingly, despite all putative similarities between *E. sakazakii*, *C. koseri*, and *E. coli* K1, and in contrast to the results obtained by Townsend et al. (2007), we were not able to isolate *E. sakazakii* from any of the neonatal rats orally inoculated with 10^9 cells, nor did the animals present any signs of infection. Although the technique used for the oral inoculation of the neonatal rats in the present study was very similar to that used by Townsend et al. (2003 and 2007), the microbiological methods used to analyze the organs may have been approached differently. In addition, the *E. sakazakii* strains chosen for each of the studies also could have caused a difference in the results.

MINIMUM INFECTIOUS DOSE

In the more than 76 documented cases of neonatal and infant *E. sakazakii* infections, the infectious dose was not determined. Iversen and Forsythe (2003) speculated that a reasonable estimate for infection might be close to that postulated for *Escherichia coli* O157:H7, *Listeria monocytogenes* 4b, or *Neisseria meningitidis*, i.e., ca. 1,000 CFU. They added that *Enterobacter* would not encounter extremely harsh pH conditions in the upper gastrointestinal tract of neonates and would pass rapidly into the small intestine. Based on the level of contamination of ca. 0.36 to 66 CFU/100 g of powder reported by Muytjens et al. (1988) and Nazarowec-White and Farber (1997) and 18 g of powder reconstituted in a single feeding, 14 generations would be needed to produce a 1,000-CFU level. This would require only 7 h at 37°C versus 17.9 h at 21°C, 1.7 days at 18°C, 7.9 days at 10°C, and nearly 9 days at 8°C (Iversen and Forsythe, 2003). The immune and microbiological status of the neonate plays an important role, as many infections are manifested early on, within the first 2 weeks of life.

Our study suggested that a large number of *E. sakazakii* cells (10^8 CFU) were required to cause infection in a suckling mice model (Pagotto et al., 2003). Extrapolating from that data, it would be tempting to suggest that a large dose of *E. sakazakii* may also be required to cause infection in human neonates. Interestingly, two of the clinical *E. sakazakii* isolates tested in that study, one being positive and the other negative in the in vitro assay using cell lines, had the lowest minimum lethal i.p. doses and were nonlethal after oral dosing. This could indicate that some *E. sakazakii* strains lack the necessary virulence factors that would allow them to cause infection. Nevertheless, analysis of *E. sakazakii* neonatal infections linked to the consumption of PIF indicates that low concentrations of the bacteria are initially present in the PIF. For example, Simmons et al. (1989) reported *E. sakazakii* and *E. cloacae* populations of 8 CFU/100 g and 48 CFU/100 g, respectively, in powdered formula linked to neonatal infection cases. Thus, it appears that low levels of *E. sakazakii* in PIF can lead to infection in neonates (Clark et al., 1990; Simmons et al., 1989). However, the mishandling of infant formula during preparation, storage, and/or feeding can facilitate the growth of *E. sakazakii* to potentially high levels (FAO-WHO, 2006). Extrapolating from the work done to date on animal models (Pagotto et al., 2003), it appears that high levels of the organism are necessary to cause illness.

HOST FACTORS IN DIFFERENT AGE GROUPS

E. sakazakii infections are primarily linked to newborns and infants, causing symptoms such as sepsis, meningitis, and necrotizing enterocolitis. Term and preterm infants are the main population of concern for this opportunistic food-borne pathogen (see chapter 4). However, while adult cases of *E. sakazakii* infection have been reported among immunocompromised adults (Dennison and Morris, 2002; Hawkins et al., 1991; Jiminez and Giminez, 1992; Lai, 2001), there have been other reports of the isolation of *E. sakazakii* from adults (Health Protection Agency, 2004). Healthy adults are, however, considered a low-risk group for *E. sakazakii* infection.

Iversen and Forsythe (2003) suggested that, similar to the pneumococci, *Haemophilus,* and meningococci, major pathogens associated with meningitis in children, *E. sakazakii* could also present a host developmental dependence on access to the central nervous system. The authors further suggested that, due to paracellular and transcellular mechanisms that may induce permeability of the blood-brain barrier, the choroid plexus may be the most likely entry site. Willis and Robinson (1988) described two cases of *E. sakazakii* infections involving cerebral infarctions followed by the development of cystic lesions, symptoms similar to those caused by *C. koseri*. Based on previous work by Foreman et al. (1984) describing possible cyst formation involving a sequence of vasculitis, necrosis, and liquefaction of the cerebral white matter possibly misdiagnosed as abscesses, Willis and Robinson (1988) suggested a similar ability of *E. sakazakii* and *C. koseri* to induce a cascade of events leading to a high rate of cyst formation. Farmer et al. (1980) had previously shown that *C. koseri* is 50% related to *E. sakazakii* by DNA-DNA hybridization, while Iversen et al. (2004) compared *Citrobacter* species and *E. sakazakii* 16S rRNA gene sequences and found that some *E. sakazakii* strains were 97.8% similar to the *C. koseri* strain assessed.

A recent review of 46 cases of invasive infections caused by *E. sakazakii* in infants focused on the identification of host risk factors and disease course (Bowen and Braden, 2006). Interestingly, using meningitis and bacteremia alone to characterize the disease, invasive infant cases were divided into two groups based on gestational age and birth weight.

FUTURE WORK

The interest in the causal relationship between PIF and this emerging food-borne pathogen has led to an interest in ascertaining its mechanism(s) of pathogenicity. This is important for the development of better treatment methods for infected infants as well as for illness prevention strategies. In addition, the

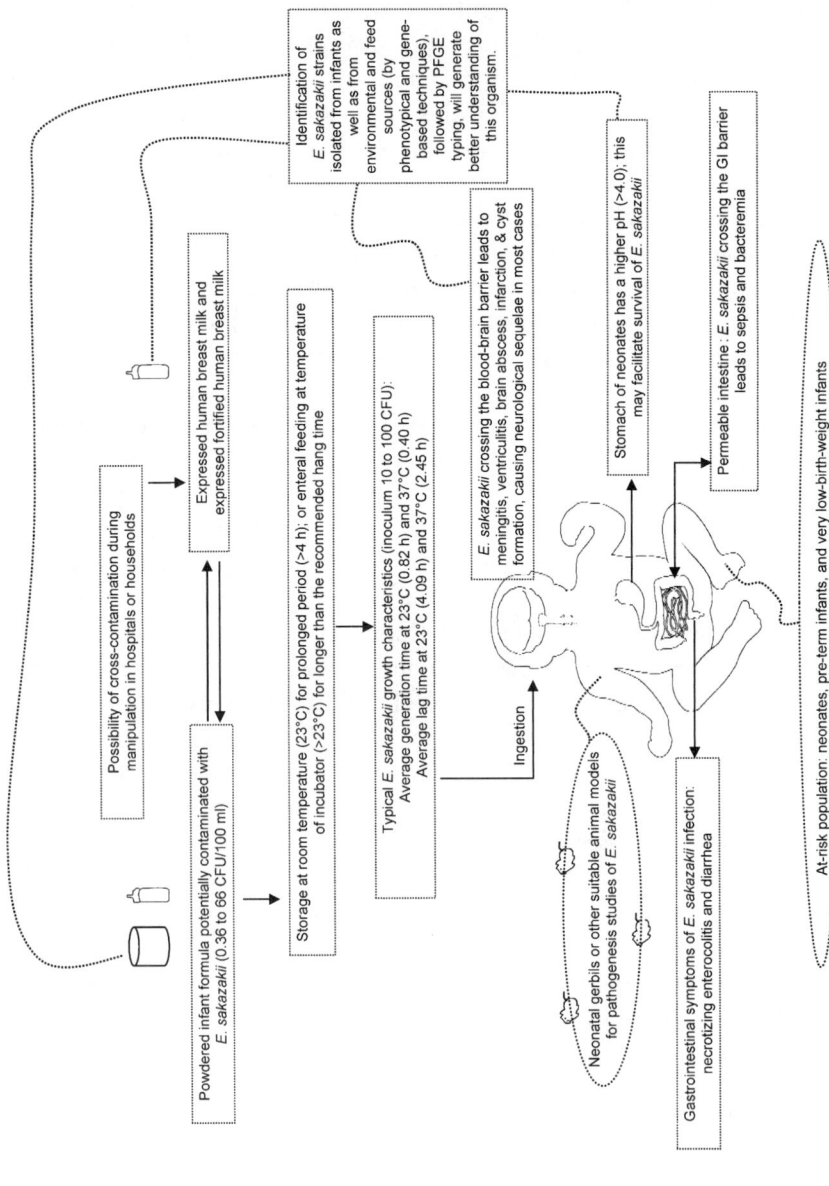

Figure 2 Proposed model for the pathogenesis of *E. sakazakii*, based on our own work and that of the literature.

present failure to arrive at a consensus on the dose response for this organism in infants limits the establishment of effective policies and regulations for *E. sakazakii* in powdered infant formula, including its preparation in hospitals and households. Further work is clearly warranted.

The research community has been joining efforts to find solutions for many of the unanswered questions that remain. The large amount of work done to increase awareness of this organism in the past 10 years is illustrated by the number of recent research publications (77 of the 116 documents published on *E. sakazakii* date from the past 6 years) as well as the increase in the number of scientific and industry meetings dedicated to this bacterium. Figure 2 summarizes the highlights of the literature and proposes a relational dependence for successful infection of the human host.

REFERENCES

Bar-Oz, B., A. Preminger, O. Peleg, C. Block, and I. Arad. 2001. *Enterobacter sakazakii* infection in the newborn. *Acta Paediatr.* **90:**356–358.

Biering, G., S. Karlsson, N. C. Clark, K. E. Jonsdottir, P. Ludvigsson, and O. Steingrimsson. 1989. Three cases of neonatal meningitis caused by *Enterobacter sakazakii* in powdered milk. *J. Clin. Microbiol.* **27:**2054–2056.

Block, C., O. Peleg, N. Minster, B. Bar-Oz, A. Simhon, I. Arad, and M. Shapiro. 2002. Cluster of neonatal infections in Jerusalem due to unusual biochemical variant of *Enterobacter sakazakii. Eur. J. Clin. Microbiol. Infect. Dis.* **21:**613–616.

Borderon, J. C., C. Lionnet, C. Rondeau, A. L. Suc, J. Laugier, and F. Gold. 1996. Current aspects of the fecal flora of the newborn without antibiotherapy during the first 7 days of life: *Enterobacteriaceae,* enterococci, staphylococci. *Pathol. Biol.* **44:**416–422.

Bowen, A. B., and C.R. Braden. 2006. Invasive *Enterobacter sakazakii* disease in infants. *Emerg. Infect. Dis.* **12:**1185–1189.

Burdette, J. H., and C. Santos. 2000. *Enterobacter sakazakii* brain abscess in the neonate: the importance of neuroradiologic imaging. *Pediatr. Radiol.* **30:**33–34.

Clark, N. C., B. C. Hill, C. M. O'Hara, O. Steingrimsson, and R. C. Cooksey. 1990. Epidemiologic typing of *Enterobacter sakazakii* in two neonatal nosocomial outbreaks. *Diagn. Microbiol. Infect. Dis.* **13:**467–472.

Collado, M. C., M. Gueimonde, M. Hernandez, Y. Sanz, and S. Salminen. 2005. Adhesion of selected *Bifidobacterium* strains to human intestinal mucus and the role of adhesion in enteropathogen exclusion. *J. Food Prot.* **68:**2672–2678.

Dennison, S. K., and J. Morris. 2002. Multiresistant *Enterobacter sakazakii* wound infection in an adult. *Infect. Med.* **19:**533–535.

Dinsmore, J. E., R. J. Jackson, and S. D. Smith. 1997. The protective role of gastric acidity in neonatal bacterial translocation. *J. Pediatr. Surg.* **32:**1014–1016.

Farmer, J. J., M. A. Asbury, F. W. Hickman, D. J. Brenner, and the *Enterobacteriaceae* Study Group. 1980. *Enterobacter sakazakii,* new species of *Enterobacteriaceae* isolated from clinical specimens. *Int. J. Syst. Bacteriol.* **30:**569–584.

Food and Agriculture Organization (FAO). 1994. Codex Alimentarius: code of hygienic practice for foods for infants and children. Food and Agriculture Organization of the United Nations, Rome, Italy.

Food and Agriculture Organization-World Health Organization (FAO-WHO). 2004. *Enterobacter sakazakii* and other microorganisms in powdered infant formula: meeting report. *Microbiological risk assessment series 6.* World Health Organization-Food and Agriculture Organization of the United Nations, Geneva and Rome. WHO Press, Geneva, Switzerland. http://www.who.int/foodsafety/publications/micro/mra6/en/index.html.

Food and Agriculture Organization-World Health Organization (FAO-WHO). 2006. *Enterobacter sakazakii* and *Salmonella* in powdered infant formula: meeting report. *Microbiological risk assessment series 10.* World Health Organization-Food and Agriculture Organization of the United Nations, Geneva and Rome. WHO Press, Geneva, Switzerland. http://www.who.int/foodsafety/publications/micro/mra10/en/index.html.

Foreman, S. D., E. E. Smith, N. J. Ryan, and G. R. Hogan. 1984. Neonatal *Citrobacter* meningitis: pathogenesis of cerebral abscess formation. *Ann. Neurol.* **16:**655–659.

Glode, M. P., A. Sutton, E. R. Moxon, and J. B. Robbins. 1977. Pathogenesis of neonatal *Escherichia coli* meningitis: induction of bacteremia and meningitis in infant rats fed *E. coli* K1. *Infect. Immun.* **16:**75–80.

Hammerman, C., and M. Kaplan. 2006. Probiotics and neonatal intestinal infection. *Curr. Opin. Infect. Dis.* **19:**277–282.

Hawkins, R. E., C. R. Lissner, and J. P. Sanford. 1991. *Enterobacter sakazakii* bacteremia in an adult. *South. Med. J.* **84:**793–795.

Health Protection Agency. 2004. *Klebsiella, Enterobacter, Serratia,* and *Citrobacter* spp. bacteraemias, England, Wales, and Northern Ireland: 2003. *Commun. Dis. Rep. CDR Wkly.* **14:**21. http://www.hpa.org.uk/cdr/archives/2004/bact_2104.pdf.

Hill, V. T., S. M. Townsend, R. S. Arias, J. M. Jenabi, I. Gomez-Gonzalez, H. Shimada, and J. L. Badger. 2004. TraJ-dependent *Escherichia coli* K1 interactions with professional phagocytes are important for early systemic dissemination of infection in the neonatal rat. *Infect. Immun.* **72:**478–88.

Hurrell, E., S. Townsend, and S. Forsythe. 2006. Comparative virulence of *Enterobacter sakazakii* with other *Enterobacteriaceae.* Poster presentation, *106th Gen. Meet. Am. Soc. Microbiol.,* 21 to 25 May, Orlando, FL.

International Commission on Microbiological Specification for Foods Micro-Organisms in Foods. 2002. *Microbiological Testing in Food Safety Management,* p. 128–130, vol. 7. Kluwer Academic/Plenum Publishers, New York, NY.

Iversen, C., and S. Forsythe. 2003. Risk profile of *Enterobacter sakazakii,* an emergent pathogen associated with infant milk formula. *Trends Food Sci. Technol.* **14:**443–454.

Iversen, C., M. Waddington, S. L. On, and S. Forsythe. 2004. Identification and phylogeny of *Enterobacter sakazakii* relative to *Enterobacter* and *Citrobacter* species. *J. Clin. Microbiol.* **42:**5368–5370.

Jimenez, E. B., and C. Gimenez. 1982. Septic shock due to *Enterobacter sakazakii. Clin. Microbiol. Newsl.* **4:**30.

Keller, R., M. A. Pedroso, R. Ritchman, and R. M. Silva. 1998. Occurrence of virulence associated properties in *Enterobacter cloacae. Infect. Immun.* **66:**645–649.

Kleiman, M. B., S. D. Allen, P. Neal, and J. Reynolds. 1981. Meningoencephalitis and com-partmentalization of the cerebral ventricles caused by *Enterobacter sakazakii. J. Clin. Microbiol.* **14:**352–354.

Kline, M. W. 1988. Pathogenesis of brain abscess caused by *Citrobacter diversus* or *Enterobacter sakazakii. Pediatr. Infect. Dis. J.* **7:**891–892.

Kodama, M., K. Murakami, R. Sato, T. Okimoto, A. Nishizono, and T. Fujioka. 2005. *Helicobacter pylori*-infected animal models are extremely suitable for the investigation of gas-tric carcinogenesis. *World J. Gastroenterol.* **11:**7063–7071.

Lai, K. K. 2001. *Enterobacter sakazakii* infections among neonates, infants, children, and adults. Case reports and a review of the literature. *Medicine* (Baltimore) **80:**113–122.

Mange, J. P., R. Stephan, N. Borel, P. Wild, K. S. Kim, A. Pospischil, and A. Lehner. 2006. Adhesive properties of *Enterobacter sakazakii* to human epithelial and brain microvascular en-dothelial cells. *BMC Microbiol.* **6:**58.

Mehall, J. R., R. Northrop, D. A. Saltzman, R. J. Jackson, and S. D. Smith. 2001. Acidification of formula reduces bacterial translocation and gut colonization in a neonatal rabbit model. *J. Pediatr. Surg.* **36:**56–62.

Mohan Nair, M. K., and K. S. Venkitanarayanan. 2006. Cloning and sequencing of the *ompA* gene of *Enterobacter sakazakii* and development of an *ompA*-targeted PCR for rapid detection of *Enterobacter sakazakii* in infant formula. *Appl. Environ. Microbiol.* **72:**2539–2546.

Monroe, P. W., and W. L. Tift. 1979. Bacteremia associated with *Enterobacter sakazakii* (yel-low, pigmented *Enterobacter cloacae*). *J. Clin. Microbiol.* **10:**850–851.

Muytjens, H. L., and L. A. Kollee. 1982. Neonatal meningitis due to *Enterobacter sakazakii. Tijdschr. Kindergeneeskd.* **50:**110–112.

Muytjens, H. L., H. C. Zanen, H. J. Sonderkamp, L. A. Kollee, I. K. Wachsmuth, and J. J. Farmer III. 1983. Analysis of eight cases of neonatal meningitis and sepsis due to *Enterobacter sakazakii. J. Clin. Microbiol.* **18:**115–120.

Muytjens, H. L., H. Roelofs-Willemse, and G. H. Jaspar. 1988. Quality of powdered substi-tutes for breast milk with regard to members of the family *Enterobacteriaceae. J. Clin. Microbiol.* **26:**743–746.

Muytjens, H. L., and L. A. Kollee. 1990. *Enterobacter sakazakii* meningitis in neonates: causative role of formula? *Pediatr. Infect. Dis. J.* **9:**372–373.

Nazarowec-White, M., and J. M. Farber. 1997. *Enterobacter sakazakii*: a review. *Int. J. Food Microbiol.* **34:**103–113.

Noriega, F. R., K. L. Kotloff, M. A. Martin, and R. S. Schwalbe. 1990. Nosocomial bac-teremia caused by *Enterobacter sakazakii* and *Leuconostoc mesenteroides* resulting from extrinsic contamination of infant formula. *Pediatr. Infect. Dis.* **9:**447–449.

Pagotto, F. J., M. Nazarowec-White, S. Bidawid, and J. M. Farber. 2003. *Enterobacter sakaza-kii*: infectivity and enterotoxin production in vitro and in vivo. *J. Food Prot.* **66:**370–375.

Simmons, B. P., M. S. Gelfand, M. Haas, L. Metts, and J. Ferguson. 1989. *Enterobacter sakazakii* infections in neonates associated with intrinsic contamination of a powdered infant formula. *Infect. Control Hosp. Epidemiol.* **10:**398–401.

Sogaard, P., and P. Kjaeldgaard. 1986. Two isolations of enteric group 69 from human clini-cal specimens. *Acta Pathol. Microbiol. Immunol. Scand. B* **94:**365–367.

Sondheimer, J., D. Clark, and E. Gervaise. 1985. Continuous gastric pH measurement in young and older healthy preterm infants receiving formula and clear liquid feedings. *J. Pediatr. Gastroenterol. Nutr.* 4:352–355.

Townsend, S., H. A. Pollack, I. Gonzalez-Gomez, H. Shimada, and J. Badger. 2003. *Citrobacter koseri* brain abscess in the neonatal rat: survival and replication within human and rat macrophages. *Infect. Immun.* 71:5871–5880.

Townsend, S., J. C. Barron, C. Loc-Carrillo, and S. Forsythe. 2007. The presence of endotoxin in powdered infant formula milk and the influence of endotoxin and *Enterobacter sakazakii* on bacterial translocation in the infant rat. *Food Microbiol.* 24:67–74.

Tift, W. L. 1977. Group B streptococcal infections in the neonate. *J. Med. Assoc. Ga.* 66:703–705.

Van Acker, J., F. de Smet, G. Muyldermans, A. Bougatef, A. Naessens, and S. Lauwers. 2001. Outbreak of necrotizing enterocolitis associated with *Enterobacter sakazakii* in powdered milk formula. *J. Clin. Microbiol.* 39:293–297.

Wang, Y., and K. S. Kim. 2002. Role of OmpA and IbeB in *Escherichia coli* K1 invasion of brain microvascular endothelial cells in vitro and in vivo. *Pediatr. Res.* 51:559–563.

Weir, E. 2002. Powdered infant formula and fatal infection with *Enterobacter sakazakii. Can. Med. Assoc. J.* 166:1570.

Willis, J., and J. E. Robinson. 1988. *Enterobacter sakazakii* meningitis in neonates. *Pediatr. Infect. Dis. J.* 7:196–199.

Xie, Y., K. J. Kim, and K. S. Kim. 2004. Current concepts on *Escherichia coli* K1 translocation of the blood-brain barrier. *FEMS Immunol. Med. Microbiol.* 42:271–279.

Xie Y., Y. Yao, V. Kolisnychenko, C. H. Teng, and K. S. Kim. 2006. KS.HbiF regulates type 1 fimbriation independently of FimB and FimE. *Infect. Immun.* 74:4039–4047.

Enterobacter sakazakii
Edited by Jeffrey M. Farber and Stephen J. Forsythe
© 2008 ASM Press, Washington, D.C.

Production of Powdered Infant Formulae and Microbiological Control Measures

6

Jean-Louis Cordier

PRODUCT CATEGORIES—DEFINITIONS

The composition, quality, and labeling requirements of powdered infant formulae are clearly laid down in either national, regional, or international regulations or standards. Examples are the Codex Alimentarius standard for infant formula (CAC, 1981), the Infant Formula Act in the United States (FDA, 2004), and Commission Directive 91/321/EEC (repealed as of 1 January 2008) and Commission Directive 2006/141/EC, established by the European Commission (EC, 1991, 1996). Many other national regulations exist, and they may differ in their definitions and requirements. However, it is not the purpose of this chapter to review them in detail.

The Commission Directive, for example, defines infant formulae as foodstuffs intended for particular nutritional use by infants during the first months of life and satisfying by themselves the nutritional requirements of such infants until the introduction of appropriate complementary feeding.

Fortifiers added to expressed human milk, as well as special formulae designed to meet the increased needs for proteins, calcium, and phosphorus of premature babies, very-low-birth-weight babies, or babies suffering from nutritional deficiencies and associated medical conditions (Schanler, 2005), would have to comply with the requirements outlined in the Commission Directive 1999/21/EC (EC, 1999, 2006, and amended in the Directive 2006/141/EC). This directive defines such dietary foods for special medical purposes as a category of foods for particular nutritional uses, especially processed or formulated and intended for the dietary management of patients and to be used under medical supervision. Infant formulae included in these

JEAN-LOUIS CORDIER, Nestlé Nutrition, Avenue Reller 22, CH-1800 Vevey, Switzerland.

two directives have been linked to outbreaks related to the presence of *Salmonella* as well as of *Enterobacter sakazakii* and therefore need to be manufactured according to the principles outlined in the section on production methods, i.e., according to very stringent hygiene measures.

A third product category is, according to Commission Directive 2006/141/EC (EC, 2006), follow-on formulae, meaning foodstuffs intended for a particular nutritional use by infants when appropriate complementary feeding is introduced and constituting the principal liquid element in a progressively diversified diet of such infants. For infants in this age group, these formulae therefore do not constitute their only food, but the formulae are given along with other foods, including fresh ones. They are manufactured under stringent control measures suitable to prevent the presence of *Salmonella*, the relevant pathogen in this case. They are frequently manufactured on the same processing lines as infant formulae and therefore also would comply with the same requirements.

SIZE OF MARKETS

Published data on volumes of infant formulae and market sizes are scarce and fragmented. Some information may be accessible through professional organizations, for example, the Danish Dairy Board (2005), which provides information on the Danish exports of infant formulae during the last 3 years. Other sources of information are reports from initiatives such as the Special Supplemental Nutrition Program for Women, Infants, and Children (WIC), in which sales in the United States are summarized (Oliveira et al., 2001). The most complete estimates and databases on the volumes manufactured are published on a regular basis by Euromonitor International, a commercial company dedicated to economical analyses of different categories of industries. Data for infant formulae over the last few years have been summarized in Table 1 for the world and for specific geographic areas. Similar tables have been published by the same company for follow-on formulae.

PATTERNS AND TRENDS

Mother's milk represents the best option to feed newborn babies. However, when a mother cannot breastfeed or chooses not to do so, infant formulae represent an appropriate alternative (WHO, 1981). In such a situation they are intended to replace mother's milk either partially or totally and are therefore formulated in order to meet the nutritional needs of infants during the first months of life. Human milk has always served as the gold standard for the

Table 1 Estimates of the worldwide production of infant formula

Location	Estimated production (vol, in 1,000 tons) in:							
	1998	1999	2000	2001	2002	2003	2004	2005[a]
World	482.6	494.5	497.2	494.6	478.2	472.1	468.1	461.2
Western Europe	67.1	67.1	67.6	68.2	67.6	68.7	70.2	71.5
Eastern Europe	12.7	12.4	12.7	12.9	13.2	13.5	14.3	15.4
North America	276.5	287.5	287.4	280.6	261.7	250.9	239.7	223.7
Latin America	23.4	22.9	23.5	22.2	21.1	21.3	21.5	22.5
Asia Pacific	87.3	88.6	91.6	94.2	97.8	100.9	105.3	110.9
Australasia	3.6	3.5	3.6	3.5	3.5	3.5	3.5	3.6
Africa and Middle East	12	12.4	12.8	13	13.3	13.4	13.5	13.8

[a]Packaged food data for 2005 is based on part-year estimates and is provisional. The copyright and database rights for these data belong to Euromonitor, 2006.

development and formulation of infant formulae, and they have evolved over time, benefiting from progresses in nutritional, medical, and pediatric sciences and processing technology. The historical development and evolution in the composition of infant formulae have been described by Barness (1987), Benson and Masor (1994), Motil (2000), Fomon (2001), Greer (2001), Carver (2003), and O'Callaghan and Wallingford (2004). The major components of infant formulae are proteins, lipids, carbohydrates, minerals, and vitamins. Their composition is based on requirements laid down in regulations and standards mentioned in the first section of this chapter or by the medical profession (Koletzko et al., 2005). New ingredients always undergo thorough evaluations according to strict criteria by independent organizations, such as the Committee on the Evaluation of the Addition of Ingredients New to Infant Formula (2004).

Powdered infant formulae can be subdivided into different product categories. Milk-based formulae are based on either casein or whey as the predominant source of proteins, with some of them being based on partially hydrolyzed proteins to reduce their allergenicity. In the case of infants suffering from cow milk protein allergies, extensively hydrolyzed formulae or soy-based products exist as alternatives. Infant formulae intended for special nutritional needs and used under medical supervision have been developed to suit such requirements. Some examples of the latter include low-birth-weight formulae high in protein, calcium, and phosphorus, formulae with a lactose concentration of less than 0.2 g in 100 g of formula in case of lactose intolerance, antiregurgitation formulae containing starch, nutrient-dense or high-calorie formulae, formulae low in phenylalanine, or formulae for chemically defined diets containing free amino acids.

PRODUCTION METHODS

Powdered infant formulae, as well as other dry products described in the first section of this chapter, are manufactured according to one of three process types.

1. **Wet-mix** processes, during which all unprocessed raw materials as well as separately processed ingredients are handled as a liquid intermediate product that is heat treated, dried, and then further handled up to the filling stage. In this process, no further additions are done after the heat treatment and, in particular, after the drying step.

2. **Dry-mix** processes, during which all separately processed ingredients are dry blended to obtain the final product, which is then further handled up to the filling stage. The process may include and combine different mixing steps to obtain the final formulation.

3. **Combined** processes, during which part of the unprocessed raw materials and part of the ingredients are processed according to process 1 to obtain a base powder. This base powder, which is considered an intermediate product, is then further used for the manufacture of different finished products. Separately processed ingredients are then added into the base powder according to the different recipes. This is usually done in batches, and different types of mixers are used, depending on the quantities and types of additions, in order to obtain homogeneous products, which are then further handled up to the filling stage. Note that, depending on the products and processing line, such separately processed ingredients also may be added to the liquid concentrate after the heat treatment or put directly into the drying chamber during drying.

An example of a processing line is provided in Fig. 1, with details of the different steps being provided in subsequent sections concerning control measures for *Salmonella* and *E. sakazakii*.

Wet-Processing Steps

Unprocessed raw materials, such as raw milk or liquid whey, are processed after reception by applying steps such as skimming, centrifugation, and standardization, as is commonly done in dairy processing; these steps are not specific to infant formulae processing. Descriptions of these general processing steps are provided by Bylund (1995), Jost (2005), and Walstra et al. (2006). After addition of some of the ingredients, as in the case of combined processes, or of all of them, as in the case of wet-mix processes, the liquid formula is submitted to a heat treatment. Depending on the process and product composition, treatments applied can range from pasteurization in a

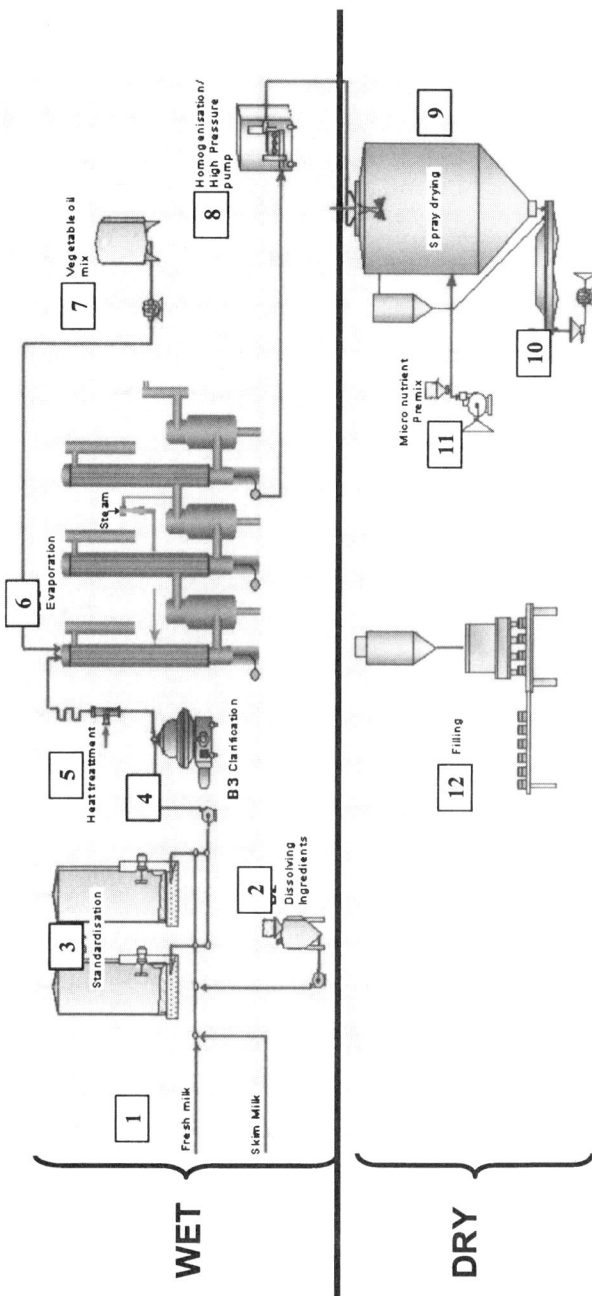

Figure 1 Example illustration of a processing line with the wet-processing and the dry-processing steps. (1) Unprocessed agricultural raw materials; (2) dissolution of processed ingredients; (3) standardization tanks; (4) clarification; (5) heat treatment (direct steam injection; CCP); (6) evaporation; (7) addition of vegetable oils (dry-mix ingredient); (8) high-pressure pumping; (9) spray drying in tower; (10) drying in after-dryer; (11) addition of vitamins (dry-mix ingredients); (12) filling. This is only an example, and different layouts and processing steps are possible (e.g., mixing and intermediate storage in the dry-processing part of the process).

plate heat exchanger (71 to 74°C for 15 to 25 s) to sterilization by means of direct steam injection (DSI) (105 to 125°C for at least 5 s).

While it is widely recognized that vegetative microorganisms such as *Salmonella* or *Listeria monocytogenes* are readily killed under such processing conditions (ICMSF, 1996), it was suggested several years ago that a high thermal resistance of *E. sakazakii* strains could account for their presence in powdered infant formulae. However, as demonstrated by numerous authors, *E. sakazakii* cells show no particular resistance and are easily killed at temperatures ranging from 60 to 70°C (Nazarowec-White and Farber, 1997a; Breeuwer et al., 2003; Edelson-Mammel and Buchanan, 2004; Iversen et al., 2004; Jung and Park, 2006). This also holds true for the distinct phenogroup identified by Edelson-Mammel and Buchanan (2004) that showed an increased heat resistance (about 20 times more resistant than the usual phenotype).

The heat-treated liquid formula is then concentrated by evaporation, most frequently using a falling-film tubular evaporator consisting of several stages ("effects") to achieve the desired level of concentration. Details on this processing step are provided by Gekas and Antelli (2004). Microbiological issues related to this step are almost exclusively due to build-up of spore formers, usually thermophilic species, reflecting the temperature profile in the different effects of the evaporators (Spillmann and Fedder, 1997; Murphy et al., 1999). This can, however, be avoided by defining appropriate run times and applying effective cleaning-in-place procedures (CIP). Depending on the processing line, the concentrate then either is pumped directly to the drying tower or stored in aseptic or refrigerated tanks before being further processed. In such cases and for technological reasons, the concentrate needs to be preheated to temperatures around 70°C before drying. This heating step allows, of course, for an additional reduction of vegetative cells; however, it is not considered a critical control point (CCP) and thus is not monitored as such.

In certain cases, the heat treatment is not applied before evaporation but to the concentrate. In such cases, an increased heat resistance of microorganisms due to the increased amount of total solids has to be accounted for. This has been shown by different authors for dairy products (milk and whey) and for other products, such as liquid eggs (Dega et al., 1972; Corry, 1974; Manas et al., 2001; Li et al., 2005). In any case, the effectiveness of the processing conditions at this step specifically applied to kill vegetative microorganisms, and thus considered a CCP, needs to be validated, taking into account the composition of the liquid formula and, in particular, the levels of total solids, which may reach approximately 40 to 45% in concentrates. The heat processes usually applied allows one, even in concentrates, to achieve killing rates of at least 6D, up to rates in excess of 100D at the higher end of the processing conditions. During validation, it is important to consider not only the reduction achieved in the liquid product (i.e., expressed as CFU/milliliter) but also the drying and concentration effect. Such considerations, in combination with figures on the quantities of powder manufactured, e.g., per hour or day, will provide information on the overall performance of this processing step, which then can be expressed per ton of product. This therefore would allow one to link the estimate of this overall performance to a performance objective, should such an objective be formulated at some stage by authorities.

Dry-Processing Steps

The objective of drying the liquid concentrate is to eliminate the water and to obtain a shelf-stable powder with a residual humidity of about 3%. Spray drying in a drying tower is the most commonly applied technology, while roller or drum drying, which also can be used for dairy products, is, to our knowledge, not used to manufacture powdered infant formulae.

Spray drying of infant formulae involves several stages, during which the preconcentrated liquid formula is pumped to the dryer by means of a high-pressure pump. At the top of the drying tower (the fourth or fifth floor of the building), the liquid is dispersed or atomized through a nozzle into a jet of fine droplets. Different types of nozzles, such as stationary ones or rotating disks or wheels, exist. The atomization pressure determines the size of the droplets, and it is necessary to alter the conditions to achieve optimal drying efficiencies. The droplets are mixed with a stream of hot air, and this leads to an immediate evaporation of the water. The temperature of the air varies between 150 and 250°C and depends on the type of dryer, as well as on the characteristics and sensitivity of the product being manufactured. The outlet temperatures at the bottom of the drying chamber usually range between 60 and 80°C. The simplest installation is a single dryer, where removal of all the moisture takes place in the drying chamber. The dryer is connected directly to a pneumatic transport system to collect the powder, to cool it, and then to transport it to the next processing step. In the two- and three-stage drying systems, the pneumatic transport equipment is replaced by a fluid bed dryer. These systems allow one to operate at lower outlet temperatures, which is important for sensitive products. With this type of equipment, the fluid bed allows for the removal of excess moisture and for the cooling of the powder by means of filtered air. The powder then is conveyed, as for single-stage systems, for further processing. Filtration of the air used to dry, cool, and convey the powder is performed using combinations of dust filters (filter classes EU4/G4 or higher) and microbiological filters (filter classes EU7/F7 or higher) to avoid airborne contamination. Details on spray dryers and principles of the spray drying are reviewed by Refstrup (2004) and Westergaard (2004).

During the drying step, a limited reduction of certain vegetative microorganisms may take place due to the high air temperature. However, due to the short and uncontrolled exposure time and, in particular, to the rapid drop in water activity, the real effect is difficult to quantify, and this step therefore is not considered a controlled killing step. The effect of different drying conditions on the survival of *Salmonella* during spray drying of skimmed milk was studied more than 30 years ago by LiCari and Potter (1970a, 1970b, and 1970c). Various other vegetative bacteria, such as *Enterobacteriaceae* (Daemen and van der Stege, 1982; Costa et al., 2002), *L. monocytogenes* (Doyle et al., 1985), *Staphylococcus aureus* (Chopin et al., 1978), and bifidobacteria (Bielecka and Majkowska, 2000; Lian et al., 2002), all have been shown to survive typical spray-drying processes.

Before further processing and depending on the manufacturing line, the powder usually is stored in tote bins, big bags, or silos. For the dry-mix or combined processes, dry mixing is performed. Such a dry mixing is done when heat-sensitive ingredients need to be added or when, for technological reasons, one needs to avoid issues such as fouling of the walls in the drying tower, which would require wet cleaning. During this mixing operation, which usually is done in a batch process, the base powder is mixed in large mixers or blenders with dry ingredients, such as vitamins, minerals, starch, carbohydrates, and others, according to the formulation. In certain cases and depending on the ingredients, in-line dosing is performed as well. The final product usually is stored again and then cans or flexible containers are filled, which are flushed with inert gas, sealed, coded, labeled, packed into shipping cartons, and palletized. Finished products typically are stored in warehouses, awaiting the results of nutritional, compositional, and microbiological analyses, which are needed, along with other processing data, for the release of the finished goods. Storage of the dry powders, even over prolonged periods, does not allow for any growth due to the low water activity but also does not lead to any die-off, as shown by Edelson-Mammel et al. (2005).

Dry-Mix Ingredients

Ingredients used in dry-mixing operations are processed ingredients that all have been submitted to some thermal inactivation steps during their manufacture. However, they all are prone to postprocess contamination with vegetative microorganisms, such as *Salmonella* and *E. sakazakii*, during further processing steps at the supplier's level. Such ingredients therefore are particularly critical, since they will not be submitted to any further "kill" steps in the manufacturing of infant formulae, and need to be selected with care. The ingredients used in such operations may vary in number depending on the type of process, e.g., all of the ingredients of the formulation in the case of pure dry-mix processes or a few ingredients in the case of combined processes. In order to ensure the compliance of finished products with the stringent requirements for *E. sakazakii,* other *Enterobacteriaceae*, and *Salmonella*, the individual ingredients must fulfill the same microbiological requirements as the final powdered infant formula. It is therefore important to perform a careful evaluation of the likely occurrence of these microorganisms and to implement thorough selection and purchase procedures, with particular attention to the most critical ones. Recent publications have indeed demonstrated that *E. sakazakii* is ubiquitous and thus is found in numerous ingredients, foods, and environments (see chapter 2 of this book; Cordier, 2006). While vitamins or minerals present a low risk of occurrence, others, such as dairy derivatives and starch, represent much higher risks (FAO-WHO, 2004). Starch already has been found at the origin of an outbreak after addition to a sterile ready-to-feed formula (FAO-WHO, 2004), and Elmadfa et al. (1999) have shown the presence of *E. sakazakii* in casein and caseinates. For certain ingredients for which data and experience are lacking, it is necessary to generate information enabling one to correctly assess the risks. One example is sucrose, which, on the basis of existing publications (Jewell et al., 2003), could be considered a low-risk ingredient in terms of microbiological contamination and the presence of *Salmonella*. Experience gained over recent years has, however, shown that the situation may vary greatly from one supplier to another. In the worst cases, contamination with *Enterobacteriaceae* can be frequent. In fact, the presence of this indicator (in 10-g samples) has been found in up to 65% of the analyzed samples in freshly manufactured sugar from certain suppliers and in 10 to 20% of sugar stored for 1 month, with a further decrease in incidence after longer periods of time. This indicates that a die-off of *Enterobacteriaceae* does occur over time. However, it is apparent that the risk of the presence of *E. sakazakii* and *Salmonella* in freshly manufactured sugar has to be considered significant.

In order to purchase dry-mix ingredients of the required microbiological quality, it therefore is extremely important to understand the potential risks, to carefully select suppliers, and to communicate clearly the reasons for the stringent requirements. It is then easier to work in a close partnership with selected suppliers to ensure improvements in preventive measures and thus the delivery of ingredients consistently fulfilling those requirements. Table 2 illustrates the effectiveness of such relationships and the improvements that can be achieved for high-risk ingredients. It should be noted, however, that such improvements cannot be achieved overnight and may require significant changes in the understanding of the particular requirements for ingredients used for this category of products, as well as for the hygiene conditions

Table 2 Number of positive samples in three major dry-mix ingredients and illustration of the improvements over 2 to 3 years

Dry-mix ingredient	Sample result in[a]:					
	2002			2003/2004		
	n	No. EB positive	No. Es positive	*n*	No. EB positive	No. Es positive
Lactose	2,219	70	0	5,354	38	0
Sucrose	1,691	28	2	18,239	28	0
Starch	1,389	155	40	1,622	34	3

[a]*n*, total number of samples tested; EB, *Enterobacteriaceae*; Es, *E. sakazakii*.

that are needed in processing facilities to ensure the delivery of ingredients which are in compliance with the requirements.

CONTROL MEASURES

As confirmed by the FAO-WHO (2004, 2006), the relevant hazards for infant formulae that require appropriate control measures are *Salmonella* and *E. sakazakii*. Other *Enterobacteriaceae* have been grouped in these two FAO-WHO reports in category B (causality plausible, but not yet demonstrated), because they are well-established causes of illness in infants. While these recent reports concluded that there is no epidemiological link to powdered infant formulae, in the efforts undertaken over the last 5 to 6 years to develop and strengthen control measures and strategies for *E. sakazakii* during processing, it was recognized very early on that it was necessary to address the whole group of *Enterobacteriaceae* in order to achieve the expected improvements. In the absence of control measures specific for *E. sakazakii*, this strategy allows one to control the *Enterobacteriaceae* to the same degree, as well any *Enterobacteriaceae* species that may potentially have the same impact on the health of infants. The implementation of such very strict measures to reduce postprocess contamination with *Enterobacteriaceae* also has been recommended by the International Association of Infant Food Manufacturers (IFM, 2004).

The safety of infant formulae is ensured through the implementation of general preventive measures such as good hygienic and good manufacturing practices (GHP and GMP, respectively), as well as of hazard analysis and critical control point (HACCP) practices, to address specific hazards through specific control measures.

As indicated in the section on production methods, the wet-mix and combined processes can be subdivided into two very distinct parts: (i) the wet part, from the reception of raw materials and ingredients up to the drying, and (ii) the dry part, from the drying up to the filling (see Fig. 1). In the dry-mix processes, all operations performed after the drying and up to the filling are exclusively dry operations.

The effect of the heat treatment as a control measure (CCP) located in the wet part of the process has been discussed in the section on wet-processing steps. In order to ensure the absence of recontamination and of growth (mainly of spore formers) in equipment located after the heat treatment, i.e., evaporators, intermediate-storage tanks, and high-pressure pumps or pipes, it is crucial that the line is designed, installed, and maintained according to the highest hygienic standards. This includes the hygienic design of the different pieces of equipment as well as the application of validated cleaning procedures, usually CIP. Additional key elements are the scheduling of run times between two CIP cycles to avoid increases in the microbial population (total viable counts) due to buildup and the definition of appropriate conditions (time and temperature) for intermediate storage, if this is done. While preheating of the concentrate before spray drying will provide a certain amount of microbial reduction, this step usually is not designed and handled as a CCP. Detailed information on the hygienic design of different types of equipment has been published by Curiel et al. (1993a, 1993b) and Lelieveld et al. (2005), and details on CIP principles and procedures are provided by DIN (1988), Chisti and Moo-Young (1994), and Bremer et al. (2006).

The presence of vegetative microorganisms such as *Salmonella* or of *Enterobacteriaceae*, including *E. sakazakii*, in finished products can be traced back to one (or both) of the following causes: (i) recontamination from the processing environment or processing line and (ii) the dry addition of contaminated ingredients. The management of ingredients is discussed in detail in the section on dry-mix ingredients. In order to implement control measures necessary to prevent contamination of powdered infant formulae with *E. sakazakii*, it is important to understand the measures implemented to control *Salmonella*, which are discussed below. They represent the basis for minimizing recontamination by *E. sakazakii*, which, as outlined below, requires measures even more stringent than those required for *Salmonella*.

Control Measures—Management of *Salmonella*

As shown by epidemiological studies, *Salmonella* is the most significant human food-borne pathogen not only for infant formulae but also for all other dry dairy-based products for all categories of consumers (ICMSF, 2005). Many targeted GHPs to control this pathogen were introduced in the late 1970s and have been further improved and refined over the last 30 years. Progress in the development and establishment of effective GMPs/GHPs has been triggered through learning from investigations of the root causes of contaminations of several outbreaks. Case reports that ultimately have led to improvements of the hygienic design of different pieces of equipment and of hygiene practices have been published by different authors. For example, investigations monitoring an outbreak in Australia have shown the insulation layer of the dryer to be the niche for *Salmonella enterica* serovar Bredeney (Forsyth et al., 2003). The presence of milk powder and moisture in the fiberglass insulation allowed not only the establishment of the pathogen but also its growth to high levels. Cracks in the inner wall of the cone (the lower part of the dryer) also allowed for direct contact of the product with contaminated residues from the insulation and allowed for the presence of salmonellae in the packed, finished product. A very similar cause was at the origin of a contamination of infant formulae with small numbers of *Salmonella enterica* serovar Ealing that caused an outbreak in the United Kingdom (Rowe et al., 1987). Today, as a consequence of these findings, dryers are operated without insulation to avoid the presence of such potential niches. Langfeldt et al. (1988) described a case of illness in Germany related to milk powder contaminated with *Salmonella enterica* serovar Mbandanka. In this instance, a wet scrubber installed to clean exhaust air was found to be the source of the pathogen, allowing for multiplication to high levels and then dissemination throughout the whole plant and, finally, into the product. Currently, wet scrubbers are rarely found and have been replaced by other systems, such as dry filters, to remove fine dust from exhaust air.

The effect of the implementation of control measures to prevent the occurrence of *Salmonella* in finished products, which have been improved and refined over the last 30 to 40 years, can be demonstrated. An example is the data from an FDA survey published by Mettler (1989), showing the effect of these improvements since the mid-1970s (Fig. 2). Today, these measures can ensure the absence of *Salmonella* in powdered infant formulae as well as in other powdered dairy products that are manufactured on similar lines and where the same or similar preventive measures are applied.

The basic measures to control *Salmonella* aim, therefore, at preventing recontamination of the product from the environment throughout the different steps and operations, from the dryer through to the filler. These GHPs and GMPs are based on the following principles:

- prevent ingress of *Salmonella* into the high-hygiene area (dry areas) where the processing lines are installed;
- in case of ingress, prevent their establishment in the premises;
- in case of ingress and establishment, prevent their multiplication, which would favor dissemination and further reinforce their establishment in favorable niches;
- when salmonellae are found, take appropriate measures to ensure their eradication.

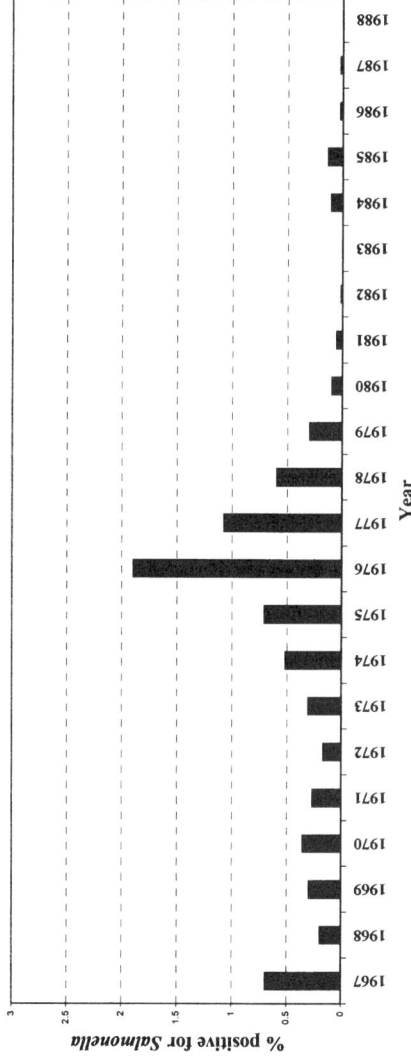

Figure 2 Impact of improved hygiene measures on the occurrence of *Salmonella* in skim-milk powder (U.S. Department of Agriculture *Salmonella* surveillance program data); adapted from Mettler (1989).

The above principles require not only the optimal design of effective preventive measures to control ingress, establishment, and multiplication of the pathogen but also having the appropriate procedures in place for their eradication, if needed. Finally, it also is necessary to have the appropriate tools in place to assess the effectiveness of the preventive measures as well as the corrective actions.

Prevention of contamination of powdered infant formulae with *Salmonella* is based on the subdivision of the factory into zones with different hygiene levels and requirements and is generally termed zoning (Duffey et al., 2003; Van Donk and Gaalman, 2004). The dry-processing and handling operations following drying (discussed in the section on dry-processing steps) always are located in high-hygiene areas. The objective here clearly is to protect the processing line and, thus, the product to avoid its recontamination before filling. To achieve this, it is necessary to physically segregate these high-hygiene areas not only from the exterior but also from other areas within the factory, such as the wet-processing areas (see Fig. 1). Areas adjacent to high-hygiene areas normally are considered medium-hygiene areas and are designed as the last line of defense to prevent contamination of the high-hygiene areas and thus the ingress of *Salmonella*. For more remote zones, such as the receiving area for raw materials or warehouses, considered basic-hygiene areas, the application of GHPs that are appropriate and adapted to the activities at these steps are considered sufficient but must be in place as a first line of defense.

The implementation of an effective zoning concept starts with consideration of the location and the situation of the plant. This has to be done with an understanding of the ecology and occurrence of *Salmonella* in the environment. While it is found more infrequently than *Enterobacteriaceae*, this pathogen can nevertheless be present in a number of niches, which can lead to a potential risk of ingress into the factory premises. Climatic conditions in a region with, for example, the risk of flooding, needs to be considered, since in such an event contaminated water easily may gain access to all areas inside the factory. Neighboring facilities, such as farms, facilities processing primary agricultural raw materials or meat, and waste disposal systems or waste-water treatment plants, represent an increased risk and are all potential sources of *Salmonella* (Espigares et al., 2006). Care also must be taken to prevent the attraction of birds, insects, or rodents, which all are known to occasionally carry *Salmonella* (Kapperud and Rosef, 1983; Tatfeng et al., 2005; Meerburg et al., 2006). This is best done by strictly keeping the surroundings of the factory clean and tidy and by implementing effective pest management measures. Unprocessed agricultural raw materials such as fresh milk or whey are known sources of the pathogen, and it is therefore necessary to carefully position and design the reception and access routes of milk tankers with respect to the rest of the factory.

The design and construction of the building also must take into account climatic conditions such as extreme temperature differences, the occurrence of earthquakes, heavy tropical rains, or dust storms. The construction of the buildings, particularly those within the high-hygiene areas such as the drying tower (a building of four to six floors), must be done in such a way to prevent any contact with the exterior, especially uncontrolled contacts such as the infiltration of water or dust. Attention must be given to the tightness of the roof, a particularly vulnerable part of the building. Openings in the external walls must be kept to a minimum, and the necessary ones, such as for the air intake or exhaust air systems, must be designed, located, and constructed in such a way that they do not weaken the integrity of the zoning principles.

The purpose of the internal zoning is to limit access routes to the high-hygiene area(s) and thus to prevent the ingress of *Salmonella*. This is achieved by the physical separation of this area from others by limiting the number of entry points and by implementing specific hygienic practices at such points. It is obvious that a single high-hygiene area encompassing all processing steps from the drying to the filling is the most effective. However, while it is possible to plan such a layout in new plants that are being constructed, it is not always possible to achieve this in existing plants. The optimal solutions therefore must be evaluated and defined according to the master plan of the factory and by taking into account all requirements necessary to ensure the production of safe powdered infant formulae.

For personnel such as operators and maintenance, engineering, and laboratory personnel or for visitors such as inspectors, it is necessary to create access zones that allow for the necessary hygiene measures, e.g., hand washing and sanitation as well as changing of shoes or of protective clothing, to be instituted. In the case of ingredients used in dry-mixing operations or for packaging material, it is necessary to implement measures to avoid ingress of *Salmonella* through, e.g., dust or soiling on packaging material. Numerous measures and procedures exist to avoid such an ingress, for example, strippable bags or protective sleeves or covers for big bags. They can be cleaned at reception, and the protective layers can be taken or stripped away in the intermediate zones (medium hygiene) before the ingredients are transferred into the high-hygiene area in clean packaging. Another example would be the spoons that are placed in the tins and that are normally packed in plastic bags and received in cardboard boxes. These, as well as an external bag, can be removed before transfer of the spoons in the inner clean bag takes place. Tins used to fill product are transported on protected conveyor belts and turned (opening down) and blown out with special devices at the entry point of the high-hygiene zone. It also is important to carefully consider the hygienic design of conveyor belts, big bags, pallets, and other transfer systems located at the interfaces between high- and medium-hygiene zones in order to avoid contacts and breaches of the zoning principles.

Air used in direct contact with the powdered infant formula, e.g., for cooling or for pneumatic transport, is filtered. Air of the same quality is used in processing areas where indirect contact with the product is possible, since equipment is not completely air tight, and where certain operations such as tipping can lead to a temporary exposure of the powder. At the same time, this air also is used to create an overpressure in the plant to avoid ingress of dust into the high-hygiene areas through openings in walls which are needed, for example, for conveyor belts, cable trays, or pipes. For this purpose, external or internal air is processed or recirculated through air-handling units (AHU) fitted with appropriate filters to ensure the removal of dust particles and microorganisms. Particular attention needs to be paid to the correct design and installation of the AHUs to ensure proper and effective filtration and drainage of condensates and to ease regular inspections, cleaning, and maintenance of the filters.

The zoning concept also must take into account the necessary transport of equipment and spare parts out of the high-hygiene area for wet cleaning and maintenance. This must be considered in terms of correct interfaces between the high-hygiene and medium-hygiene zones, as well as procedures to avoid recontamination when transferring them back.

While the occurrence of *Salmonella* is a rare event and thus the pathogen is infrequently found in factories, it nevertheless may happen that a strain gains access into the high-hygiene area(s). This may occur due to weaknesses in or to a breakdown of the preventive measures, for example, when urgent and/or unforeseen repairs or construction work is done without the necessary care and preventive measures. Other reasons may be infiltrations of water or dust, the use of contaminated dry ingredients, etc. This will increase the risk and ultimately put the processing line and finally the product at risk of being contaminated. In such situations, it is necessary to prevent the multiplication and spread of *Salmonella* within the high-hygiene area. In the case of *Salmonella*, two types of situations can be observed: (i) transient strains that will disappear rapidly, and (ii) strains that will establish themselves and thus represent a risk for prolonged periods of time. Such strains or serotypes are often termed "house strains," and their spread and establishment are greatly favored by growth. Growth is possible in the presence of water, which may originate from wet cleaning, from the formation of condensation, from infiltrations, or, in the case of fire alarms, from the unforeseen release of sprinkling water. The establishment of *Salmonella* in environmental niches in the processing environment is particularly favored by the presence of residual humidity and product residues in cracks and crevices in floors or walls. Additionally, (i) interfaces between floor and equipment, (ii) hollow structures such as insulation material, (iii) metallic structures supporting equipment, and (iv) equipment itself or wet residues accumulating in places difficult to access, such as in cable trays, etc., also can further support the establishment of salmonellae. Investigations to understand the effect of different parameters on the survival of pathogens under such conditions have been performed by various authors, e.g., Allan et al. (2004). Once established, these persistent strains of salmonellae become extremely difficult to eradicate, and they may reappear occasionally over the years without any possibility of finding the source and definitively eliminating them. Further detailed discussions on such niches are provided by the ICMSF (2002).

Prevention of such situations is achieved by minimizing or, if possible, by eliminating the use and thus the presence of water in such processing zones. In addition, the hygienic design of processing rooms and of the equipment and its installation represent an additional preventive measure to avoid condensation and buildup of residues, and it facilitates dry-cleaning operations to eliminate powder residues (Duffey et al., 2001, 2003). Where wet cleaning is nevertheless performed, particular precautions must be taken in order to minimize the consequences. This means that cleaning, sanitation, and drying procedures must be established in such a way that traces of water are eliminated as rapidly as possible.

The design and effectiveness of the preventive measures outlined above need to be assessed through appropriately designed surveillance and monitoring plans. Regular visual inspections are important to detect weaknesses in the zoning and hygiene practices. However, in addition, it is essential to carry out sampling of the processing environment and of the processing line to demonstrate the absence of *Salmonella*. Such sampling plans are performed (i) to establish a baseline of the processing line for a plant considered to be under control and then to assess whether this is maintained over time, (ii) to rapidly detect breaches, and (iii) to investigate the sources of contamination and the effectiveness of the corrective actions put in place in case of such an event. Effective monitoring plans are based on the analysis of different types of samples, reflecting the relevant elements that have an impact on the microbiological quality of the powdered infant formulae. The types of samples which should be considered include:

1. samples from the processing environment;
2. samples from the processing lines, i.e., from the product contact surfaces; and
3. samples of semifinished and finished products.

The first element, environmental samples, mainly encompasses samples taken from the processing environment in the high-hygiene areas. The sampling points are defined according to their closeness to the processing equipment/processing line and to the risk of pathogen contamination. Preference therefore is given to points close to openings in equipment such as (i) hatches, (ii) access doors in the drying chamber, and (iii) connections between pieces of equipment that may be flexible and that are opened or dismantled occasionally by personnel for inspection or when problems such as stoppages occur.

Environmental samples are not taken exclusively in the high-hygiene areas. It also is important to focus on interfaces between medium- and high-hygiene zones, such as access locks for personnel or transfer areas for ingredients. These samples are important in assessing the effectiveness of the barriers and the adherence to the defined hygienic practices. Sampling of more remote areas as well as the exterior of the factory buildings really does not provide meaningful and easily interpretable information, and their numbers should be very limited.

Sampling plans for the processing environment are established to achieve an appropriate balance between routine samples taken at regular intervals from the same locations and investigational samples taken from different locations and points according to the local situation and conditions of the factory. The quality of the samples often is more important than their number or the quantity of sampled material, the purpose really being to find *Salmonella*, if it is present in specific niches. Therefore, small quantities of residues from an interface between the floor and equipment may provide more useful and relevant information than large quantities of spilled clean powder found on the floor. It is extremely important that sampling be performed by skilled and trained personnel using appropriate tools adapted to the sampling points. Samples from vacuum cleaners equipped with appropriate filters, for example, provide general information on the status of larger areas (Trakumas et al., 2001; Haysom and Sharp, 2003). On the other hand, scrapers or spatulas of different sizes, shovels and brushes, slightly humidified sponges, and swabs are ideal either to sample residues accumulated in cracks or crevices, external surfaces of equipment, electrical boxes or cable trays, etc., or to sample larger surfaces on floors and walls where traces of past or present humidity have been identified. In the case of dry-processing environments, the application of tools such as contact plates and small cotton/alginate swabs as described in the International Organization for Standardization (ISO) standard 18593 (ISO, 2004), which may be useful in wet-processing plants, really are not appropriate (Deberghes et al., 1995).

Shut-down periods, during which equipment and machines are stopped for maintenance or replacement or when construction or installation work is performed, are very appropriate periods to perform additional investigative sampling. On such occasions, sites and points that normally are not accessible during normal operations are sampled. Care should be taken during sampling to avoid disseminating contamination from dislodged residues; this is of particular importance during shut-down periods in which numerous exceptional activities are taking place.

The frequency of sampling needs to be adapted to the analytical findings, as well as to the overall hygienic situation. Both the number of samples and sampling frequency are increased in cases of positive results. These would include, for example, the presence of *Salmonella* in one or more samples, deviations in hygienic status based on visual inspections, results of process hygiene indicator testing or after specific events such as major wet cleaning procedures, or the startup of processing lines.

The second element of the monitoring plans is line samples taken directly from equipment surfaces in contact with the manufactured powder. Such sampling points specifically are chosen to represent places where contaminating microorganisms could gain access from the surrounding environment or where particular conditions, e.g., the presence of condensation, would favor a buildup of contaminating salmonellae. These points are, of course, specific to a production line and to the type of equipment and processing conditions applied. Examples of representative line samples are the first powder manufactured or filled, residues from sifter tailings in after-dryers or fillers, and residues from other equipment, such as the ones used for storage and/or mixing. Additional information and details on environmental sampling plans and sampling techniques are provided by Holah (1999) and the ICMSF (2002). The third element in the monitoring plan is the sampling and testing of finished products. This is done as a final and overall verification and is usually used as one of the elements of the release procedure. This element needs to take into account the disposition, number, and interconnections of the processing lines from the drying to the filling lines to ensure appropriate coverage of all manufactured lots.

As in the case of environmental monitoring, sampling plans for line samples and finished products need to be adapted to the analytical findings. Both the frequency and number of samples or quantities analyzed need to be increased if problems are identified. In the worst case, extensive sampling and testing of finished product codes, even beyond the usual 60 samples of 25 g each required by several regulations, need to be conducted. In addition, thorough investigations must be undertaken to trace the source(s) of *Salmonella* and to take appropriate corrective actions. Such periods, during which the finished product is tested at the highest level, should last as long as the problem remains and positives are found, or for a prolonged period of time if all additional results are negative.

In the case of issues and investigations, quantitative determinations (e.g., most-probable-number testing) can provide useful information on the extent of the contamination and can assist in the determination of the root cause. In addition, serological or molecular identification of *Salmonella* isolates can be performed to assist in the tracing of the sources of contamination (Threlfall and Frost, 1990; Winokur, 2003; Vivanco et al., 2004).

In the case of dry dairy products, including powdered infant formulae, *Enterobacteriaceae* play a key role as indicators of process hygiene (Cox et al., 1988; Eyles and Davey, 1989; Mossel and Struijk, 1995; ICMSF, 2002; Tortorello, 2003). Traditionally, *Enterobacteriaceae* have been used as indicators for processing environments, while coliforms have been used for line samples and finished products, according to the requirements of most national legislation and of the Codex Alimentarius (CAC, 1979). However, coliforms gradually are being replaced by *Enterobacteriaceae*, thus reflecting the evolution of regulations. The use of the *Enterobacteriaceae* allows one to simply and cheaply assess the hygienic status of dry-processing areas; analytical results usually are obtained within 24 h after sampling (compared to 3 to 5 days for *Salmonella*). It is important to point out here that in the types of samples discussed above, the determination of both *Salmonella* and *Enterobacteriaceae* is necessary. Increasing trends or strongly fluctuating levels of *Enterobacteriaceae* are indicative of deviations from the hygienic conditions and, as a consequence, point to an increased risk for the presence of *Salmonella*; low levels of the indicators do not, however, guarantee the absence of *Salmonella*.

Despite these improvements, sporadic outbreaks have occurred recently (Park et al., 2004; Espié et al., 2005). The published case studies often show that these outbreaks were not caused by systemic weaknesses of the outlined principles but rather by errors or breaches of the underlying principles. In the case of an outbreak described in France (Anonymous, 2005), the incriminating product was contract manufactured in a processing environment harboring *Salmonella* and then packed in another facility. It is interesting that products from that particular contract manufacturer already had been at the origin of a previous outbreak with the same serotype (Espié et al., 2005), thus confirming the common environmental origin in both products; in total, 143 infants were affected in the two outbreaks. The presence of atypical salmonellae, such as lactose-fermenting strains, has been reported (about 1% of isolates) in this type of facility (Blackburn and Ellis, 1973), and this may lead to difficulties in detection and identification. Thus, the presence of atypical salmonellae in samples may be overlooked and thus jeopardize the effectiveness of the monitoring procedures. Such particular strains have indeed been at the origin of at least two outbreaks (Louie et al., 1993; Ustera et al., 1996).

Thus, laboratories need to use appropriate methodology, such as the use of selective media able to correctly identify atypical salmonellae, e.g., lactose-positive strains.

Control Measures—Management of *E. sakazakii*

What are the differences between the management and control of *E. sakazakii* and that of *Salmonella*? As indicated above, the preventive measures implemented to control *Salmonella* form the basis of the control measures and also are a prerequisite to address and control *E. sakazakii*. However, contrary to *Salmonella*, *E. sakazakii* cannot, at least today, be singled out and specifically controlled to the same extent. In the course of investigating the occurrence of *E. sakazakii* in processing environments of powdered infant formulae and other dairy products to assist in defining additional or more stringent measures, it rapidly has become clear that this organism is ubiquitous and already was present and well established in high-hygiene areas of such plants. This was later confirmed by Kandhai et al. (2004), who demonstrated the presence of this organism in dry-processing environments of very diverse foods as well as in households. This widespread occurrence in high-hygiene areas is in fact not specific to *E. sakazakii* and is also observed for the whole group of *Enterobacteriaceae*, which have been successfully used for decades as indicators of process hygiene. Based on these observations, the most appropriate approach to define more stringent control measures than the existing ones is to address the whole group of *Enterobacteriaceae*, including *E. sakazakii*, and thus to reduce globally the occurrence of postprocess contamination.

In considering the existing preventive measures, the ingress of *Enterobacteriaceae* into the high-hygiene area can, due to their ubiquitous occurrence, only be minimized and not completely suppressed. Minimizing can only be achieved through the stringent and consequent application of existing measures, as presented in the first section of this chapter, and their reinforcement where necessary and feasible. The most important impact in reducing the levels of *Enterobacteriaceae* and thus the probability of contamination is through the very strict management of water in the high-hygiene area, the target being the complete elimination of the presence of any water.

In the case of *Salmonella* management, the presence of water, even if undesired, will not lead to multiplication as long as no ingress into the high-hygiene area takes place, and, under these conditions, there is no immediate threat of contamination. However, in the case of *Enterobacteriaceae*, water, even if present only in traces, will lead to an immediate increase of the population. An illustration of how such a situation would appear in trend analyses performed at the factory level is shown in Fig. 3.

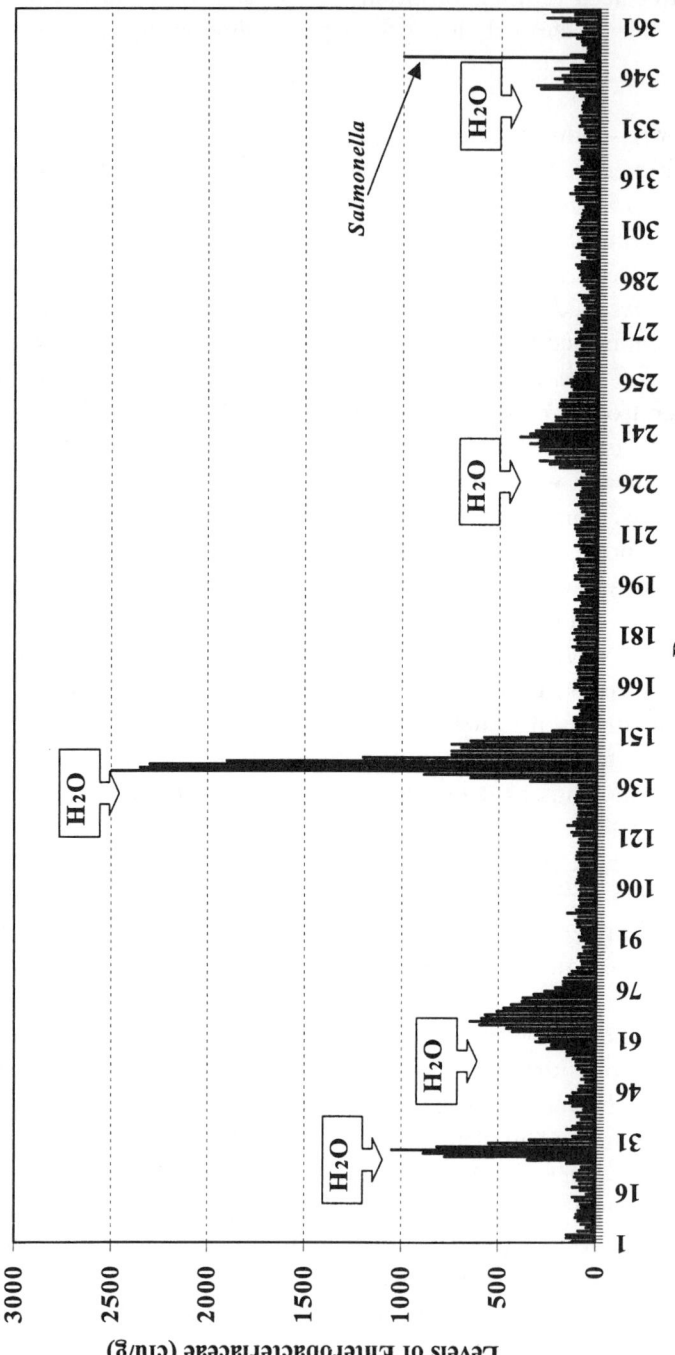

Figure 3 Illustration of the fluctuations in the levels of *Enterobacteriaceae* in environmental samples from high-hygiene areas as a function of the presence of water. In case of an ingress of *Salmonella* at very low levels (which would have occurred between days 270 and 330) remaining undetected during monitoring, the presence of water would invariably lead to an increase of the pathogen, putting the production lines and the product at risk (a hypothetical example is used for illustration).

For simplicity reasons and for clarity of the figure, the data for *Enterobacteriaceae* presented here are derived from real representative data, and the peak corresponding to the detection of *Salmonella* is included only for illustration. In such a situation, ingress of *Salmonella* may have occurred in the period from about day 270 to 330 without being detected, and the presence of water would have led to an increase in the numbers of salmonellae and their detection. Levels of *Enterobacteriaceae,* including *E. sakazakii,* on the other hand, will fluctuate according to the presence of water. This increase may be local and contained as, for example, in the presence of points of condensation, or it may be more widespread, as in the case of wet cleaning. While a decline in the levels of *Enterobacteriaceae* is observed after drying, part of the population will survive for prolonged periods of time. This is most likely linked with the resistance to desiccation of several members of the group, with *E. sakazakii* being one of the most resistant species, as shown by Breeuwer et al. (2003) and Caubilla-Barron and Forsythe (2006). The presence of water thus will constantly lead to a refilling of the reservoir, therefore maintaining a high environmental pressure and, thus, a significant risk of contamination of the powder during handling and processing.

In order to achieve consistent reductions in the levels of *Enterobacteriaceae* and to maintain these very low levels over long periods, it is necessary to eliminate the presence of any water, and this can be achieved through different measures. The exclusive application of dry-cleaning procedures to the processing environment and processing equipment is an essential measure in achieving the desired outcome. While, for example, the International Dairy Federation (1991) refers to dry cleaning in its hygiene recommendations, wet cleaning of installations also is described in detail (IDF, 1991), a practice which represents a high risk in the case of the control and management of *E. sakazakii*, where exclusive dry cleaning is the practice of choice. Therefore, it is important that equipment be designed in such a way as to facilitate an effective dry cleaning; recommendations have been published by Duffey et al. (2001, 2003).

Where wet cleaning is still needed, e.g., in the management of allergens when different products are manufactured on the same line, alternative solutions must be sought and implemented as appropriate. This may include the transport of parts of equipment out of the high-hygiene area for wet cleaning, sanitation, and drying, with subsequent transport back. This implies, as indicated above, the implementation of well-designed interfaces and the application of specific procedures to avoid recontamination of the equipment or carryover from the medium-hygiene zones. An alternative that is more effective in controlling *Enterobacteriaceae*, including *E. sakazakii,* is the manufacture of such products on dedicated lines, allowing the implementation of dry-cleaning procedures. However, these types of solutions may require time, for example, to build a new production line. In the meantime, alternative management solutions need to be implemented, e.g., very tightly controlled wet-cleaning and sanitation procedures followed by rapid and careful drying.

All measures taken in addition to the ones to control *Salmonella* have been shown to be effective in significantly reducing the levels of *Enterobacteriaceae* in processing environments of the high-hygiene areas. As a consequence, the levels of *E. sakazakii* are reduced as well. These measures allow one, in principle, to reach and consistently maintain levels of *Enterobacteriaceae* below 10 CFU per g or per surface unit (depending on the sampling point) in such processing areas. The improvements over time and the consolidation at low levels are illustrated by a representative example in Fig. 4.

The results of specific *E. sakazakii* monitoring in such environments show, however, that despite the low levels of *Enterobacteriaceae* reached, sporadic detection of this organism is still possible when thorough sampling techniques as well as appropriate analytical methods are applied (i.e., after enrichment). The incidence may vary from factory to factory or even within different areas of a single factory, but complete absence in processing environments is unlikely. The detection of *E. sakazakii* obviously depends on the sampling points and the quality and volume of the sample. In the case of environmental samples and due to the presence of competitive microorganisms, in particular, other *Enterobacteriaceae*, it also is important to use appropriate analytical methods, such as the one developed by Guillaume-Gentil et al. (2005), which has served as the basis for the development of a new ISO technical specification for powdered infant formula (ISO, 2006). See chapter 2 for further details on analytical methods.

In the case of the processing line illustrated in Fig. 4, for example, the presence of *E. sakazakii* was found in 5 to 10% of the samples analyzed by enrichment techniques. This highlights the fact that eradication is not possible, as it is for *Salmonella*. Thus, the mere presence of *E. sakazakii* in the processing environment signifies that there is always a certain risk of end product contamination occurring, which can be kept to a minimum only as long as no increase takes place. The thorough investigation of the recent outbreak in France (Coignard and Vaillant, 2006), for example, has shown that lots contaminated with *E. sakazakii* (present in samples of 50 or 100 g, with the same molecular type) were manufactured over a period of about 6 months, from April to September 2004, during which wet cleaning had taken place on several occasions. Even though no environmental data are published in the case report, this confirms that the persistence of *E. sakazakii* in the processing environment of a manufacturing facility can lead to a significant source of recontamination over prolonged periods of time.

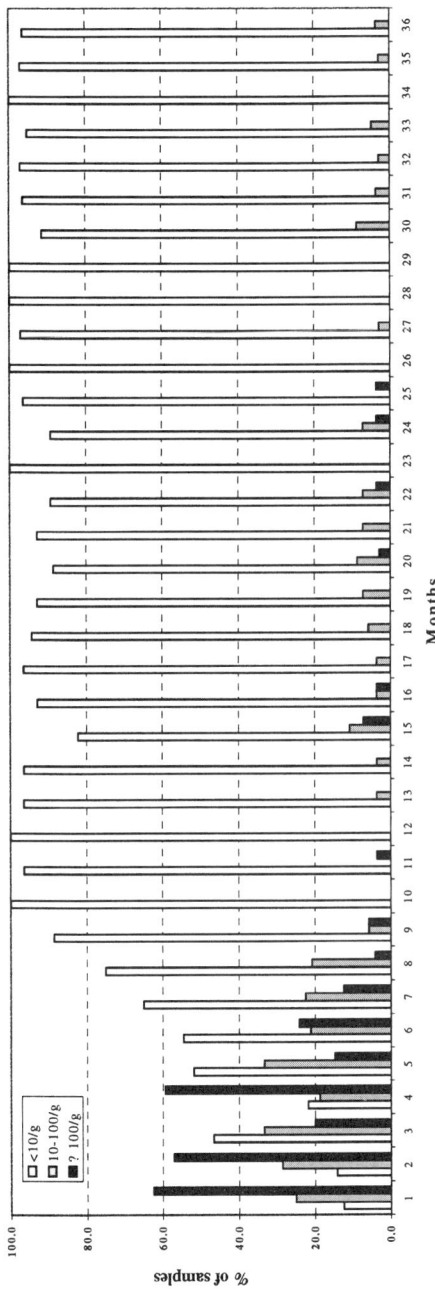

Figure 4 Evolution of the levels of environmental *Enterobacteriaceae* following the implementation of the stringent hygiene measures described in the text at around the fourth month. The target level is <10 CFU/g. Values of >10 CFU/g, <100 CFU/g, and >100 CFU/g automatically trigger an increase of the testing frequency of finished product for both *E. sakazakii* and other *Enterobacteriaceae*. The question mark in the key at upper left indicates ">" (greater than).

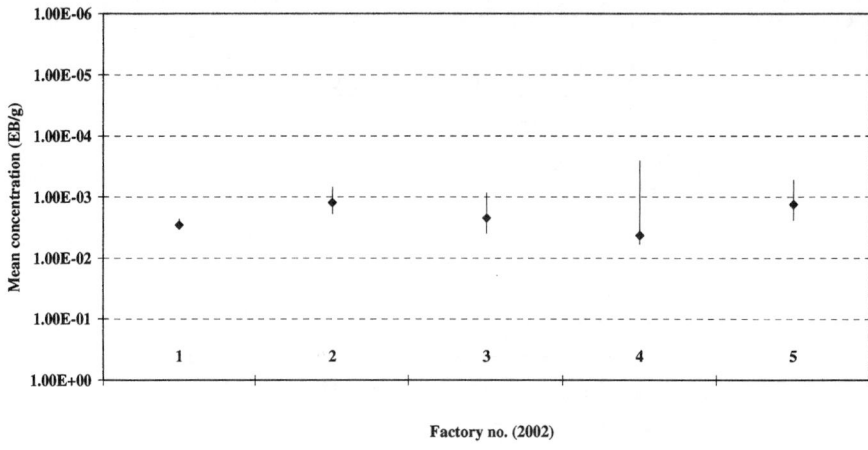

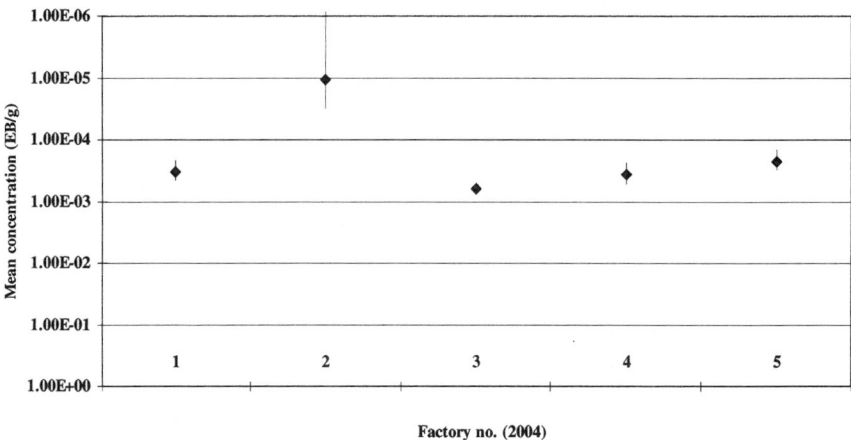

Figure 5 Comparison of the mean concentrations of *Enterobacteriaceae* (EB) (90% credibility interval) in powdered infant formulae manufactured in five different factories before (2002) and after (2004) the introduction of the more stringent control measures described in the text.

The impact of the significant and consistent reductions of *Enterobacteriaceae* in the processing environment is reflected in the results of the finished products. This can be seen in Fig. 5, which shows the results of testing for the mean concentration of *Enterobacteriaceae* in five factories, as determined using the analytical data accumulated for the products manufactured in 2002 (about 6 months) and 2004, i.e., between about 5,000 and 10,000 sample results per year and per factory, for *Enterobacteriaceae*. The mean concentrations are shown using the analytical data accumulated for the products manufactured in five factories in 2002 (about 6 months) and 2004.

The data for 2002 show the mean concentration of *Enterobacteriaceae* before introduction of a systematic dry cleaning, while the data for 2004 show the mean concentration 1 to 2 years after such a systematic implementation. The data show a significant decrease in the numbers of *Enterobacteriaceae* during this period of time, i.e., a factor ranging between about 10 and about 100, depending on the factory. While the distribution of *Enterobacteriaceae* in the product may be log normal, for these calculations and in the absence of suitable data to determine a standard deviation a Poisson distribution of the contamination was assumed, as this also allows for comparisons of trends. The mean concentrations for *E. sakazakii* (which are not shown in the figure) were, in general, 2 to 10 times lower than those for other *Enterobacteriaceae* and followed the same trends between 2002 and 2004.

Increases in numbers of both *E. sakazakii* and other *Enterobacteriaceae* in the processing environment due, for example, to the uncontrolled presence of water following an emergency shower are unavoidable. As a consequence, this may increase the probability of their presence at low levels in the finished product, which may lead to blockage and rework or destruction of the products if they are out of compliance from a regulatory standpoint. An additional example of the effect of the enhanced control measures is shown in Fig. 6. In this case, the opposite phenomenon can be observed in a factory switching from the manufacture of powdered infant formulae to less sensitive dry dairy products, for which only control measures for *Salmonella* are necessary. In this situation, implementation of the more relaxed control measures mentioned in the first section of this chapter (and therefore effective to control *Salmonella*) led to an immediate and steady increase in the levels of *Enterobacteriaceae* in the processing environment. Criteria for the hygiene indicators in such products are usually based on analytical volumes of 1 g, i.e., the absence of *Enterobacteriaceae* in 1-g samples.

Currently, the risk of contamination from the processing environment can be assessed only in a qualitative way. However, more data and information to understand the precise routes of contamination and factors influencing it may allow one to determine this risk in a more quantitative way. Certain tools such as the molecular identification of strains of *E. sakazakii* isolated from different types of samples, such as ribotyping or PCR-based assays (Nazarowec-White and Farber, 1999; Lehner et al., 2006), or mathematical models to determine the transmission or fate of microorganisms from the environment and during subsequent processing steps, such as mixing (Den Aantrekker et al., 2003; Nauta, 2005), may be of assistance in the future. For the time being and based on practical and historical experience over many

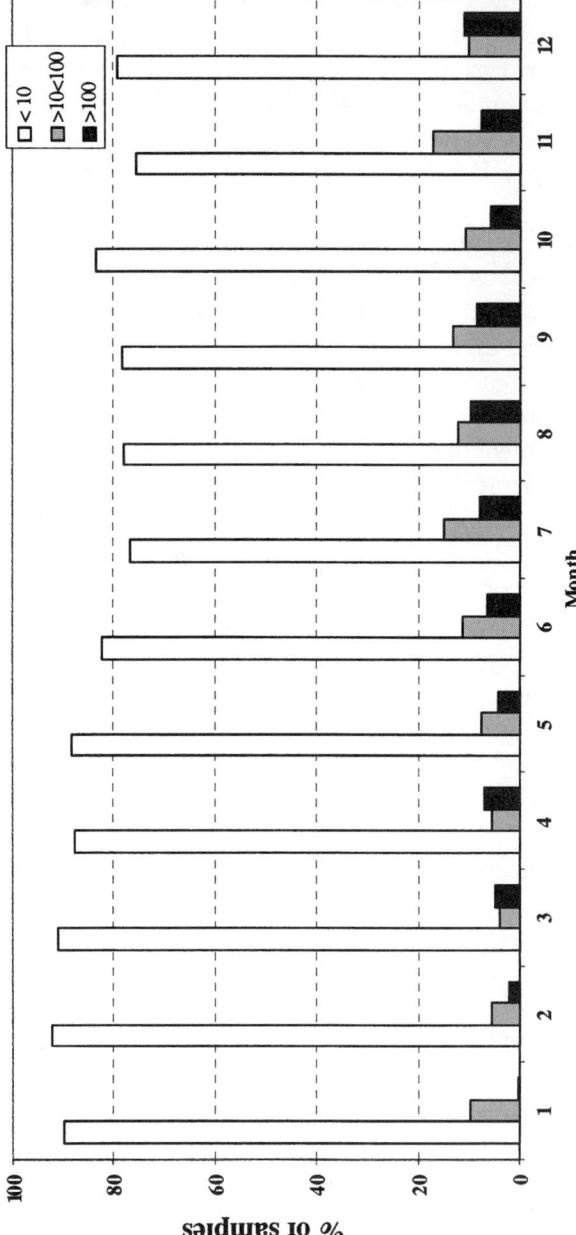

Figure 6 Evolution of the levels of environmental *Enterobacteriaceae* in processing environments in a factory switching from the production of powdered infant formulae to other dry dairy products with different microbiological requirements (*Enterobacteriaceae* must be absent from 1-g samples) but equivalent requirements for *Salmonella*. The key at upper right indicates numbers per gram.

years, it has been shown that levels of *Enterobacteriaceae* that are >100 CFU/g represent an increased risk of product recontamination with *E. sakazakii*. This value has been established as a threshold to trigger investigations of why this is occurring and to apply corrective actions, which may include modifications of the release procedures and an increased sampling and testing plan.

As in the case of *Salmonella*, release of products is based on the outcome of testing programs for *E. sakazakii* and other *Enterobacteriaceae* and includes environmental and line samples (food contact surfaces) as well as finished products. Testing of finished products and line samples is performed on the basis of sampling plans encompassing several samples of 10 g (analytical volumes), which is equivalent to the sampling plans that have been put into place by the EC or that have been discussed at the Codex Alimentarius level. Complete monitoring programs are applied to reject noncompliant products, but the aim, of course, is to detect signs of deviations at an early stage and to take corrective actions before contamination of the final product occurs.

Such integrated programs and release procedures can be very complex and need to take into account all elements of the processing lines and environment, the flow of products, and the production schedules. While for a factory operating with a single dryer and one filling line the procedure may be straightforward, for factories with several dryers, several mixing units, and numerous filling lines running in parallel, the design of the monitoring and release procedure is much more complex. In addition, the impact of a positive result and the implementation of the necessary measures need to be clearly assessed and defined in anticipation of such possible issues. A very important element of the management is the use of the data to perform trend analyses, allowing one to effectively use the amount of data generated and to make appropriate management decisions.

MICROBIOLOGICAL CRITERIA

Microbiological specifications for powdered infant formulae have been established by the Codex Alimentarius Commission (CAC, 1979) and are presented in Table 3. They were based on the epidemiological data and knowledge available at that time and have proven effective in the management of *Salmonella*. Based on new epidemiological evidence and the emergence of *E. sakazakii* as an opportunistic pathogen for a specific group of infants, the need for revision has become evident.

A number of years ago, certain manufacturers introduced strengthened release criteria based on their internal assessments of the emerging issue and in anticipation of evolving regulatory requirements. Following the first expert

Table 3 Current situation (2007) with respect to microbiological criteria for infant formulae at Codex Alimentarius and EC levels[c]

Standard and criteria (sample size)	n	c	m	M
CAC/RCP 21—1979 (CAC, 1979)				
Mesophilic aerobic counts	5	2	10^3	10^4
Coliforms	5	1	<3	20
Salmonella (25 g)	60	0	0	0
CAC/RCP 21—1979, proposed revision (CCFH, 2007)				
Mesophilic aerobic counts (1 g)	5	2	500	5,000
Enterobacteriaceae (10 g)	10	2	0	NA[a]
Enterobacter sakazakii (10 g)	30	0	0	
Salmonella (25 g)	60	0	0	
EC 2073/2005 (EC, 2005)				
Enterobacteriaceae (10 g)	10	0	0	
Enterobacter sakazakii (10 g)[b]	30	0	0	
Salmonella (25 g)[b]	30	0	0	

[a]NA, not applicable.
[b]These criteria were observed only if one or more samples were positive for *Enterobacteriaceae*.
[c]n is the number of samples to be analyzed per lot; c is the number of samples allowed between "*m*" and "*M*" values; m is the microbiological limit that separates good from marginally acceptable quality; and M is the microbiological limit that separates marginally acceptable from unacceptable quality.

meeting on *E. sakazakii* and other microorganisms in powdered infant formulae (WHO, 2 to 5 February 2004, Geneva, Switzerland), the need for modifications of the original specifications was identified. At this meeting, the switch from coliforms to *Enterobacteriaceae* was identified as a necessary step, as well as the need for strengthening specifications as one of the risk mitigation measures (FAO-WHO, 2004). The initial risk assessment performed during this meeting and published in Appendix C of the meeting report (FAO-WHO, 2004) estimated the effect of a reduction of the frequency/extent of contamination of powdered infant formulae on the relative risk of *E. sakazakii* infection. This scenario has shown that while a reduction of the relative risk was possible by reducing the prevalence of contaminated products, it could go only so far. A more significant reduction was possible only in combination with additional measures, such as the hygienic preparation and handling of the reconstituted formulae.

The risk assessment and available published data on the incidence and levels of *E. sakazakii* and other *Enterobacteriaceae* have been used by the ICMSF to propose modified criteria to the Codex Alimentarius Working Group performing the revision of the relevant Code of Hygienic Practice (CAC, 1979; see Table 3). During the meeting of the Codex Committee on Food Hygiene (CCFH) in Buenos Aires in 2005, it was decided, based on the most recent data, that the proposal should be revisited (Codex Committee on Food Hygiene, 2004) and was then confirmed during a recent meeting of the Working Group (June 2007) (CCFH, 2007). Further discussions and calculations on the impact of microbiological criteria on the risk reduction took place during a recent ICMSF meeting in 2005 as well as during the second expert meeting organized by the FAO-WHO in Rome, Italy (2006). The outcomes of these discussions have been presented and were discussed during the further revision of the Code of Hygienic Practices (CAC/RCP 21-1979) in 2006 and 2007 (CCFH, 2007). While the proposal for *E. sakazakii* has been maintained, a slightly less stringent proposal has been put forward for *Enterobacteriaceae* (i.e., a c = 2).

In parallel to the Codex Alimentarius activities, the European Food Safety Authority (EFSA) instructed the Biohazard Panel to review the microbiological risks related to infant formulae. The scientific panel concluded that *E. sakazakii* presented the greatest risk for neonates up to ca. 4 to 6 weeks of age, for premature neonates, for low-birth-weight infants, and for immunocompromised infants (EFSA, 2004). This report also recommended the possibility of introducing a performance objective (PO) aiming at very low levels of *Salmonella* and *E. sakazakii* (e.g., absence in 1, 10, or 100 kg of sample) and to use *Enterobacteriaceae* (testing for its absence in 10 g) to verify compliance. While these recommendations have not resulted in the establishment of a PO, the opinion has been used by the EC to establish microbiological criteria for powdered infant formulae, which came into force in January 2006 (EC, 2005). These criteria (summarized in Table 3) outline a two-step approach, with initial testing for *Enterobacteriaceae* considered an indicator of process hygiene, and, in the case of positives, testing for *E. sakazakii* and *Salmonella* as food safety parameters. Very recently, the EC requested that the EFSA review its initial position (EFSA, 2004) to provide a scientific opinion on the possible correlation between *E. sakazakii,* other *Enterobacteriaceae,* and *Salmonella* in infant formulae and on the correlation between *Enterobacteriaceae* and *Salmonella* in the case of follow-on formulae. The Biohazard Panel concluded that because *Salmonella* is so rarely present, there is no correlation between the hygiene indicator and *Salmonella* in both types of products. In contrast, the panel concluded that there is a relationship between the presence of *E. sakazakii* and other *Enterobacteriaceae,* but that in the absence of sufficient data no universal correlation can be established. The panel recommended the generation of additional data using common protocols to draw more definitive conclusions (EFSA, 2007).

In July 2007, the Standing Committee on Food Chain and Animal Health (SCOFCAH) approved several amendments to the EC regulation 2073/2005 by deleting the link (in terms of analyses) between the hygiene parameter *Enterobacteriaceae* and the safety parameters *Salmonella* and *E. sakazakii* for powdered infant formulae, i.e., now both parameters are being considered at the same level. In addition, *Salmonella* ($M = 30$, $c = 0$, $m = 0$ [in 25 g]) have been introduced for follow-on formulae as well as a stringent hygiene criterion, i.e., *Enterobacteriaceae* with $n = 5$, $c = 0$, $m = 0$ (in 10 g). This technical vote will be confirmed later in 2007, and amendments are thus likely to be implemented by the end of 2007 or the beginning of 2008.

Other regulations or action levels are being used by different authorities and are discussed in chapter 9 of this book. It is important for manufacturers to have clearly defined regulatory criteria, consistent between each other if possible, in order to design the necessary preventive measures in a correct and appropriate way. Such references also are necessary to establish appropriate monitoring and release procedures, as well as to validate the complex internal procedures outlined in the section on control measures for management of *E. sakazakii* to ensure the release of safe products that comply with regulatory requirements.

PREPARATION AND HANDLING OF THE PRODUCT

As discussed in the previous sections, it is possible to reduce the levels of *E. sakazakii* and other *Enterobacteriaceae* in powdered infant formulae to comply with the stringent criteria outlined in the section on microbiological criteria. However, as acknowledged by FAO-WHO (2004, 2006), the products are not sterile, and even products complying with the most stringent microbiological criteria sporadically may be contaminated at very low levels, as shown by Iversen and Forsythe (2004). As indicated in the risk assessment (FAO-WHO, 2004, 2006), the introduction of even more stringent criteria than the ones proposed would not significantly contribute to the reduction of the relative risk. A significant risk reduction can be achieved only in combination with other risk management options. One of these measures is the appropriate preparation and reconstitution of the powdered infant formulae, as well as the handling of the bottles up to their final consumption. Recommendations to users, health professionals, or parents on the preparation, handling, and storage of these products play an essential role in ensuring the safety of the products. Recommendations made by manufacturers on the safe use of the products are related to the storage and handling conditions after reconstitution of the powders.

These recommendations are based on well-known microbiological principles governing the growth behavior of microorganisms and have been confirmed in studies specific to *E. sakazakii* (Nazarowec-White and Farber, 1997b; Iversen et al., 2004). The importance of a rapid refrigeration of the reconstituted bottles, in the case of refrigerated intermediate storage, has been highlighted in a recent publication (Kandhai et al., 2006). Recommendations of manufacturers on the immediate consumption of reconstituted infant formulae take into account the growth characteristics of *E. sakazakii* and, in particular, the apparently rapid recovery of dry-stressed cells, as well as the rapid growth at body temperatures. In hospitals with appropriate control measures and infrastructure, such as bottle kitchens with the correct layout, refrigerated storage capabilities, etc., one can store the product for limited periods of time (in general, for a maximum of 24 h at 4°C). Information with respect to such practices has been prepared by different manufacturers and is shared with health care professionals. As illustrated in several publications and reports, the application and assessment of good hygiene measures are critical and may point out deficiencies (Rowan et al., 1997; Schnebelen, 2005).

Several organizations have provided recommendations on the reconstitution of powdered infant formulae in order to ensure the safety of the products up to consumption. Examples are the European Association of Paediatricians (Agostoni et. al., 2004), as well as authorities such as the FDA (2002), the Agence Française de Sécurité Sanitaire des Aliments (AFSSA, 2005), and the Food Standards Agency UK (FSA, 2006). The recommendations take into account the issues outlined in the FAO-WHO reports (2004, 2006) as well as other elements, such as the potential loss of nutrients and the potential risk of the scalding of infants as described by Jeffery et al. (2000). While the ultimate aim of these documents is the safe feeding of reconstituted formulae, they contain contradictory recommendations, for example, for the reconstitution temperatures, which range from <20°C to ≥70°C. There is certainly a need for a more harmonized approach or for different options depending on the type of product being commercialized. In the case of infant formulae containing probiotics, for example, reconstitution at 70°C would lead to a killing of the beneficial flora. In addition, formula containing starch may not dissolve properly at very low temperatures. The use of a model as described in FAO-WHO (2004), once available, will certainly allow one to compare different options and evaluate their impact on the relative risk reduction.

REFERENCES

Agence Française de Sécurité Sanitaire des Aliments (AFSSA). 2005. Recommandations d'hygiène pour la préparation et la conservation des biberons, Juillet 2005. http://www.sante. gouv.fr/htm/actu/biberon/rapport_afssa.pdf.

Agostoni, C., J. Axelson, O. Goulet, B. Koletzko, K. F. Michaelsen, J. W. L. Puntis, J. Rigo, R. Shamir, H. Saajewska, D. Turck, Y. Vandenplas, and L. T. Neaves. 2004. Preparation and handling of powdered infant formula: a commentary by the ESPGHAN Committee on Nutrition. *J. Pediatr. Gastroenterol. Nutr.* **39**:320–322.

Allan, J. T., Z. Yan, and J. L. Kornacki. 2004. Surface material, temperature, and soil effects on the survival of selected foodborne pathogens in the presence of condensate. *J. Food Prot.* 67:2666–2670.

Anonymous. 2005. Epidémie de salmonellose à *Salmonella enterica* sérotype Agona chez des nourrissons, France, Janvier–Avril 2005. Point final de l'investigation au 10 Juin 2005. www.invs.sante.fr/presse/2005/le_point_sur/salmonella_agona_150605/index.html.

Barness, L. A. 1987. History of infant feeding practices. *Am. J. Clin. Nutr.* 46:168–170.

Benson, J. D., and M. L. Masor. 1994. Infant formula development: past, present and future. *Endocrin. Regul.* 28:8–16.

Bielecka, M., and A. Majkowska. 2000. Effect of spray drying temperature of yoghurt on the survival of starter cultures, moisture content and sensoric properties of yoghurt powder. *Nahrung* 44:257–260.

Blackburn, B. O., and E. M. Ellis. 1973. Lactose-fermenting *Salmonella* from dried milk and milk drying plants. *Appl. Microbiol.* 26:672–674.

Breeuwer, P., A. Lardeau, M. Peterz, and H. M. Joosten. 2003. Dessication and heat tolerance of *Enterobacter sakazakii*. *J. Appl. Microbiol.* 95:967–973.

Bremer, P. J., S. Fillery, and A. J. McQuillan. 2006. Laboratory scale clean-in-place (CIP) studies on the effectiveness of different caustic and acid wash steps on the removal of dairy biofilms. *Int. J. Food Microbiol.* 106:254–262.

Bylund, G. 1995. *Dairy Processing Handbook*. Tetra Pak Processing Systems, Lund, Sweden.

Carver, J. D. 2003. Advances in nutritional modifications of infant formulas. *Am. J. Clin. Nutr.* 77:1550S–1554S.

Caubilla-Barron, J., and S. J. Forsythe. 2006. Long-term persistence and recovery of *Enterobacter sakazakii* and other *Enterobacteriaceae* from powdered infant milk formula, P-019. *Abstr. 106th Gen. Meet. Am. Soc. Microbiol.* American Society for Microbiology, Washington, DC.

Chisti, Y., and M. Moo-Young. 1994. Clean-in-place systems for industrial bioreactors: design, validation and operation. *J. Ind. Microbiol.* 13:201–207.

Chopin, A., S. Tesone, J. P. Vila, Y. Le Groet, and G. Mocquot. 1978. Survival of *Staphylococcus aureus* during preparation and conservation of dried skim milk. Problems of survivor enumeration. *Can. J. Microbiol.* 24:1371–1380.

Codex Alimentarius Commission (CAC). 1979. Recommended international code of hygienic practice for foods for infants and children. CAC/RCP 21-1979. Food and Agriculture Organization, Rome, Italy.

Codex Alimentarius Commission (CAC). 1981. Codex Standard for infant formula, Codex Stan 72-1981 and amendments in 1983, 1985, 1987 and 1997. Food and Agriculture Organization, Rome, Italy.

Codex Committee on Food Hygiene (CCFH). 2004. CX/FH 05/37/4. Codex Committee on Food Hygiene, 37th session, Buenos Aires, Argentina. Proposed Draft Revision of the Recommended International Code of Practice for Foods for Infants and Children. ftp://ftp.fao.org/codex/ccfh37/fh37_04e.pdf.

Codex Committee on Food Hygiene (CCFH). 2007. CX/FH 07/39/4. Codex Committee on Food Hygiene, 39th session, New Delhi, India. Proposed Draft Code of Hygienic Practice for Powdered Formulae for Infants and Young Children at Step 3. ftp://ftp.fao.org/codex/ccfh39/fh39_04e.pdf.

Coignard, B., and V. Vaillant. 2006. Infections à *Enterobacter sakazakii* associées à la consommation d'une préparation en poudre pour nourrissons, p. 88. Rapport d'Investigation. Institut de Veille Sanitaire, Saint-Maurice, France.

Committee on the Evaluation of the Addition of Ingredients New to Infant Formula. 2004. Infant formula: evaluating the safety of new ingredients. The National Academics Press, Washington, DC.

Cordier, J. L. 2006. *Enterobacteriaceae. In* Y. Motarjemi and M. Adams (ed.), *Emerging Foodborne Pathogens.* Woodhead Publishing Ltd., Cambridge, United Kingdom.

Corry, J. E. 1974. The effect of sugars and polyols on the heat resistance of salmonellae. *J. Appl. Bacteriol.* **36**:31–43.

Costa, E., N. Teixido, J. Usall, E. Tons, V. Gimeno, J. Delgado, and I. Vinas. 2002. Survival of *Pantoea agglomerans* strain CPA-2 in a spray-drying process. *J. Food Prot.* **65**:185–191.

Cox, L. J., N. Keller, and M. van Schothorst. 1988. Use and misuse of quantitative determinations of *Enterobacteriaceae* in food microbiology. *Soc. Appl. Bacteriol. Symp. Ser.* **17**:237S–249S.

Curiel, G. J., G. Hauser, P. Peschel, and D. A. Temperley. 1993a. Hygienic equipment design criteria. Document 10. European Hygienic Equipment Design Group, Brussels, Belgium.

Curiel, G. J., G. Hauser, P. Peschel, and D. A. Temperley. 1993b. Hygienic design of closed equipment for the processing of liquid food. Document 8. European Hygienic Equipment Design Group, Brussels, Belgium.

Daemen, A. L. M., and H. J. van der Stege. 1982. The destruction of enzymes and bacteria during the spray drying of milk and whey. 2. The effect of the drying conditions. *Neth. Milk Dairy J.* **36**:211–229.

Danish Dairy Board. 2005. Danish dairy statistics: 8f exports of infant formula by market areas. http://www.mejeri.dk/smcms/danishdairyboard_dk/Facts_figures/Danish_dairy/8__Preserved_milk/8_f__Exports_of/Index.htm?ID=5189.

Deberghes, P., T. Cordier, J. P. Vincent, J. P. Hornez, and M. Catteau. 1995. Amélioration de l'efficacité du prélèvement de surface par utilisation d'éponges. *Microbiologie-Aliments-Nutrition* **13**:409–412.

Dega, C. A., J. M. Goepfert, and C. H. Amundson. 1972. Heat resistance of salmonellae in concentrated milk. *Appl. Microbiol.* **23**:415–420.

Den Aantrekker, E. D., R. M. Boom, M. H. Zwietering, and M. van Schothorst. 2003. Quantifying recontamination through factory environments—a review. *Int. J. Food Microbiol.* **80**:117–130.

Deutsches Institut für Normung (DIN). 1988. DIN-Fachbericht 18: Milchwirtschaftiche Anlagen, Reinigung und Desinfektion nach dem CIP-Verfahren. Beuth Verlag, Berlin, Germany.

Doyle, M. P., L. M. Meske, and E. H. Marth. 1985. Survival of *Listeria monocytogenes* during the manufacture and storage of nonfat dry milk. *J. Food Prot.* **48**:740–742.

Duffey, J. L., G. Hauser, H. Hutten, K. Mager, R. R. Maller, K. Masters, G. M. H. Meesters, W. Rumpf, and G. Schleining. 2001. General hygienic design criteria for the safe processing of dry particulate materials. Document 22. European Hygienic Equipment Design Group, Brussels, Belgium. CCFRA Technology Ltd., Chipping Campden, United Kingdom.

Duffey, J. L., G. Hauser, H. Hutten, K. Mager, K. Masters, G. M. H. Meesters, J. Ossterom, W. Rumpf, G. Schleining, and J. Roels. 2003. Hygienic engineering of plants for the processing of dry particulate materials, November 2003. Document 26. European Hygienic Equipment Design Group, Brussels, Belgium. CCFRA Technology Ltd., Chipping Campden, United Kingdom.

Edelson-Mammel, S. G., and R. I. Buchanan. 2004. Thermal inactivation of *Enterobacter sakazakii* in rehydrated infant formula. *J. Food Prot.* **67**:60–63.

Edelson-Mammel, S. G., M. K. Porteous, and R. I. Buchanan. 2005. Survival of *Enterobacter sakazakii* in a dehydrated powdered infant formula. *J. Food Prot.* **68**:1900–1902.

Elmadfa, I., A. Titz, and P. Burger. 1999. Expertengutachten zur Lebensmittelsicherheit – Lebensmittelbestrahlung. Bericht im Auftrag des Bundeskanzleiamts. Institut für Ernährungswissenschaften der Universität Wien, Vienna, Austria.

Espié, E., F. X. Weill, C. Brouard, J. Capek, G. Delmas, A. M. Forgues, F. Grimont, and H. de Valk. 2005. Nationwide outbreak of *Salmonella enterica* serotype Agona infections in infants in France, linked to infant milk formula, investigations ongoing. *Eurosurveillance* **10**:54.

Espigares, E., A. Bueno, M. Espigares, and R. Galvez. 2006. Isolation of *Salmonella* serotypes in waste water effluent: effect of treatment and potential risk. *Int. J. Hyg. Environ. Health* **209**:103–107.

European Commission (EC). 1991. Commission Directive 91/321/EEC of 14 May 1991 on infant formulae and follow-on formula. *Official J. Eur. Union* L175, corrigendum *Official J. Eur. Union L101* and amending acts 96/4/EC, 1999/50/EC, 2003/14/EC, and 2003/5/EC.

European Commission (EC). 1999. Commission Directive 1999/21/EC of 25 March 1999 on dietary foods for special medical purposes. *Official J. Eur. Union* L91/29–36.

European Commission (EC). 2005. Commission regulation (EC) no. 2073/2005 of 15 November 2005 on microbiological criteria for foodstuffs. *Official J. Eur. Union* L338/1–26.

European Commission (EC). 2006. Commission Directive 2006/141/EC of 22 December 2006 on infant formulae and follow-on formulae and amending Directive 1999/21/EC. *Official J. Eur. Union* L401.

European Food Safety Authority (EFSA). 2004. Opinion of the Scientific Panel on Biological Hazards on the request from the Commission related to the microbiological risks in infant formulae and follow-on formulae. *EFSA J.* **113**:1–35.

European Food Safety Authority (EFSA). 2007. Scientific opinion of BIOHAZ Panel on the request from the Commission for review of the opinion on microbiological risks in infant formulae and follow-on formulae with regard to *Enterobacteriaceae* as indicators. *EFSA J.* **444**:1–14.

Eyles, M. J., and J. A. Davey. 1989. Enteric indicator organisms in foods. *In* K. A. Buckle, J. A. Davey, M. J. Eyles, A. D. Hocking, K. G. Newton, and E. J. Stuttard (ed.), *Foodborne Microorganisms of Public Health Significance,* 4th ed. AIFST Food Microbiology Group, Pymble, Australia.

Food and Agriculture Organization-World Health Organization (FAO-WHO). 2004. *Enterobacter sakazakii* and other microorganisms in powdered infant formula: meeting report. *Microbiological risk assessment series 6.* World Health Organization-Food and Agriculture

Organization of the United Nations, Geneva and Rome. WHO Press, Geneva, Switzerland. http://www.who.int/foodsafety/publications/micro/mra6/en/index.html.

Food and Agriculture Organization-World Health Organization (FAO-WHO). 2006. *Enterobacter sakazakii* and *Salmonella* in powdered infant formula: meeting report. *Microbiological risk assessment series 10.* World Health Organization-Food and Agriculture Organization of the United Nations, Geneva and Rome. WHO Press, Geneva, Switzerland. http://www.who.int/foodsafety/publications/micro/mra10/en/index.html.

Food and Drug Administration (FDA). 2004. Federal Food Drug and Cosmetic Act as amended through December 31, 2004, Chapter IV, Section 412—Requirements for infant formula. http://www.fda.gov/opacom/laws/fdcact/fdctoc.htm.

Food and Drug Administration (FDA). 2002. Health Professional Letter on *Enterobacter sakazakii* infections associated with use of powdered (dry) infant formulas in neonatal intensive care units. http://www.cfsan.fda.gov/~dms/inf-ltr3.html

Food Standards Agency UK (FSA). 2006. Guidance on preparing infant formula. http://www.food.gov.uk/news/newsarchive/2005/nov/infantformulastatementnov05.

Fomon, S. 2001. Infant feeding in the 20th century: formula and beikost. *J. Nutr.* **131:** 409S–420S.

Forsyth, J. R., N. M. Bennett, S. Hogben, E. M. Hutchinson, G. Rouch, A. Tan, and J. Taplin. 2003. The year of the *Salmonella* seekers—1977. *Aust. N.Z. J. Public Health* **27:**385–389.

Gekas, V., and K. Antelli. 2004. Evaporators. *In* H. Roginski, J. Fuquay, and P. Fox (ed.), *Encyclopedia of Dairy Sciences.* Elsevier Ltd., Philadelphia, PA.

Greer, F. R. 2001. Feeding the premature infant in the 20th century. *J. Nutr.* **131:**426S–430S.

Guillaume-Gentil, O., V. Sonnard, M. C. Kandhai, J. D. Marugg, and H. Joosten. 2005. A simple and rapid cultural method for detection of *Enterobacter sakazakii* in environmental samples. *J. Food Prot.* **68:**64–69.

Haysom, I. W., and K. Sharp. 2003. The survival and recovery of bacteria in vacuum cleaner dust. *J. R. Soc. Health* **123:**39–45.

Holah, J. 1999. Effective microbiological sampling of food processing environments. Guideline no. 20. Campden & Chorleywood Food Research Association, Chipping Campden, United Kingdom.

Infant Food Manufacturers (IFM). 2004. ISDI position on *Enterobacter sakazakii* in powdered infant formula. http://www.ifm.net/issues/esakazakii_position.htm.

International Commission on Microbiological Specifications for Foods (ICMSF). 1996. *Microorganisms in foods,* vol. 5. *Characteristics of microbial pathogens.* Blackie Academic & Professional, London, United Kingdom.

International Commission on Microbiological Specifications for Foods (ICMSF). 2002. *Microorganisms in foods,* vol. 7. *Microbiological testing in food safety management.* Kluwer Academic/Plenum Publishers, New York, NY.

International Commission on Microbiological Specifications for Foods (ICMSF). 2005. *Microorganisms in foods,* vol. 6. *Microbial ecology of food commodities.* Kluwer Academic/Plenum Publishers, New York, NY.

International Dairy Federation (IDF). 1991. IDF recommendations for the hygienic manufacture of spray dried milk powders. *Bull. IDF* **267**:1–24.

International Organization for Standardization (ISO). 2004. Microbiology of food and animal feeding stuffs—horizontal methods for sampling techniques from surfaces using contact plates and swabs. Specification ISO 18593. ISO, Geneva, Switzerland.

International Organization for Standardization (ISO). 2006. Milk and milk products—detection of *Enterobacter sakazakii*. Technical specification ISO/TS 22964//IDF/RM 210. ISO, Geneva, Switzerland.

Iversen, C., and S. J. Forsythe. 2004. Isolation of *Enterobacter sakazakii* and other *Enterobacteriaceae* from powdered infant formula milk and related products. *Food Microbiol.* **21**:771–776.

Iversen, C., M. Lane, and S. J. Forsythe. 2004. The growth profile, thermotolerance and biofilm formation of *Enterobacter sakazakii* grown in infant formula milk. *Lett. Appl. Microbiol.* **38**:378–382.

Jeffery, S. L. A., T. C. S. Cubison, C. Greenaway, P. M. Gilbert, and N. Parkhouse. 2000. Warming milk—a preventable cause of scalds in children. *Br. Med. J.* **520**:235.

Jewell, K., P. Voysey, and K. Hon-Yin Hau. 2003. An industrial microbiological risk assessment of salmonellae in dried powders. R&D report no. 185. Campden and Chorleywood Food Research Association, Chipping Campden, United Kingdom.

Jost, R. 15 January 2005, posting date. Milk and dairy products. *In Ullmann's Encyclopedia of Industrial Chemistry.* http://www.mrw.interscience.wiley.com/ueic/articles/a16_589/frame.html.

Jung, M. K., and J. H. Park. 2006. Prevalence and thermal stability of *Enterobacter sakazakii* from unprocessed ready-to-eat agricultural products and powdered infant formulas. *Food Sci. Biotechnol.* **15**:152–157.

Kandhai, M. C., M. W. Reij, L. G. Gorris, O. Guillaume-Gentil, and M. van Schothorst. 2004. Occurrence of *Enterobacter sakazakii* in food production environments and households. *Lancet* **363**:39–40.

Kandhai, M. C., M. W. Reij, C. Grognou, M. van Schothorst, L. G. Gorris, and M. H. Zwietering. 2006. Effects of preculturing conditions on lag time and specific growth rate of *Enterobacter sakazakii* in reconstituted powdered infant formula. *Appl. Environ. Microbiol.* **72**:2721–2729.

Kapperud, G., and O. Rosef. 1983. Avian wildlife reservoir of *Campylobacter fetus* subsp. *jejuni, Yersinia* spp., and *Salmonella* spp. in Norway. *Appl. Environ. Microbiol.* **45**:375–380.

Koletzko, B., S. Baker, G. Cleghorn, U. Fagundes Neto, S. Gopalan, O. Hennell, Q. S. Hock, P. Jirapinyo, B. Lennerdal, P. Penharz, H. Pzyrembel, J. Ramirez-Mayans, R. Shamir, D. Turck, Y. Yamashiro, and D. Zong-Yi. 2005. Global standard for the composition of infant formula: recommendations of an ESPGHAN coordinated international expert group. *J. Pediatr. Gastroenterol. Nutr.* **41**:584–599.

Langfeldt, N., W. Heeschen, and G. Hahn. 1988. Zum Vorkommen von Salmonellen in Milchpulver: Untersuchungen zur Kontamination durch Analyse kritischer Punkte. *Kieler Milchwirtschftl. Forschungsber.* **40**:81–90.

Lehner, A., S. Nitzsche, P. Breeuwer, B. Diep, K. Thelen, and R. Stephan. 2006. Comparison of two chromogenic media and evaluation of two molecular based identification systems for *Enterobacter sakazakii* detection. *BMC Microbiol.* **6**:15–30. [Epub ahead of print.]

Lelieveld, H. L., M. A. Mostert, and J. Holah. 2005. *Handbook of hygiene control in the food industry.* Woodhead Publishing Ltd., Cambridge, United Kingdom.

Li, X., B. W. Sheldon, and H. R. Ball. 2005. Thermal resistance of *Salmonella enterica* serotypes, *Listeria monocytogenes*, and *Staphylococcus aureus* in high solids liquid egg mixes. *J. Food Prot.* **68**:703–710.

Lian, W. C., H. C. Hsiao, and C. C. Chou. 2002. Survival of bifidobacteria after spray drying. *Int. J. Food Microbiol.* **74**:79–86.

LiCari, J. J., and N. N. Potter. 1970a. *Salmonella* survival during spray drying and subsequent handling of skimmilk powder. I. *Salmonella* enumeration. *J. Dairy Sci.* **53**:865–870.

LiCari, J. J., and N. N. Potter. 1970b. *Salmonella* survival during spray drying and subsequent handling of skimmilk powder. II. Effect of drying conditions. *J. Dairy Sci.* **53**:871–876.

LiCari, J. J., and N. N. Potter. 1970c. *Salmonella* survival differences in heated skimmilk and in spray drying of evaporated milk. *J. Dairy Sci.* **53**:1287–1289.

Louie, K. K., A. M. Paccagnella, H. Lior, B. J. Francis, and M. T. Osterholm. 1993. *Salmonella* serotype Tennessee in powdered milk and infant formula—Canada and United States, 1993. *Morb. Mortal. Wkly. Rep.* **42**:516–517.

Manas, P., R. Pagan, F. J. Sala, and S. Condon. 2001. Low molecular weight milk whey components protects *Salmonella senftenberg* 775W against heat by a mechanism involving divalent cations. *J. Appl. Micobiol.* **91**:871–877.

Meerburg, B. G., W. F. Jacobs-Reitsma, J. A. Wagenaar, and A. Kijlstra. 2006. Presence of *Salmonella* and *Campylobacter* spp. in wild small mammals on organic farms. *Appl. Environ. Microbiol.* **72**:960–962.

Mettler, A. E. 1989. Pathogens in milk powder—have we learned the lesson? *J. Soc. Dairy Technol.* **42**:48–53.

Mossel, D. A., and C. B Struijk. 1995. *Escherichia coli*, other Enterobacteriaceae and additional indicators as markers of microbiologic quality of food: advantages and limitations. *Microbiologia* **11**:75–90.

Motil, K. J. 2000. Infant feeding: a critical look at infant formulas. *Curr. Opinion Pediatr.* **12**:469–476.

Murphy, P. M., D. Lynch, and P. M. Kelly. 1999. Growth of thermophilic spore forming bacilli in milk during the manufacture of low heat powders. *Int. J. Dairy Technol.* **52**:45–50.

Nauta, M. J. 2005. Microbiological risk assessment models for partitioning and mixing during food handling. *Int. J. Food Microbiol.* **100**:311–322.

Nazarowec-White, M., and J. M. Farber. 1997a. Thermal resistance of *Enterobacter sakazakii* in reconstituted dried infant formula. *Lett. Appl. Microbiol.* **24**:9–13.

Nazarowec-White, M., and J. M. Farber. 1997b. Incidence, survival, and growth of *Enterobacter sakazakii* in infant formula. *J. Food Prot.* **60**:226–230.

Nazarowec-White, M., and J. M. Farber. 1999. Phenotypic and genotypic typing of food and clinical isolates of *Enterobacter sakazakii. J. Med. Microbiol.* **48**:559–567.

O'Callaghan, D. M., and J. C. Wallingford. 2004. Infant formulae—new developments, pp. 1384–1392. *In* H. Roginski, J. Fuquay, and P. Fox (ed.), *Encyclopedia of Dairy Sciences.* Elsevier Ltd., Philadelphia, PA.

Oliveira, V., M. Prell, D. Smallwood, and E. Frazão. 2001. Infant formula prices and availability—final report to Congress. Economic Research Service, U.S. Department of Agriculture, Washington, DC.

Park, J. K., W. S. Seok, B. J. Choi, H. M. Kim, B. K. Lim, S. S. Yoon, S. Kim, Y. S. Kim, and J. Y. Park. 2004. *Salmonella enterica* serovar London infections associated with consumption of infant formula. *Yonsei Med. J.* **29**:43–48.

Refstrup, E. 2004. Drying principles, p. 860–871. *In* H. Roginski, J. Fuquay, and P. Fox (ed.), *Encyclopedia of Dairy Sciences*. Elsevier Ltd., Philadelphia, PA.

Rowan, N. J., J. G. Anderson, and A. Anderton. 1997. The bacteriological quality of hospital prepared infant feeds. *J. Hosp. Infect.* **35**:259–267.

Rowe, B., N. T. Begg, D. N. Hutchinson, H. C. Dawkins, R. J. Gilbert, M. Jacob, B. H. Hales, F. A. Rae, and M. Jepson. 1987. *Salmonella ealing* infections associated with consumption of infant dried milk. *Lancet* **ii**:900–903.

Schanler, R. J. 2005. Human milk supplementation for preterm infants. *Acta Paediatr. Suppl.* **94**:64–67.

Schnebelen, C. 2005. Audit des pratiques professionelles dans les biberonneries: conduite d'une étude dans la région Rhône-Alpes. http://cclin-sudest.chu-lyon.fr/Audit/Biberonnerie/Audit_biberon.pdf.

Spillmann, H., and I. Fedder. 1997. Thermophile Sporen in Milchpulver. *DMZ Lebensmittelind. Milchwirtschaft.* **118**:66–75.

Tatfeng, Y. M., M. U. Usuanlele, A. Orupke, A. K. Digban, M. Okodua, F. Oviasogie, and A. A. Turay. 2005. Mechanical transmission of pathogenic organisms: the role of cockroaches. *J. Vector Borne Dis.* **42**:129–134.

Threlfall, E. J., and J. A. Frost. 1990. The identification, typing and fingerprinting of *Salmonella*: laboratory aspects and epidemiological applications. *J. Appl. Bacteriol.* **68**:5–16.

Tortorello, M. L. 2003. Indicator organisms for safety and quality—uses and methods for detection: minireview. *J. AOAC Int.* **86**:1208–1217.

Trakumas, S., K. Willeke, S. A. Grinshpun, T. Reponen, G. Mainelis, and W. Friedman. 2001. Particle emission characteristics of filter-equipped vacuum cleaners. *AIHAJ* **62**:482–493.

Ustera, M. A., A. Echeita, A. Aladueña, M. C. Blanco, R. Reymundo, M. I. Prieto, O. Tello, R. Cano, D. Herrera, and F. Martinez-Navarro. 1996. Interregional foodborne salmonellosis outbreak due to powdered infant formula contaminated with lactose-fermenting *Salmonella virchow*. *Eur. J. Epidemiol.* **12**:377–381.

Van Donk, D. P., and G. Gaalman. 2004. Food safety and hygiene: systematic layout planning of food processes. *Chem. Eng. Res. Design* **82**:1485–1493.

Vivanco, A. B., J. Alvarez, I. Laconcha, N. Lopez-Molina, A. Rementeria, and J. Garaizar. 2004. Molecular genotyping and methods and computerized analysis for the study of *Salmonella enterica*. *Methods Mol. Biol.* **268**:49–58.

Walstra, P., J. T. Wouters, and T. J. Geurt. 2006. *Dairy Science and Technology*. Taylor and Francis, London, United Kingdom.

Westergaard, V. 2004. Drying design, p. 871–889. *In* H. Roginski, J. Fuquay, and P. Fox (ed.), *Encyclopedia of Dairy Sciences*. Elsevier Ltd., Philadelphia, PA.

Winokur, P. L. 2003. Molecular epidemiological techniques for *Salmonella* strain discrimination. *Front. Biosci.* 8:c14–c24.

World Health Organization (WHO). 1981. International code of marketing breast-milk substitutes. WHO, Geneva, Switzerland.

Enterobacter sakazakii
Edited by Jeffrey M. Farber and Stephen J. Forsythe
© 2008 ASM Press, Washington, D.C.

Use of Nonsterile Nutritionals for Neonates in the Hospital and after Hospital Discharge: Control Measures Currently Instituted at One Tertiary Care Institution

7

Deborah L. O'Connor, Joan Brennan, Susan Dello, and Laurie Streitenberger

INTRODUCTION

Powdered infant formulas have been implicated as the cause of a number of infectious *Enterobacter sakazakii* outbreaks presenting as meningitis, sepsis, and/or necrotizing enterocolitis (FAO-WHO, 2004; Food and Drug Administration, 2002; Health Canada, 2002; Himelright et al., 2002). Reported death rates among infected infants range from 20% to more than 50%, with many survivors exhibiting severe developmental sequelae (Gurtler et al., 2005). *E. sakazakii* is ubiquitous in nature; however, only powdered infant formula and preparation equipment have been linked to *E. sakazakii* outbreaks among infants. As a result, a number of regulatory agencies have issued warnings about the safety of powdered formulas for use in infants, most notably for use in infants that are of low birth weight, are born prematurely, or are immunocompromised. While accredited health care institutions in North America had infection prevention and control policies and procedures in place at the time of these warnings, they prompted reevaluation of the rigor of these processes as they applied to feeding, and a flurry of activity related to the relative merit versus the risk of using powdered infant formula where a sterilized liquid counterpart was absent.

As recommended by a number of expert committees, at The Hospital for Sick Children (SickKids) we encourage the use of human milk over infant formula whenever possible (American Academy of Pediatrics, 2005; Canadian

DEBORAH L. O'CONNOR, Department of Clinical Dietetics, The Hospital for Sick Children, and Department of Nutritional Sciences, The University of Toronto, Toronto, Ontario, Canada. JOAN BRENNAN, Department of Clinical Dietetics, The Hospital for Sick Children, Toronto, Ontario, Canada. SUSAN DELLO, Department of Food and Nutritional Services, The Hospital for Sick Children, Toronto, Ontario, Canada. LAURIE STREITENBERGER, Department of Infection Prevention and Control, The Hospital for Sick Children, Toronto, Ontario, Canada.

Paediatric Society, Dietitians of Canada, Health Canada, 1998; Health Canada, 2004; World Health Organization, 2001). Due to the elevated nutritional requirements of many of our hospitalized infants, however, human milk may be fortified with energy and nutrient modules that are not sterile, including specially designed powdered human milk multinutrient fortifiers (human milk fortifier). Sterile nutritional counterparts for many of these products are unavailable. Early regulatory advice concerning *E. sakazakii* provided little guidance on how to support the nutritional status of human-milk-fed infants without compromising the volume of human milk delivered.

Much remains to be done to establish the best evidence-based practices to maximize the benefit of available nutritional products and to minimize the risk of microbial contamination. The purpose of this chapter is to outline some of the strategies we have adopted to date at SickKids to optimize this balance in the context of infants (birth to 12 months of age). We caution the reader that our policies and procedures with regard to infection prevention and control continue to evolve, and what we suggest herein needs to be evaluated in the context of the environment in which they are applied. In this chapter we will discuss the guidelines for optimal feeding of infants, both term born and born prematurely. We will address the approaches we have taken to enrich human milk and formula with nutrients in the event that this is required to improve the nutritional status of a small and/or sick infant. As much of the confusion in responding to initial safety warnings about *E. sakazakii* surrounded a lack of knowledge about appropriate substitutions for specific categories of powdered formula, in this chapter we will review the general categories of infant formulas (cow milk, soy, lactose free, hydrolysate and amino acid based, etc.) and indications for use. We will then review guidelines and specific procedures we have implemented to collect breast milk, inventory formula and expressed breast milk, and store them centrally. We also will review guidelines and procedures we have implemented to prepare specialized feedings for infants and review basic bedside management of delivery and administration of these feedings (called hang and hold times). Finally, we will outline how we approach the issue of infants going home on specialized feedings.

DEFINITIONS

There are two terms that are worth defining before we begin. First, in this chapter, when we refer to an infant being breast-fed we are, in fact, referring to a child receiving his or her mother's own milk at the breast, by bottle, by feeding device, or by tube. Second, when we refer to enteral feeding, we are referring to any nutrient-containing substance provided via the gastrointestinal or alimentary tract or any route that connects to this system. Enteral feeding could include breast milk or formula administered orally or via an enterostomy tube (i.e., nasogastric, orogastric, gastrostomy, nasojejunum, or jejunostomy tube). Enteral feeding does not include nutrition provided by an intravenous route.

GUIDELINES FOR OPTIMAL INFANT FEEDING

Breast-feeding is the gold standard and strongly preferred method of feeding healthy term-born and prematurely born infants. The World Health Organization, Health Canada, and the American Academy of Pediatrics recommend human milk as the exclusive nutrient source for full-term infants for the first 6 month of life and indicate that breast-feeding be continued at least through the first 12 months of life and thereafter as long as mother and baby mutually desire (American Academy of Pediatrics, 2005; Canadian Paediatric Society, Dietitians of Canada, Health Canada, 1998; Health Canada, 2004; World Health Organization, 2001).

The scientific rationale for recommending breast-feeding as the preferred feeding choice for both term-born and prematurely born infants stems from its acknowledged benefits to infant nutrition, gastrointestinal function, host defense, neurodevelopment, and psychological, economic, and environmental well-being (American Academy of Pediatrics, 2005; American Academy of Pediatrics, 2004b; American Academy of Pediatrics, 2004c). The American Academy of Pediatrics Policy Statement entitled "Breastfeeding and the Use of Human Milk" is an excellent resource, and it includes a succinct discussion of the specific benefits of breast-feeding (American Academy of Pediatrics, 2005). Briefly, the policy acknowledges that research provides good evidence that breast-feeding decreases the rate of postneonatal infant mortality (~21%) and reduces the incidence of a wide range of infectious diseases, including bacterial meningitis, bacteremia, diarrhea, respiratory tract infection, necrotizing enterocolitis, otitis media, urinary tract infection, and late-onset sepsis in preterm infants. Breast-feeding also is associated with slight improvements in cognitive development in both term-born and prematurely born infants, although the benefits appear to be greatest for low-birth-weight infants (Anderson et al., 1999). We showed a positive association between duration of human milk feeding and cognitive development among prematurely born infants ($n = 463$), even after controlling for maternal and family characteristics known to influence these outcomes (O'Connor et al., 2003). Further, we reported that premature infants with chronic lung disease for whom more than 50% of their feeds were human milk had, on average, an ~11-point higher Bayley Motor Index score at 12 months corrected age than the corresponding subset of exclusively formula-fed infants.

Available data suggest that approximately 70% of North American women currently initiate breast-feeding (Ryan et al., 2002; Statistics Canada, 2001). At 6 months postpartum, however, only 32.5% of American women are still breast-feeding (Ryan et al., 2002). The duration of breast-feeding appears to be significantly shorter among mothers delivering babies born prematurely. Collectively, these data suggest additional effort (education, support to families, etc.) needs to go into closing the gap between recommended and actual infant feeding practices; also, anti-infective hospital and home infant feeding guidelines need to address both the breast-fed and formula-fed infant.

In the event that breast-feeding is contraindicated or a mother chooses not to breast-feed, a commercially prepared cow-milk-based iron-fortified formula is the next best option under most circumstances. There are, in fact, very few contraindications to breast-feeding, even among hospitalized infants (American Academy of Pediatrics, 2004b). Infants with galactosemia (1 out of 40,000 newborns) are unable to metabolize galactose, one of the monosaccharides found in the principal sugar (lactose) in breast milk, and as such, these infants should not breast-feed. Infants with other inborn errors of metabolism, such as phenylketonuria (1 out of 15,000 newborns), may ingest some human milk, but the amount needs to be closely monitored in relation to the baby's protein requirement. Women in North America who are infected with human immunodeficiency virus (HIV) and those with human T-cell lymphotropic virus types 1 and 2 should not breast-feed. Women with active herpes simplex lesions on their breast(s) should not breast-feed, nor should women who have active pulmonary tuberculosis infection, until they have been treated (Association for Professionals in Infection Control and Epidemiology, 2005).

Most medications are compatible with breast-feeding, or a substitute medication may exist. Depending on the therapy, women receiving antimetabolite chemotherapy should not breast-feed. Women consuming drugs of abuse should not breast-feed until they cease using the abused drug. Good resources to identify those medications that are incompatible with breast-feeding include Appendix B of the 5th edition of the American Academy of *Pediatrics Nutrition Handbook* (American Academy of Pediatrics, 2004a) or the Motherisk Helpline at The Hospital for Sick Children [(416) 813-6780].

SPECIAL CONSIDERATIONS FOR THE SICK AND/OR LOW-BIRTH-WEIGHT INFANT

Often, sick and/or low-birth-weight infants ($<$1,800 g) have nutrient requirements that exceed what reasonably can be met by the usual enteral intakes. In these instances, we strongly encourage continued breast-feeding and nutrient enrichment of some or all of the human milk provided (O'Connor et al., 2004). Unfortunately, there is not a "one size fits all" approach to addressing the nutrient needs of these babies, but we will provide some general principles that will allow the reader to gain an appreciation of competing priorities. Regardless of the configuration of the health care institution, some mechanism should be in place to screen hospitalized babies for their nutritional risk(s). Plotting an infant's weight, length, and head circumference on growth charts can most expeditiously accomplish this task (de Onis et al., 2006; Gibson, 2005; Dietitians of Canada, Canadian Paediatric Society, The College of Family Physicians of Canada and Community Health Nurses Association of Canada, 2004; Kuczmarski et al., 2000). Percentiles derived from plotting a child's weight, length, and/or head circumference for his or her age are interpreted based on the following standards: (i) 10th to 90th percentile, healthy for most pediatric patients; (ii) 3rd to 10th or 90 to 97th percentiles, possibly at risk of poor nutrition, worth further evaluation; and (iii) <3rd or >97th percentiles, at risk of malnutrition or overnutrition, needs further investigation (American Dietetic Association, 2000). Further nutritional evaluation may include examination of a child's weight in relation to his or her length/height for that age (weight-for-length percentile) and/or calculation of ideal body weight (IBW) (IBW is determined as follows: [i] plot the child's current length-for-age on a growth chart to identify the length-for-age percentile; [ii] the ideal weight is then the weight at the same percentile as the length for the same age and gender), and percent ideal body weight (%IBW) (%IBW is determined by the following operation: [actual weight/IBW] × 100). %IBW is interpreted as follows: (i) 90 to 109%, normal weight; (ii) 80 to 89%, possibly or mildly underweight; (iii) <80%, moderately or severely underweight; (iv) 110 to 119%, overweight; and (v) >120%, obese (American Dietetic Association, 2000). For children older than 24 months of age, a body mass index (BMI) (weight [kg]/height [m]^2) also may be calculated. BMI is interpreted as follows: (i) <5th percentile, underweight; (ii) 85th to 94th percentile, risk of being overweight; (c) >95th percentile, overweight (Dietitians of Canada, Canadian Paediatric Society, The College of Family Physicians of Canada, and Community Health Nurses Association of Canada, 2004).

A complete nutritional assessment should be conducted on babies determined to be at possible risk or at risk of poor nutrition. In addition, there will be populations of hospitalized infants that, by nature of their medical condition, are at nutritional risk, and hence a complete nutritional assessment should be completed. A complete nutritional assessment includes a longitudinal examination of the aforementioned growth parameters on growth charts, assessment of energy and nutrient intakes in relation to nutrient requirements, review of blood indices indicative of protein malnutrition and/or electrolyte imbalance, and an appraisal of clinical signs indicative of inadequate nutritional status. We strongly encourage that qualified individuals, such as a clinical dietitian, physician, or advanced nurse practitioner with a background in pediatrics, conduct this nutritional assessment. Table 1 provides a selective list of resources for those new to this field.

At all stages of the life cycle, overnutrition can be as harmful as undernutrition, so it is imperative that infants are carefully monitored and nutrient intakes, particularly fat-soluble vitamin intakes, are regularly assessed to ensure that levels consumed are within safe intake limits. Multivitamin/mineral preparations containing fat-soluble vitamins may need to be altered or discontinued while nutrient-enriched formula or nutrient-enriched human milk is being fed. The duration of time that infants are nutrient enriched should be kept to a minimum and adjusted or discontinued as infants are able to consume larger volumes of milk and are able to meet energy and nutrient requirements from human milk or standard-term formula alone.

THE BREAST-FED INFANT REQUIRING NUTRIENT ENRICHMENT

A number of different approaches can be used to increase the energy and nutrient intake of breast-fed infants. One approach is to replace a number of human milk feedings each day with nutrient-enriched ready-to-feed formula, liquid human milk fortifier (for the premature infant), or a mixture of human milk with undiluted liquid formula concentrate. These approaches work very well at times when mother's milk supply is limited. Of course, the main disadvantage of this approach is that it substantially reduces the amount of human milk fed to the infant whose mother's milk supply is adequate. In order to maximize the use of human milk and to increase the provision of nutrients over and above what can be achieved using this latter approach, specially designed powdered human milk fortifiers (e.g., Similac Human Milk Fortifier [Ross Laboratories] or Enfamil Human Milk Fortifier [Mead Johnson]) for the premature infant may be used. These powdered fortifiers provide a very concentrated source of energy and protein, as well as vitamins and minerals, and were specifically designed to meet the needs of the prematurely born infant. Approximately 1 g of powdered human milk fortifier, provided in individually sealed packets, is mixed with ~25 ml of human milk. Disadvantages of this approach include the following: (i) the products are not sterile; (ii) the products are not always available after hospital discharge without regulatory approval; and (iii) babies must be closely monitored to ensure that the combination of enteral feedings and

Table 1 Nutritional assessment of infants—key resources to get you started[a]

Type of assessment and reference

Anthropometric

  Growth charts for term-born infants

    de Onis et al., 2006. WHO child growth standards

    Kuczmarski et al., 2000. CDC growth charts: United States

    Dietitians of Canada, Canadian Paediatric Society, The College of Family Physicians of Canada and Community Health Nurses Association of Canada, 2004. The use of growth charts for assessing and monitoring growth in Canadian infants and children

  Growth charts for prematurely born infants

    Babson and Benda, 1976. Growth graphs for the clinical assessment of infants of varying gestational age

    Fenton, 2003. A new growth chart for premature babies: Babson and Benda's chart updated with recent data in a new format

  Infant health and development growth charts

    Casey et al., 1990. Growth patterns of low birth weight premature infants: a longitudinal analyses of a large varied sample

    Casey et al., 1991. Growth status and growth rates of varied sample of low-birth-weight, premature infants: a longitudinal cohort from birth to 3 years of age

    Guo et al., 1996. Weight-for-length reference data for preterm, low-birth-weight infants

Dietary

  Term-born infants

    Institute of Medicine, 1997. Dietary reference intakes for calcium, phosphorus, magnesium, vitamin D, and fluoride

    Institute of Medicine, 1998. Dietary reference intakes for thiamin, riboflavin, niacin, vitamin B6, folate, vitamin B12, pantothenic acid, biotin, and choline

    Institute of Medicine, 2000. Dietary reference intakes for vitamin C, vitamin E, selenium, and carotenoids

    Institute of Medicine, 2001. Dietary reference intakes for vitamin A, vitamin K, arsenic, boron, chromium, copper, iodine, iron, manganese, molybdenum, nickel, silicon, vanadium, and zinc

    Institute of Medicine, 2005a. Dietary reference intakes for energy, carbohydrate, fiber, fat, fatty acids, cholesterol, protein, and amino acids

    Institute of Medicine, 2005b. Dietary reference intakes for water, potassium, sodium, chloride, and sulfate

    Raiten et al., 1998. Assessment of nutrient requirements for infant formulas

  Prematurely born infants

    Nutrition Committee of the Canadian Paediatric Society, 1995. Nutrient needs and feeding of preterm infants

    Klein, 2002. Nutrient requirements for preterm infant formulas

    Klein and Heird, 2005. Summary and comparison of recommendations for nutrient contents of low-birth-weight infant formulas

(Continued)

Table 1 (*continued*)

Type of assessment and reference
Biochemical
Gibson, 2005. Principles of nutritional assessment
Soldin et al., 2003. Pediatric reference ranges
Clinical
Gibson, 2005. Principles of nutritional assessment

^aReferences cited in this table can be found in the reference list at the end of this chapter.

vitamin/mineral drops do not result in unsafe nutrient levels, particularly fat-soluble vitamins.

A third approach to increase the energy and nutrient intakes of breast-fed infants would be to mix unmodified human milk with a powdered infant formula (standard or nutrient enriched). As sterile nutritional equivalents are available as are single-dose packets of powdered human milk fortifier in North America, we usually do not need to use powdered formulas from a can to nutrient enrich feedings of infants at our institution. The exception is if an infant has a unique medical condition that requires a very specialized feeding available only as a powder (infants with inborn errors of metabolism, for example), in which case we would need to use powdered formulas. This option may be used as babies are readied for transition home or after hospital discharge, as it is often the least expensive option for families. Further, once a can of powder is opened, it can be resealed and kept up to 1 month, unlike liquid varieties, which need to discarded within 48 h. Given the small volume of formula required for enrichment of human milk, most of the liquid concentrate would be discarded.

IF FORMULA FEEDING, WHAT FORMULA?

As summarized in Fig. 1 there are numerous infant formulas on the market. The purpose of this next section is to provide the reader with a working knowledge of the major categories of formula available in North America and their indications for use. We begin with a discussion of why unmodified cow milk is not an appropriate option for infant feeding.

Unmodified Cow Milk

In the event that breast-feeding is contraindicated or a mother chooses not to breast-feed, a commercially prepared cow-milk-based iron-fortified formula is recommended for the first 9 to 12 months of life (American Academy of Pediatrics, 2004c). Whole (full-fat), 2%, 1%, or skim cow milk,

Breastfeeding is the preferred feeding for preterm and term infants—if breastmilk is not available, or when weaning, choose:

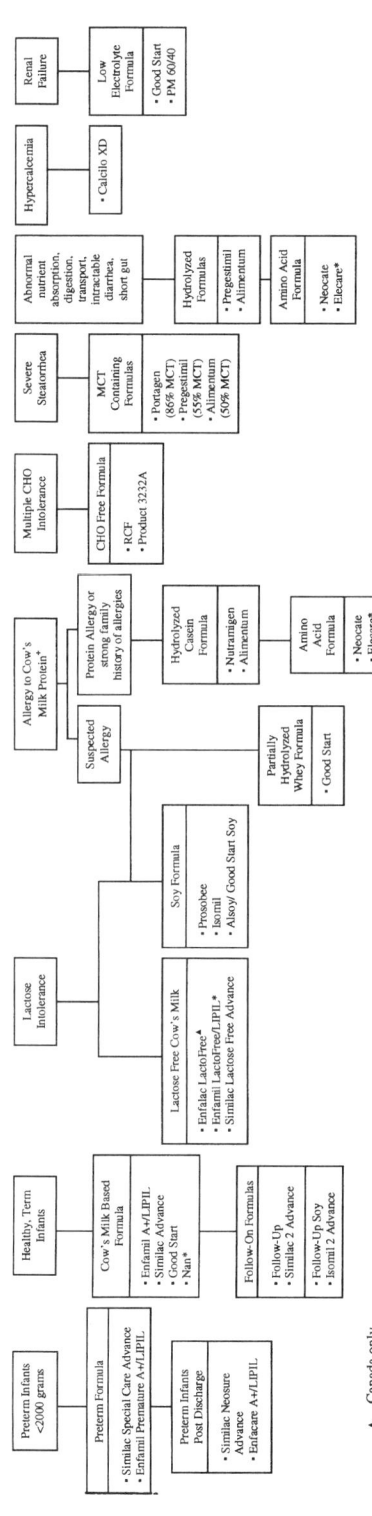

Figure 1 Suggested enteral feeding selection for infants. This list is not exclusive; additional products exist that may be equally suitable to an infant's needs.

▲ Canada only.

* United States only.

+ Infants with documented cow's milk protein enteropathy or enterocolitis may be as sensitive to soy protein and should not be given soy formula. They should be given an extensively hydrolyzed formula. Many infants with documented IgE-mediated allergy to cow's milk protein will tolerate a soy formula.

goat milk, or evaporated milk is not recommended during this time. The nutrient composition of unmodified cow and goat milks differ quite markedly from that of human milk. For a comprehensive discussion of these differences, the reader is referred to the text entitled *Handbook of Milk Composition* (Jensen, 1995). The protein content of unmodified cow milk (3.5 g/dl) is significantly greater than that of human milk (1.0 g/dl), and the types of and concentrations of protein and nonprotein nitrogen differ considerably. Further, the fatty acid profiles of cow and human milks differ, as does the distribution of these fatty acids on the glycerol backbone of the triglyceride molecule where most fatty acids are bound. It is well known that cow milk contains limited amounts of the essential fatty acids linoleic and alpha-linolenic acid.

The predominant source of carbohydrate in human milk is lactose. The concentration of lactose in human milk (~6.1 g/dl) is significantly higher than that of cow milk (~4.8 g/dl) (Jensen, 1995). Lactose, a disaccharide composed of the monosaccharides galactose and glucose, is thought to play an important role in mineral bioavailability. Other prominent milk sugars in human milk include neutral and acidic oligosaccharides. Oligosaccharides are the largest constituent of human milk after water, lipid, and lactose, and they are found in concentrations 10 times higher than those in cow milk. A significant fraction of oligosaccharides escape digestion and are fermented in the colon, promoting colonization of the gastrointestinal tract with nonpathogenic bacteria that can inhibit the binding of pathogens to the epithelial surface of the gut.

The mineral content, including sodium and potassium concentrations, of unmodified cow milk is significantly higher than that of human milk and is thought to reflect the higher growth rate of the calf compared to that of the human infant (Ziegler and Fomon, 1989). The high concentrations of protein, sodium, and potassium together increase the renal solute load of cow milk compared to that of human milk or commercially available formula (American Academy of Pediatrics, 2004c). Renal solute load, by definition, reflects the solutes excreted per liter of milk or formula consumed. A higher renal solute load puts an unnecessarily high demand on the immature kidney relative to that of human milk and doubles urine osmolality (Fuchs et al., 1992). While the kidney of the healthy term infant usually can handle this increase, this might not be the case at the times of increased water loss that occur during a bout of acute febrile illness, diarrhea, and/or vomiting (Fomon, 1993). Consumption of unmodified cow milk has been associated with blood loss in the stool of infants less than 6 months of age (Fuchs et al., 1992; Raiten et al., 1998). The mechanism for this loss is presumably mediated via an allergic reaction between some component of cow milk protein and the enterocytes of the gastrointestinal tract. Blood losses, in combination with the low concentration and bioavailability of iron in cow milk, may predispose infants weaned to cow milk to iron deficiency anemia.

Commercially Available Infant Formulas
The Food and Drug Administration in the United States (FDA), under the provisions of the Federal Food, Drug, and Cosmetic Act (commonly known as the Infant Formula Act), as amended, is responsible for ensuring the safety and nutritional quality of infant formulas marketed in the United States (21 Code of Federal Regulations 107) (*Federal Register*, 1985). The Infant Formula Act specifies the minimum concentrations of 29 nutrients and the maximum concentrations of 9 of these nutrients. Likewise, in Canada, Health Canada, under provisions of the Federal Food and Drug Regulations (Part B, Division 25), has set out extensive nutritional requirements for infant formulas (Food and Drug Directorate, 1990). At the request of the FDA, expert panels in 1998 and again in 2002 made recommendations for revision of the Infant Formula Act as it applied to the nutrient content of formula designated for term-born infants (Raiten et al., 1998) and prematurely born infants (Klein, 2002). The latter document was expanded upon in 2005 (Klein and Heird, 2005). The reader is encouraged to refer to these comprehensive resources to study the research literature supporting specific nutrient requirements for formulas designed for infants.

Cow-Milk-Based Formulas
Cow-milk-based formulas designed for term-born infants currently on the market in North America contain 67 kcal/dl or 20 kcal per fluid oz (formulas are generally described in terms of their energy contribution per volume). However, note that formulas with a higher energy contribution typically contain increased concentrations of protein, calcium, phosphorus, and other macro- and micronutrients per unit of volume as well.

The protein concentration of cow-milk-based standard-term formula is significantly lower than that of unmodified cow milk and is similar to that of human milk. Different formulas vary in their whey-to-casein ratio, with some having a whey-to-casein ratio of 18:82 or 100:1 and others having a ratio of 60:40, similar to that of human milk. The terms casein and whey were developed by the dairy industry as a way to describe the physical properties of milk. Casein precipitates into curds when it is acidified, while the whey proteins remain in solution. A number of studies, though not all, suggest that whey-based formulas accelerate gastric emptying faster than casein-predominant milk formulas and may be associated with fewer episodes of emesis and gastroesophageal reflux (Billeaud et al., 1990; Fried et al., 1992; Khoshoo et al.,

1996; Thorkelsson et al., 1994; Tolia et al., 1992). These observations may be important in the selection of an infant formula for those infants so affected.

Because the proteins in the whey fraction of cow milk differ from those in human milk, it is not surprising that increasing the whey:casein ratio of infant formula does not necessarily produce an essential amino acid profile in the blood of the formula-fed infant that better matches that of the exclusively breast-fed infant. While much research needs to be done in this area, we do know from the adult and infant literature that the dietary amino acid composition can affect plasma amino acid patterns in blood and, in turn, may affect maturation of behavior (Heine, 1999; O'Connor et al., 1997; Oberlander et al., 1992; Rassin, 1994; Steinberg et al., 1992; Yogman and Zeisel, 1983).

The carbohydrate in cow-milk-based standard-term formula is composed either of lactose alone or of a combination of lactose and maltodextrin. Maltodextrins are moderately sweet, easily digestible polysaccharides of 5 to 10 glucose units, and they are produced from the hydrolysis of starch. Like human milk, standard-term formulas are designed to derive 50% of their energy from lipid. The lipid fraction or the oil blend is composed of a combination of vegetable oils, including coconut (a source of short- and medium-chain saturated fatty acids), palm (a source of long-chain saturated fatty acids), sunflower, safflower, and soybean oils. The latter three oils provide a source of polyunsaturated fatty acids, in particular, the essential fatty acids linoleic acid and alpha-linolenic acid. Genetic variants of safflower and sunflower oils have been used to increase the monounsaturated fatty acid and specifically the oleic content of the fat blend. While formula manufacturers have done a relatively good job of incorporating the most prevalent fatty acids in human milk into formula, the oil blends do not approach mimicking the presence or concentration of the ~167 fatty acids found in human milk or its triacylglycerol profile and sterol, phospholipid, sphingomyelin, neutral glycosylceramides, acidic glycosphingolipid, or ganglioside content. Recently, the long-chain polyunsaturated fatty acids arachidonic acid (ARA) and docosahexaenoic acid (DHA) have been added to many term formulas. These fatty acids can be synthesized by infants from the precursor essential fatty acids linoleic and alpha-linolenic acid, respectively. While data are conflicting on this point, some clinical trials have demonstrated improved short-term performance on tests of visual and cognitive function among infants fed formulas containing ARA and DHA (American Academy of Pediatrics, 2004c).

Both brand names (e.g., Enfamil Good Start Supreme and Similac Advance) and generics (e.g., President's Choice) are widely available. Cow-milk-based term formulas are provided as concentrated liquids, ready-to-use liquid, and powdered options. In addition, cow-milk-based formulas designed for older infants and toddlers now are available in North America and are called follow-up or toddler formula. The rationale for the development of the latter class of formulas was to more closely meet the nutritional needs of older infants and toddlers who are consuming solid foods and to deter parents from the early introduction of cow milk, as these products are generally less expensive than starter formulas. They tend to be higher in calcium and phosphorus than starter milk-based formulas. While nutritionally complete, there is no clear evidence to suggest they produce better nutritional and health outcomes than those of starter formulas.

Powdered formulas are frequently more attractive than are liquid concentrate and ready-to-use forms, as they usually are the least expensive (by ~110 to 150 Canadian dollars a month) and, once opened, can be used for up to 1 month before being discarded. Powdered formulas are particularly attractive for mothers who use formula to supplement breast-feeding. They are lightweight and generally more portable than liquid options. Powdered formulas usually are reconstituted using 1 scoop of powder in 60 ml of clean potable water; scoop sizes do differ among products. For all hospitalized infants, immunocompromised infants at home, and all healthy term infants less than 4 months of age, tap water should be boiled or commercially prepared sterilized water should be purchased. The length of time suggested to boil tap water varies from 2 to 5 min depending on the authoritative body or regulatory agency (Canadian Paediatric Society, Dietitians of Canada, Health Canada, 1998; American Academy of Pediatrics, 2004c; American Dietetic Association, Pediatric Nutrition Practice Group, 2004) . A recent joint Food and Agriculture Organization-World Health Organization workshop recommended either using boiling water to reconstitute powdered formula or heating reconstituted powdered formula (FAO-WHO, 2004). The FDA withdrew a previous recommendation to use boiling water to reconstitute powdered formula over concerns with the resultant physical stability of the formula (clumping and/or separation) and nutrient degradation (American Dietetic Association, Pediatric Nutrition Practice Group, 2004). Using boiling water to reconstitute powdered formula also may present a safety issue for staff. More recently still, it was shown that reconstituting powdered cow-milk-based infant formula with hot (not less than 70°C) but not boiling water produced significant inactivation of *E. sakazakii* (FAO-WHO, 2006). While there was some destruction of vitamins, particularly vitamin C, the levels of all vitamins remained above those claimed on the product label. As recommended by the American Dietetic Association, at SickKids we use cooled, commercially prepared, sterilized water to reconstitute powdered formula in a laminar-flow hood located in our enteral feeding preparation room (American Dietetic Association, Pediatric Nutrition Practice Group, 2004).

Additional details on these procedures are provided below. We are investigating, as are others, the feasibility of using hot water (not less than 70°C) to reconstitute powdered infant formula in the hospital in cases where a nutritionally equivalent sterile product is unavailable.

Liquid concentrates are sterile and are available in 13-oz (385-ml) cans or aseptic containers in the United States. They require dilution with clean water (the water must be prepared as described above) in a 1:1 ratio. Mixing and measurement are somewhat easier than they are with powdered formula, but liquid concentrates typically are 30 to 40% more expensive and must be used within 48 h of opening and within 24 h of being mixed with water. Ready-to-feed formula is the most convenient, since it requires no mixing and does not depend on a clean water supply; however, it is by far the most expensive way to buy formula—up to four times the cost of powdered formula.

Lactose-Free Cow-Milk-Based Formulas

Lactose-free formulas (67 kcal/dl), as the name suggests, are free of the milk sugar lactose but contain a cow-milk-based protein isolate as their protein source. Current commercially available products are designed to meet the nutritional requirements for term-born infants only, though they may be appropriate for premature infants after hospital discharge or later in their first year of life if they are growing well. The carbohydrate fraction of lactose-free formulas is composed of corn syrup solids (Enfamil LactoFree LIPIL) or malodextrin and sucrose (Similac Lactose Free Advance). Their fat blends tend to be relatively similar to their respective lactose-containing parent brands. Lactose-free formulas in North America are available as liquid concentrates, ready-to-feed liquids, and powder.

Lactose-free formulas are appropriate for use in infants with congenital lactase deficiency, a relatively rare disorder inherited as an autosomal recessive gene. Though somewhat controversial, lactose-free formulas also may be useful during periods of secondary disaccharidase deficiency due to acute enteritis or chronic conditions affecting the integrity of the small intestine, such as diarrhea, enteropathies, and Crohn's disease (Canadian Paediatric Society, Dietitians of Canada, Health Canada, 1998; American Academy of Pediatrics, 2004c). Many families self-prescribe a lactose-free formula in response to a baby that is excessively fussy or gassy and/or a suspicion that the baby is sensitive to lactose. Due to lactose's role in mineral absorption and in nonpathogenic bacterial colonization of the gastrointestinal tract, before switching to a lactose-free formula there should be some consideration of the relative merits of a lactose-containing versus a lactose-free formula for the particular infant in question.

Soy-Protein-Based Formulas

Soy-based formulas (67 kcal/dl) are currently available in North America as liquid concentrates, ready-to-use liquids, and powder and are free of both cow milk protein and lactose. They are designed to meet the nutritional requirements of healthy term-born infants. Historically, soy-based formulas have accounted for ~20% of the infant formula market in the United States (Strom et al., 2001). The protein fraction of current formulations is made up of soy protein isolate and not soy flour. This modification is thought to account, at least in part, for the improved tolerance and increased mineral bioavailability of soy-based products. Due to the fact that soy protein tends to be limiting in the essential amino acid methionine, this amino acid is added. Carnitine, generally at low levels in plants, is added to soy formula at levels found in breast milk, as is taurine, an antioxidant and a major conjugate of bile acids in early infancy. Starch, corn syrup solids, corn maltodextrin, and sucrose, or some combination of these, forms the carbohydrate fraction of soy formula. Minerals (i.e., calcium and phosphorus) are added in concentrations greater than those of cow-milk-based formula to account for an anticipated reduction in mineral bioavailability caused by the presence of phytates found in the soy protein isolate. The fat blend of these products generally parallels that of the cow-milk-based flagship brand for each manufacturer.

Concern has been expressed recently about using these formulas without clear indication due to their phytoestrogen, specifically isoflavone, content (Anthony et al., 1996; Chen and Rogan, 2004; Klein, 1998; Setchell et al., 1997; Sharpe et al., 2002; Sheehan, 1995; Strom et al., 2001). Exposure of laboratory animals to phytoestrogens (also called dietary or plant estrogens) has produced altered sexual development in some studies (Anthony et al., 1996; Chen and Rogan, 2004; Klein, 1998). Controlled prospective human studies are lacking, but a retrospective cohort study based on telephone interviews reported no overt reproductive problems in young adults fed soy-based infant formulas as infants (Strom et al., 2001). Likewise, no adverse effects of feeding soy-based formulas to infants have come to light after decades of use.

Because these products are lactose free, they are appropriate for use in term infants with hereditary lactose deficiency, galactosemia (some products are preferred over others for infants with galactosemia), or demonstrated lactose intolerance due to acute gastroenteritis (American Academy of Pediatrics, Committee on Nutrition, 1998). Most (>85%) infants with documented immunoglobulin E (IgE)-mediated allergy to cow milk will do well on isolated soy-based formula (American Academy of Pediatrics, Committee on Nutrition, 1998; Bock and Atkins, 1990; Ladodo and Borovick, 1992). These formulas are considerably less expensive and more palatable than extensively hydrolyzed cow-milk-based products. Infants with documented cow milk protein-induced enteropathy or enterocolitis (non-immunoglobulin E-mediated allergy) are frequently sensitive to soy protein as well, and therefore they should be provided with formula for which the protein fraction is made up of extensively hydrolyzed protein or 100% free amino acids (Burks et al., 1994; Eastham, 1989; Whitington and Gibson, 1977).

Soy formulas may be suitable for term infants of families that wish to follow a vegetarian-based diet and require a supplement for breast-feeding. Soy formulas are not designed to meet the needs of prematurely born infants (American Academy of Pediatrics, Committee on Nutrition, 1998).

Protein-Hydrolysate- and Amino-Acid-Based Formulas

Protein hydrolysate formulas currently on the market can be categorized loosely according to the extent that the protein component is hydrolyzed: (i) 100% free-amino-acid-containing formula (SHS Neocate and Similac EleCare); (ii) extensively hydrolyzed protein-containing formula (Enfamil Nutramigen, Enfamil Pregestimil, and Similac Alimentum); and (iii) partially hydrolyzed protein-containing formula (Good Start and Enfamil Gentlease). All provide 67 to 68 kcal/dl, are iron fortified, and are designed for term-born infants. Not all products are available as sterile ready-to-feed liquids or liquid concentrates. The nitrogen fraction of the extensively and partially hydrolyzed products consists of the following, either individually or in combination: nonfat cow milk, cow milk casein, and/or cow milk whey. Milk proteins are heat treated and enzymatically hydrolyzed. Various amino acids are added to extensively hydrolyzed products to compensate for losses that occur in processing (e.g., L-cystine, L-tryptophan, and L-tyrosine). Most of the nitrogen in extensively hydrolyzed products is in the form of free amino acids and peptides that are <1,500 kDa. As a rule, the more extensively hydrolyzed the formula, the less antigenic it is, the poorer the taste, and the higher the price. Partially hydrolyzed formulas have a greater proportion of their total nitrogen content made up of larger peptides. As the name implies, the nitrogen fraction of amino-acid-based formulas is composed of 100% free amino acids.

The fat contents of protein-hydrolysate- and amino-acid-based formulas contain a combination of the following oils: corn oil, soy oil, high oleic safflower or sunflower oil, coconut oil, palm olein, and a single-cell oil blend rich in DHA and ARA. Alimentum, EleCare, and Pregestimil contain approximately one-third to one-half of their total fatty acids as medium-chain triglycerides (MCT; fatty acids containing 6 to 12 carbon atoms). In contrast, Neocate contains 5% of total fatty acids as MCT oil, and Nutramigen, Good Start, and Enfamil Gentlease are devoid of MCT oil. MCT are more easily and rapidly hydrolyzed and absorbed than long-chain fatty acids by infants with fat malabsorption. Unlike long-chain fatty acids, MCT do not require pancreatic enzymes or bile salts for digestion and absorption and can directly enter portal circulation for transport to the liver.

The carbohydrate fractions of extensively hydrolyzed protein- and free-amino-acid-based formulas are devoid of lactose and are made up of a combination of the following: modified cornstarch, modified tapioca starch, maltodextrins, corn syrup solids, and sucrose.

Protein-hydrolysate- and amino-acid-based formulas are used for (i) infants with severe milk protein allergy, (ii) infants at high risk of developing milk allergy, and (iii) infants with significant malabsorption due to gastrointestinal or hepatobiliary disease (e.g., cystic fibrosis, short gut syndrome, biliary atresia, cholestasis, and protracted diarrhea) (American Academy of Pediatrics, Committee on Nutrition, 2000; American Academy of Pediatrics, 2004c). In many instances, these formulas are lifesaving and are preferable over the only available option—parenteral nutrition. In general, hydrolyzed products with a significant fraction of their fat blend as MCT are used for malabsorption disorders.

Table 2 summarizes the American Academy of Pediatrics' feeding recommendations for infants at risk of or exhibiting symptoms of food allergy. Hypoallergenic formulas, by definition, are those that have been shown in randomized controlled trials not to provoke an allergic reaction in infants with confirmed cow milk protein allergy. Given this definition, only extensively protein-hydrolyzed and 100% amino-acid-containing formulas can be called hypoallergenic. There is some evidence suggesting that partially hydrolyzed formulas also may be efficacious in allergy prevention, especially atopic dermatitis, among infants with a family history of allergy (Baumgartner et al., 1998; Marini et al., 1996; von Berg et al., 2003). As the members of this latter class of formula are considerably cheaper and better tasting than their extensively hydrolyzed and amino-acid-based counterparts, they may be a suitable alternative for some families. Partially hydrolyzed protein-based formulas should not be used for infants at risk of severe milk protein allergy or infants with confirmed cow milk allergy.

Formulas Designed for Infants Born Prematurely

Formulas designed for premature infants during their initial in-hospital course are commonly called premature formulas (Similac Special Care Advance or Enfamil Premature LIPIL [Enfamil A+ Premature in Canada]) (American Academy of Pediatrics, 2004e). These products are available in a commercially sterile ready-to-use form (68 or 81 kcal/dl; 20 or 24 kcal/oz); they are not available in North America as a powder or liquid concentrate.

Table 2 Guidelines for preventing and addressing symptoms of food allergy in infants[a]

Category and guideline

Prevention

> Mothers of infants at a high risk of developing allergy (e.g., strong family history) are encouraged to breast-feed for the first year of life or longer. Mothers should eliminate peanuts and tree nuts (e.g., almonds and walnuts) from their diet and consider eliminating or reducing consumption of eggs, cow milk, and fish.

> As a restrictive diet may not meet all nutritional requirements of a breastfeeding woman, appropriate dietary counseling, as well as supplements, should be provided.

> In the event that a mother wishes to supplement breast-feeding, a hypoallergenic formula should be used.

> Introduction of solids to an at-risk infant should be delayed until 6 months of age, with introduction of dairy products delayed until 1 year, eggs until 2 years, and peanuts, nuts, and fish until 3 years.

Addressing symptoms of allergy

> Maternal restriction of cow milk, egg, fish, peanuts, and tree nuts may be beneficial to the breast-fed infant who develops symptoms of food allergy.

> If maternal restriction is unsuccessful, use an extensively hydrolyzed protein-containing formula. If symptoms persist, try a 100% free amino acid-based formula. Infants with IgE-associated symptoms of allergy may do well on a soy-based formula, as the risk of concomitant cow and soy allergy is low. For infants with non-IgE-associated symptoms, such as enterocolitis, proctocolitis, malabsorption syndrome, or esophagitis, do not use a soy-based product, as the risk of concomitant cow and soy allergy is relatively high.

> Formula-fed infants with confirmed cow milk allergy may benefit from the use of a hypoallergenic or soy formula as described above.

[a]Adapted from the American Academy of Pediatrics, Committee on Nutrition (2000).

These formulas are higher in protein, vitamins, and minerals than standard-term formula in order to facilitate extrauterine growth of prematurely born infants similar to that of a fetus of the same postconceptional age. For example, current premature formulas on the market in North America contain 3.0 g protein/100 kcal, compared to 2.1 to 2.2 g/100 kcal in standard-term formula. To facilitate skeletal growth at in utero rates, current premature formulas contain 165 to 180 and 83 to 100 mg/100 kcal of calcium and phosphorus, respectively, compared to the 78 and 42 mg/100 kcal, respectively, found in their standard-term parent brand. Premature formulas are designed to improve absorption, and as such, the fat blend is 40 to 50% MCT. Standard-term formulas designed for healthy infants do not contain MCT. The fatty acids DHA and ARA have been added to the fat blend of all premature formulas available in North America. Clinical trials with prematurely born infants demonstrate at least a short-term developmental advantage (visual, cognitive, and language) to supplementing premature infant formulas with DHA and ARA (American Academy of Pediatrics, 2004e).

The carbohydrate fraction of premature formulas is composed of lactose and glucose polymers. Glucose polymers are used to reduce the lactose content and osmolality of the formula. Premature infants may not be able to fully digest lactose in a formula matrix, because lactase activity is not fully achieved in the fetal intestine until quite late in gestation (36 to 40 weeks). Excessive amounts of undigested carbohydrates entering the large intestine could result in gas production and intestinal distention. While the main milk sugar in human milk is lactose, human milk generally is better tolerated than infant formula, suggesting that the relationship between lactose and tolerance is quite complex.

Postdischarge Formulas
More recently, cow-milk-based formulas designed for premature infants after hospital discharge have become commercially available (Similac Neosure Advance and EnfaCare A+/LIPIL) (American Academy of Pediatrics, 2004e). They were designed to address the fact that, after hospital discharge, the nutrient needs of prematurely born infants (particularly those born at less than 1,500 g) may continue to exceed that provided by a standard-term formula. In general, the nutrient levels in postdischarge formulas are between those of premature formulas designed for in-hospital use and standard-term formulas. They contain 75 kcal/dl (22 kcal/oz) of energy and 2.8 g, 105 to 120 mg, and 62 to 66 mg/100 kcal of protein, calcium, and phosphorus, respectively. EnfaCare A+/LIPIL and NeoSure Advance have a whey:casein ratio of 60:40 and 48:52, respectively. Both formulations contain a combination of lactose and glucose polymers, and the fat blend is made up of vegetable, MCT, and single-cell oils, the latter providing a source of DHA and ARA. In North America, these products are manufactured in a commercially sterile ready-to-use form and as a nonsterile powder.

Metabolic Formulas
Several different formulas or medical foods have been developed for infants, children, and adults with inborn errors of metabolism (American Academy of Pediatrics, 2004d). Collectively, the total incidence of inborn errors in metabolism is low (~1:1,000 births); however, timely availability of a specific metabolic formula could be life-saving and frequently will prevent severe developmental sequelae. Given the low turnover of many of these products, they are typically available only as powders. Powders have a longer shelf life than concentrates and ready-to-use options. Metabolic formulas typically contain 67 kcal/dl. The nitrogen component of formulas designed for those with inborn errors of amino acid metabolism are made up of 100% free amino acids devoid of the amino acid or its precursor(s) that cannot be adequately metabolized. For example, metabolic formulas designed for infants

with phenylketonuria are prepared without the amino acid phenylalanine. Phenylketonuria is an inherited error of metabolism caused by a deficiency of the enzyme phenylalanine hydroxylase. This enzyme is responsible for the conversion of phenylalanine to tyrosine. For inborn errors of metabolism not involving amino acid or protein metabolism, components of cow milk protein or soy protein isolate may comprise the nitrogen fraction of the metabolic formula.

Whereas the carbohydrate fractions of most metabolic formulas are composed of corn syrup solids or a combination of corn syrup solids, sugar, and cornstarch, some metabolic formulas have been developed to contain no source of carbohydrate. The latter formulation may be used as part of a diet designed to induce ketosis (ketogenic diet) to manage seizure disorders that are not responsive to pharmacotherapy. The fat blend of metabolic formulas usually consists of vegetable oils.

Other Specialty Formulas, Nutrient Modulars, and Dietary Supplements

Hundreds of other specialty formulas, nutrient modulars, and dietary supplements, not covered above, are used to feed infants with very specialized nutritional requirements. Specialty formulas include Portagen and Monogen, cow-milk-based powders with most of their fat blend being MCT. These are the only available formula options for infants with chylothorax, a condition where there is a tear or leak in the thoracic duct, causing chylous fluid to collect in the pleural cavity. Nutrient modulars such as Polycose powder (glucose polymer), Moducal (maltodextrin), and microlipid (safflower oil emulsion) are used for energy boosting. Dietary supplements generally refer to vitamins and minerals but can be interpreted more broadly to include any natural product consumed enterally, including thickening agents, herbs, enzymes, metabolites, extracts, or concentrates. Some specialty formulas, nutrient modulars, and dietary supplements are sterile; most are not. Many of these specialty products are made in manufacturing environments not as closely monitored as a manufacturing facility for infant formula. Careful consideration must be given to the relative nutritional benefits versus the risks of using these products.

PREPARATION OF ENTERAL FEEDINGS IN THE HOSPITAL

We highly recommend that the reader either interested in setting up an enteral feeding preparation room or wishing to understand more details about the configuration of such a facility refer to the "Guidelines for Preparation of Formula and Breast Milk in Health Care Facilities," authored by the Pediatric Nutrition Practice Group of the American Dietetic Association (2004). These

Table 3 Guidelines for preparation and handling of enteral feedings designed for infants in the hospital setting[a]

Guideline
Human milk feeding should be encouraged.
Formula products, including nutrient modulars, should be selected based on nutritional need.
Whenever possible, an alternative to powdered infant formulas or nonsterile nutrition modulars, such as a sterile ready-to-feed formula, concentrated liquid formula, or a sterile modular, should be chosen.
Trained personnel should prepare infant formulas, nutrient-enriched human milk, and other specialty enteral feeds using aseptic techniques in a designated space and, ideally, under a laminar-flow hood.
Sterilized water should be used for reconstitution of infant formula.
After reconstitution, enteral feedings including fortified human milk should be refrigerated immediately and discarded if not used within 24 h after preparation.
The administration, or "hang," time for continuous enteral feedings, including fortified human milk feedings, should not exceed 4 h.
Written hospital guidelines should be available in the event of a manufacturer product recall, including notification of health care providers, a system for reporting and follow-up of specific formula products used, and retention of recall records.

[a]Adapted from the following sources: Food and Agriculture Organization of the United Nations (2004); Food and Drug Administration (2002); Health Canada (2002); and American Dietetic Association, Pediatric Nutrition Practice Group (2004).

are the guidelines that we at SickKids have patterned our policies, procedures, and practices after. A brief synopsis of the important elements in the preparation and handling of enteral products in a health care institution to reduce the risk of microbial contamination is provided in Table 3. Institutions adhering to these guidelines prior to the regulatory warnings about *E. sakazakii* likely had to make relatively minor adjustments to policies and procedures regarding preparation, delivery, and administration of enteral feeds. Data do not currently exist to indicate how many institutions in developed countries adhere to these guidelines.

Physical Facilities and Personnel

In a tertiary pediatric health care center, a significant proportion of infants, children, and youth may be immunologically at risk. By definition, a tertiary-care institution is one that provides highly specialized skills, technology, and support services. Many infants and children who are immunocompromised require specialized enteral feeding. As summarized above, many of these specialized feedings are not sterile. Immunologically at-risk children in tertiary-care pediatric settings are not confined to a single patient care unit, although most infants in the neonatal intensive care and critical care units are clearly immunocompromised. This would differ, of course, in hospitals where neonatology and/or pediatrics are subspecialties occupying but one of many patient care areas. As recommended by the American Dietetic Association's guidelines, we operate a centrally located facility for preparation of enteral feedings, whether infant formula based or breast milk based. The entire area is a secured space. A registered dietitian supervises this facility, with input from a qualified infection control professional.

In addition, as advised, within the enteral preparation area we operate a specially designated milk preparation room used solely for the purpose of preparing enteral feedings using aseptic techniques. The milk preparation room should be designed to facilitate efficient, safe workflow and must include product storage facilities that meet published standards for the safe storage of food products (American Dietetic Association, Pediatric Nutrition Practice Group, 2004). The milk preparation room is accessible only to those individuals preparing or supervising the preparation of enteral feeds. It is held at positive air pressure, has a clean air supply, and is equipped with a laminar-flow hood fitted with a high-efficiency bacteria-retentive filter (a HEPA filter). Preparation of enteral feedings in the laminar-flow hood helps to ensure a clean source of air to the work area. Airflow out of the hood will facilitate suspension and removal of contaminants introduced to the work area by staff. A preventive maintenance program is in place to ensure proper functioning of the laminar-flow hood. A separate hand-washing sink is available in the milk preparation room with controls for water that do not require hand operation.

The importance of hiring and retaining high-quality staff responsible for enteral feeding preparation should not be underestimated. Enteral feeding room technicians should be well trained. They should understand the importance of their role in providing nutrition support to hospitalized infants, the importance of nutrition in the positive health outcomes of these infants, and the consequences of contamination (i.e., microbial, viral, and allergenic) in negative health outcomes. Technicians should have reading, writing, and mathematical skills at the high school level or beyond. They need to be in good health, adhere to good hygienic practices, and observe a dress code consistent with aseptic technique. Personnel should have access to written policies and procedures regarding enteral feeding preparation and should participate in their development and revision. Care should be taken to ensure that their workload is conducive to minimizing errors and that sufficient staff are available to provide coverage 7 days a week. In order to promote focus on the task at hand, telephone calls into the milk preparation area should be restricted. Our milk preparation area is equipped with a speaker phone, which allows staff to receive important information with minimal interruption.

Equipment, Utensils, and Supplies

All equipment in the milk preparation room is sanitized after each type of enteral feeding is prepared. A high-temperature dishwasher in the milk preparation area is dedicated to this purpose. The milk preparation room is well stocked with supplies (e.g., gowns, bottles, nipples, sanitizing solutions, utensils, and equipment) for the day to ensure aseptic technique and to minimize trips in and out of the room. As recommended, we use single-use nipples and Grad-U-Feeds to reduce the risk of microbial and food allergen cross-contamination. Plastic bottles are washed, sterilized, and reused due to cost considerations. We purchase sterilized water and keep it chilled in the milk preparation room refrigerator for reconstitution of liquid concentrate and powdered formula. It is noteworthy that using chilled water does reduce the solubility of powdered nutritionals and hence makes them more difficult to reconstitute, but it can help to ensure that the temperature of the final product is within the desired range ($<4°C$) at the time of delivery. Prepared enteral feedings are labeled, refrigerated, and transported to patient care units. Labels include the patient's name, medical record identification number, patient care unit, type of enteral feeding, energy density, and date of order. Labels are prepared outside the milk preparation room prior to staff entering the milk preparation area. In epidemiological investigations where rehydrated powdered formula was implicated as the vector for *E. sakazakii*-related meningitis, equipment and utensils used to prepare formula also tested positive for this organism (Gurtler et al., 2005). These observations underscore the importance of taking the appropriate precautions with equipment and supplies to ensure that, in the event of formula contamination with *E. sakazakii*, they do not serve as vectors for contamination of other enteral feeds.

HANDLING OF EXPRESSED BREAST MILK

Given the many benefits of human milk feeding, women should be encouraged by health care providers to supply breast milk for their hospitalized infants. Two excellent and detailed resources for understanding the key considerations for expression, storage, and thawing of human milk can be found in the "Guidelines for the Preparation of Formula and Breastmilk" and in a report written by Ruth Lawrence in an issue of *Advances in Nutritional Research* (Lawrence, 2001; American Dietetic Association, Pediatric Nutrition Practice Group, 2004). Table 4 briefly summarizes the key recommendations from these documents. Our processes with regard to handling breast milk have evolved significantly at SickKids over the past 5 years in response to these documents, our increased use of human milk, and new information on processes to prevent breast milk mix-ups.

Mothers of hospitalized infants are strongly encouraged to use mechanical expression to establish and/or maintain their milk supply if infants are

Table 4 Guidelines for the collection and storage of human milk for hospitalized infants[a]

Guideline
Mechanical expression needs to begin as soon as possible after giving birth, ideally with the use of a hospital-grade electric breast pump.
Personal collection kits should be cleaned after each use and sterilized at least daily. Mothers must be instructed in writing and/or orally regarding appropriate hand-washing, pumping, labeling, storage, and transport techniques.
Human milk must be stored in food-grade plastic containers or glass.
To prevent errors in breast milk delivery, human milk supplied to the facility must be labeled with complete and accurate information, including the infant's name, medical record number, and date and time of pumping.
Human milk transported to and from the hospital should be maintained at a temperature (2–6°C) that will prevent the loss of nutrients and to minimize bacterial growth.
Human milk should be stored in separate labeled bins or zippered bags to prevent misadministration of breast milk and to prevent cross-contamination of that milk with milk from other feedings.
For breast milk storage, refrigerator temperatures should be maintained at 2 to 4°C, and freezer temperatures should be at least −20°C. Infants should receive fresh breast milk whenever possible because of the enhanced activity of anti-infective cellular components.
Frozen breast milk should be used in the order in which it was expressed (oldest milk first) and placed in the refrigerator to thaw gradually. Alternatively, thaw it under cool running water or in a disposable cup containing warm water.

[a]Adapted from the American Dietetic Association, Pediatric Nutrition Practice Group (2004).

unable to nurse at the breast. We advise mothers to collect milk from both breasts simultaneously to stimulate prolactin production, a hormone that promotes milk synthesis. Pumping should begin as soon as possible after birth (i.e., in the first 24 h). At SickKids, we have around 32 hospital-grade mechanical pumps to facilitate onsite milk collection. Mothers are encouraged to use these pumps, and they are provided personal collection kits free of charge, which are washed and steam autoclaved after each use. We discourage the use of manual pumps with a rubber bulb, as the suction is difficult to control (and can damage the breast), milk frequently gets up into the bulb, and they are difficult to clean. Mothers may pump at their infant's bedside or in designated private rooms for this purpose. Many mothers opt to collect milk in the privacy of their own homes. Mothers are asked to wash pumps they are using at home with hot soapy water and rinse them between each use. Electric pumps are available for rent on site if a mother wishes to take a hospital-grade pump home. It is hospital policy that mothers are taught the importance of hand hygiene before expressing milk, and the operation of the electric breast pump is reviewed. Sterile plastic containers (30 or 90 ml, depending on 24-h infant milk intake), as well as computer-generated labels, are provided to mothers to identify their breast milk with the baby's name, medical record identification number, and date the milk was expressed. It is not unusual to have infants with the same name in the same patient area, hence the need for a unique medical record number.

Breast milk is picked up from a designated locked freezer compartment on each patient care unit and stored centrally in the enteral preparation area. In instances when breast milk does not need to be modified and will be used within 48 h, it is kept in a locked refrigerator on the units. A locked, walk-in commercial freezer ($-25°C$) in the enteral preparation area is designated solely for the storage of breast milk with a first-in, first-out inventory system. Each mother's milk is kept in a separate bin. Milk to be used the next day is thawed overnight in the refrigerator in the milk preparation room. Any nutrient additions are made under a laminar-flow hood in the milk preparation room.

While the focus of this book is *E. sakazakii*, it should be noted that expressed breast milk is a nonsterile body fluid and may contain any number of nonpathogenic and pathogenic bacteria, including *Staphylococcus epidermidis*, lactobacilli, *Bacillus coliformis*, enterococci, and *Staphylococcus aureus*. We do not screen the mothers for the aforementioned bacteria or for cytomegalovirus, HIV, or any other pathogens. For these reasons and because breast milk modified to meet the individual nutrient needs of one particular infant could be harmful to another, it is critical that the appropriate breast milk get to the right infant. To reduce the risk of breast milk mix-ups, a number of policies and procedures have been put into place recently at SickKids. When breast milk is delivered to patient care units, it is locked in a designated refrigerator. Before milk is fed to a baby, the patient name and medical record number on the container must be checked against the same identifiers on the patient's arm band. Two signatures (i.e., one health care professional and a parent or two health care professionals) must sign the patient flow sheet to verify that this double-check process was completed. A copy of the label also is affixed to the patient flow sheet. We have policies and procedures in place to address a breast milk mix-up and possible inadvertent transmission of an infectious agent. This includes testing both the donor and recipient infant's mother for key infectious diseases such as HIV, human T-lymphotropic virus type 1, and cytomegalovirus. In the event the donor mother tests positive for an infectious disease, prophylaxis will be delivered to the recipient infant and subsequent testing of the recipient infant will take place as appropriate.

DELIVERY AND BEDSIDE MANAGEMENT OF INFANT FEEDING

At our institution, as in others, nursing staff primarily are the personnel responsible for ensuring that proper handling of enteral products occurs once they have been delivered to patient units. Standard ready-to-feed formulas also are stocked in the pantries on patient units. We actively discourage any mixing of enteral feeds outside the milk preparation room unless there is a strong rationale for doing so. For example, thickening agents added to breast milk to prevent aspiration during a swallow must be added immediately prior to feeding. Personnel in the milk preparation room may add this thickener to breast milk, or they may provide the thickener in a preweighed single-unit dose with patient identifiers to the patient unit. Ongoing staff and parent education regarding hand hygiene as well as handling enteral feeds, bottles, and other feeding devices is an absolute must, as staff turns over and vigilance regarding hygienic practices wane.

At SickKids, the preferred method for thawing frozen breast milk is to place it in a locked refrigerator in the milk preparation room overnight. For immediate thawing of breast milk, it is placed in tepid water. It is recommended that if formula or breast milk is to be warmed prior to patient feeding, the bottle is placed in a plastic bag and warmed in a patient-specific cup containing warm water. We do not use microwaves to either thaw or warm feedings due to concerns about overheating and hotspots. We have discontinued the use of electronic water baths for communal warming of enteral feeds due to concern about the consistency of their cleanliness and the risk of cross-contamination.

In order to reduce confusion and enhance compliance with enteral feeding policies, we use the same bedside enteral feeding guidelines (including timelines) throughout the hospital. Once enteral feeding products are removed from the refrigerator, they must be used within 4 h, whether the enteral feeding is fed by tube or mouth. Tube feedings either are fed over a relatively short period of time (called bolus or gavage feeding) or are fed at a constant rate over a longer period of time, usually using a pump (continuous feeding). Nutrients, particularly minerals, tend to come out of solution over time, and fat eventually will separate from the feeding and adhere to the feeding device and tubing. From a nutritional point of view, then, bolus feeding is desirable. Likewise, minimizing the time that enteral feedings are held above 4°C reduces the risk of bacterial growth. Nonetheless, there are occasions when continuous feeding is necessary due to feeding intolerance, aspiration, reflux, and so on. In these instances, no more than 4 h of enteral feeding should be hung, and every 4 h the feeding set needs to be completely emptied and refilled with fresh human milk or formula as appropriate. The feeding systems for bolus-fed patients are thoroughly cleaned with hot water and rinsed with sterile water after each feeding and are placed to dry away from sinks. Feeding bags for continuous feeds are flushed with sterile water every 8 h. All enteral feeding setups are discarded and replaced with a new setup every 24 h. The setups are labeled with the date and time of changing. A recent study by Telang et al. (2005) concluded that bacterial growth rates are low in human milk and infant formula prepared from powder when held up to 6 h at room temperature (22°C).

PREPARATION OF ENTERAL FEEDINGS AT HOME

Once sick infants are stable, there is considerable interest in discharging them home or back to the nontertiary referring institution. It is very unusual in 21st century North American medicine to keep a baby hospitalized for a specialized feeding issue alone. Needless to say, babies discharged home are generally less sick and immunocompromised than hospitalized infants, yet they remain at relatively higher risk following exposure to infectious agents than healthy term-born babies. The principles of formula preparation, as provided on the can by manufacturers, are the same for premature and healthy term newborns (Table 5). Generally it is always a good idea to switch babies over to the feeding that they are going to go home on a few days before hospital discharge in order to assess tolerance and reduce parental anxiety after hospital discharge. Formula switches for some babies will result in a few days of gassiness, fussiness, and changes in stool color and consistency. In general, it is advisable to make formula switches over multiple days. The relative benefits versus risks of using a powdered formula after hospital discharge will need to be evaluated. For example, it is advisable that immuno-compromised infants discharged home on overnight continuous feeds be fed a sterile product, if available, for their specialized feeding requirements. On the other hand, it is much more expensive for families of exclusively human-milk-fed infants to nutrient enrich with a formula concentrate versus a powdered formula, as only a small amount of either would be required on a daily basis. A concentrate would need to be discarded 48 h after opening, whereas a powder could be resealed and kept for up to a month.

In addition to written instructions on how to prepare and feed standard formula and/or specialized feedings, it is often a good idea to provide a hands-on demonstration with parents. This is especially important when English or French is not the parents' first language or their literacy level is low.

CONCLUSION

Breast-feeding is the preferred feeding method for both term and prematurely born infants. On the rare occasions when feeding human milk is contraindicated or when a mother chooses not to breast-feed, an iron-fortified cow-milk-based formula is the next best option. For some sick and/or

Table 5 Guidelines for preparation and handling of enteral feedings designed for infants at home[a]

Guideline
Breastfeeding should be encouraged.
Advise families that proper hygiene, preparation, dilution, use, and storage are important in preparing infant formula. Unless there is a specific medical reason to do otherwise, families should be ecouraged to follow carefully all directions provided on the formula container.
As powdered infant formulas and many nutrient modulars are not sterile, they should not be fed to prematurely born or immunocompromised infants if a nutritionally equivalent alternative is available. Parents of term-born infants should be advised of the risks associated with the use of powdered infant formulas and made aware of sterile options.
For healthy term infants under 4 months of age, premature infants after hospital discharge, and immunocompromised infants, water used to reconstitute formula should be brought to a rolling boil for 2 min and cooled to room temperature. Likewise, bottles, nipples, caps, supplemental nursers, utensils, and mixing equipment should be boiled for 2 min.
Potable water from a cold-water tap should be used instead of a hot-water tap to reduce lead exposure. In homes constructed prior to 1990, standing water should be removed each morning by flushing the toilet prior to preparing formula.
Hands and the enteral feeding preparation area should be washed and dried carefully.
Carefully wipe the tops of containers before opening them.
Several bottles can be prepared at one time and stored in the refrigerator (at 2–4°C) for up to 24 h if reconstituted from a powder and 48 h if reconstituted from a concentrate or ready to feed.
Never use a microwave oven to warm formula. It could produce hot spots that may affect the nutrient concentration and anti-infective properties (in the case of human milk) of the enteral feeding and could burn the baby. Placing the bottle in a container of warm water or holding the bottle under a warm running tap can warm enteral feedings.
Once at room temperature, bottles should be discarded after 2 h; i.e., 1 h after being warmed or feeding has commenced. Most formulas can be fed cold, at room temperature, or warmed.
For tube-fed infants, unless medically contraindicated, use gavage feeding instead of continuous feeding.

[a]Adapted from the Canadian Paediatric Society, Dietitians of Canada, Health Canada (1998).

very-low-birth-weight infants, neither human milk alone nor standard-term formula will meet the elevated nutritional requirements necessary to promote growth and optimal health. For other infants, there may be particular constituents in human milk or cow-milk-based formulas that are allergenic, create food intolerance, or cannot be adequately metabolized. In these instances a specialized formula may be required. For both hospitalized infants and infants discharged home, formula selection should be guided in the first instance by nutritional need. Whenever possible, a sterile feeding should be chosen over a powdered nonsterile option; however, frequently the most specialized formulas do not have a sterile option. Policies and procedures need to be put into place to address the preparation of formula and breast milk in health care institutions and to advise families on how to feed their infants after hospital discharge. The authors hope that this chapter provides a preliminary framework for the various considerations required in this effort. As we did at the beginning of the chapter, we acknowledge that our policy and procedures with regard to infection prevention and control and feeding continue to evolve with new information and regulatory guidance. For example, we are evaluating the feasibility and nutrient integrity of reconstituting powdered formulas with hot water (at least 70°C), and as this chapter goes to press, we have received the new "Safe Preparation, Storage and Handling of Powdered Infant Formula Guidelines" from the World Health Organization (2007). In addition to reconstituting powdered formulas with hot water at >70°C, the WHO guidelines suggest reducing enteral feeding hang times to 2 h. Both recommendations will need to be carefully evaluated. Finally, while many of the policies and procedures reviewed in this chapter may work well at other health care institutions, they need to be evaluated in the context of the environment in which they are applied.

REFERENCES

American Academy of Pediatrics. 2004a. Appendix B, p. 885–917. *In* R. E. Kleinman (ed.), *Pediatric Nutrition Handbook*, 5th ed. American Academy of Pediatrics, Elk Grove Village, IL.

American Academy of Pediatrics. 2004b. Breastfeeding, p. 55–86. *In* R. E. Kleinman (ed.), *Pediatric Nutrition Handbook*, 5th ed. American Academy of Pediatrics, Elk Grove Village, IL.

American Academy of Pediatrics. 2004c. Formula feeding of term infants, p. 87–97. *In* R. E. Kleinman (ed.), *Pediatric Nutrition Handbook*, 5th ed. American Academy of Pediatrics, Elk Grove Village, IL.

American Academy of Pediatrics. 2004d. Inborn errors of metabolism, p. 481–490. *In* R. E. Kleinman (ed.), *Pediatric Nutrition Handbook*, 5th ed. American Academy of Pediatrics, Elk Grove Village, IL.

American Academy of Pediatrics. 2004e. Nutritional needs of the preterm infant, p. 23–54. *In* R. E. Kleinman (ed.), *Pediatric Nutrition Handbook*, 5th ed. American Academy of Pediatrics, Elk Grove Village, IL.

American Academy of Pediatrics. 2005. Breastfeeding and the use of human milk. *Pediatrics* **115:**496–506.

American Academy of Pediatrics, Committee on Nutrition. 2000. Hypoallergenic infant formulas. *Pediatrics* **106:**346–349.

American Academy of Pediatrics, Committee on Nutrition. 1998. Soy protein-based formulas: recommendations for use in infant feeding. *Pediatrics* **101:**148–153.

American Dietetic Association, Pediatric Nutrition Practice Group. 2004. Infant feedings: guidelines for preparation of formula and breastmilk in health care facilities. American Dietetic Association, Chicago, IL.

American Dietetic Association. 2000. *Manual of Clinical Dietetics,* 6th ed. American Dietetic Association, Chicago, IL.

Anderson, J. W., B. M. Johnstone, and D. T. Remley. 1999. Breast-feeding and cognitive development: a meta-analysis. *Am. J. Clin. Nutr.* **70:**525–535.

Anthony, M. S., T. B. Clarkson, C. L. Hughes, Jr., T. M. Morgan, and G. L. Burke. 1996. Soybean isoflavones improve cardiovascular risk factors without affecting the reproductive system of peripubertal rhesus monkeys. *J. Nutr.* **126**:43–50.

Association for Professionals in Infection Control and Epidemiology. 2005. *APIC text of infection control and epidemiology*, 2nd ed. Association for Professionals in Infection Control and Epidemiology, Washington, DC.

Babson, S. G., and G. I. Benda. 1976. Growth graphs for the clinical assessment of infants of varying gestational age. *J. Pediatr.* **89**:814–820.

Baumgartner, M., C. M. Brown, M. C. Secretin, M. Van't Hof, and F. Haschke. 1998. Controlled trials investigating the use of one partially hydrolyzed whey formula for dietary prevention of atopic manifestations until 60 months of age: an overview using meta-analytical techniques. *Nutr. Res.* **18**:1425–1442.

Billeaud, C., J. Guillet, and B. Sandler. 1990. Gastric emptying in infants with or without gastro-oesophageal reflux according to the type of milk. *Eur. J. Clin. Nutr.* **44**:577–583.

Bock, S. A., and F. M. Atkins. 1990. Patterns of food hypersensitivity during sixteen years of double-blind, placebo-controlled food challenges. *J. Pediatr.* **117**:561–567.

Burks, A. W., H. B. Casteel, S. C. Fiedorek, L. W. Williams, and C. L. Pumphrey. 1994. Prospective oral food challenge study of two soybean protein isolates in patients with possible milk or soy protein enterocolitis. *Pediatr. Allergy Immunol.* **5**:40–45.

Canadian Paediatric Society, Dietitians of Canada, Health Canada. 1998. *Nutrition for healthy term infants.* Minister of Public Works and Government Services, Ottawa, Canada.

Casey, P. H., H. C. Kraemer, J. Bernbaum, J. E. Tyson, C. J. Sells, M. W. Yogman, and C. R. Bauer. 1990. Growth patterns of low birth weight premature infants. A longitudinal analysis of a large varied sample. *J. Pediatr.* **117**:298–307.

Casey, P. H., H. C. Kraemer, J. Bernbaum, M. L. O. Yogman, and J. C. Sells. 1991. Growth status and growth rates of a varied sample of low birth weight, premature infants: a longitudinal cohort from birth to 3 years of age. *J. Pediatr.* **119**:599–605.

Chen, A., and W. J. Rogan. 2004. Isoflavones in soy infant formula: a review of evidence for endocrine and other activity in infants. *Annu. Rev. Nutr.* **24**:33–54.

de Onis, M., C. Garza, A. Onyango, and R. Martorell. 2006. WHO child growth standards based on length/height, weight and age. *Acta Paediatr. Suppl.* **450**:76–85.

Dietitians of Canada, Canadian Paediatric Society, The College of Family Physicians of Canada, and Community Health Nurses Association of Canada. 2004. The use of growth charts for assessing and monitoring growth in Canadian infants and children. *Can. J. Diet. Pract. Res.* **65**:22–32.

Eastham, E. J. 1989. Soy protein allergy, p. 223–236. *In* R. N. Hamburger (ed.), *Food intolerance in infancy: allergology, immunology, and gastroenterology.* Carnation Nutrition Series. Raven Press, New York, NY.

Federal Register. 1985. Rules and regulations. Nutrient requirements for infant formulas. *Fed. Regist.* **50**:45106–45108.

Fenton, T. R. 2003. A new growth chart for premature babies: Babson and Benda's chart updated with recent data in a new format. *BMC Pediatr.* **3**:13.

Fomon, S. J. 1993. *Nutrition of normal infants.* Mosby, St. Louis, MO.

Food and Agriculture Organization-World Health Organization (FAO-WHO). 2004. *Enterobacter sakazakii* and other microorganisms in powdered infant formula: meeting report. *Microbiological risk assessment series 6.* World Health Organization-Food and Agriculture Organization of the United Nations, Geneva and Rome. WHO Press, Geneva, Switzerland. http://www.who.int/foodsafety/publications/micro/mra6/en/index.html.

Food and Agriculture Organization-World Health Organization (FAO-WHO). 2006. *Enterobacter sakazakii* and *Salmonella* in powdered infant formula: meeting report. *Microbiological risk assessment series 10.* World Health Organization-Food and Agriculture Organization of the United Nations, Geneva and Rome. WHO Press, Geneva, Switzerland. http://www.who.int/foodsafety/publications/micro/mra10/en/index.html.

Food and Drug Administration. 2002. Health professionals letter on *Enterobacter sakazakii* infections associated with use of powdered (dry) infant formulas in neonatal intensive care units. http://www.cfsan.fda.gov/~dms/inf-ltr3.html.

Food and Drug Directorate. 1990. Food and drug regulations, division 25. *Canada Gazette* 124:73E–73H.

Fried, M. D., V. Khoshoo, D. J. Secker, D. L. Gilday, J. M. Ash, and P. B. Pencharz. 1992. Decrease in gastric emptying time and episodes of regurgitation in children with spastic quadriplegia fed a whey-based formula. *J. Pediatr.* **120:**569–572.

Fuchs, G. J., A. S. Gastanaduy, and R. M. Suskind. 1992. Comparative metabolic study of older infants fed infant formula, transition formula, or whole cow's milk. *Nutr. Res.* **12:**1467–1478.

Gibson, R. S. (ed.). 2005. *Principles of nutritional assessment,* 2nd ed., p. 245–272. Oxford University Press, New York, NY.

Guo, S. S., K. Wholihan, A. F. Roche, W. C. Chumlea, and P. H. Casey. 1996. Weight-for-length reference data for preterm, low-birth-weight infants. *Arch. Pediatr. Adolesc. Med.* 150:964–970.

Gurtler, J. B., J. L. Kornacki, and L. R. Beuchat. 2005. *Enterobacter sakazakii*: a coliform of increased concern to infant health. *Int. J. Food Microbiol.* **104:**1–34.

Health Canada. 2002. Health professional advisory: *Enterobacter sakazakii* infection and powdered infant formulas. http://www.hc-sc.gc.ca/fn-an/securit/ill-intox/esakazakii/enterobacter_sakazakii_e.html.

Health Canada. 2004. Exclusive breastfeeding duration. http://www.hc-sc.gc.ca/fn-an/nutrition/child-enfant/infant-nourisson/excl_bf_dur-dur_am_excl_e.html.

Heine, W. E. 1999. The significance of tryptophan in infant nutrition. *Adv. Exp. Med. Biol.* **467:**705–710.

Himelright, I., E. Harris, V. Lorch, and M. Anderson. 2002. *Enterobacter sakazakii* infections associated with the use of powdered infant formula—Tennessee 2001. *Morb. Mortal. Wkly. Rep.* **51:**298–300.

Institute of Medicine. 1997. *Dietary reference intakes for calcium, phosphorus, magnesium, vitamin D and fluoride.* The National Academy Press, Washington, DC.

Institute of Medicine. 1998. *Dietary reference intakes for thiamin, riboflavin, niacin, vitamin B6, folate, vitamin B$_{12}$, pantothenic acid, biotin, and choline.* The National Academy Press, Washington, DC.

Institute of Medicine. 2000. *Dietary reference intakes for vitamin C, vitamin E, selenium and carotenoids.* The National Academy Press, Washington, DC.

Institute of Medicine. 2001. *Dietary reference intakes for vitamin A, vitamin K, arsenic, boron, chromium, copper, iodine, iron, manganese, molybdenum, nickel, silicon, vanadium and zinc.* The National Academy Press, Washington, DC.

Institute of Medicine. 2005a. *Dietary reference intakes for energy, carbohydrate, fiber, fat, fatty acids, cholesterol, protein, and amino acids.* The National Academy Press, Washington, DC.

Institute of Medicine. 2005b. *Dietary reference intakes for water, potassium, sodium, chloride and sulfate.* The National Academy Press, Washington, DC.

Jensen, R. G., ed. 1995. *Handbook of milk composition.* Academic Press, San Diego, CA.

Khoshoo, V., M. Zembo, A. King, M. Dhar, R. Reifen, and P. Pencharz. 1996. Incidence of gastroesophageal reflux with whey- and casein-based formulas in infants and in children with severe neurological impairment. *J. Pediatr. Gastroenterol. Nutr.* **22:**48–55.

Klein, C. J. 2002. Nutrient requirements for preterm infant formulas. *J. Nutr.* **132:** 1395S–1577S.

Klein, K. O. 1998. Isoflavones, soy-based infant formulas, and relevance to endocrine function. *Nutr. Rev.* **56:**193–204.

Klein, C. J., and W. C. Heird. 2005. Summary and comparison of recommendations for nutrient contents of low-birth-weight infant formulas. Life Sciences Research Office, Bethesda, MD. http://www.lsro.org/articles/lowbirthweight_rpt.pdf.

Kuczmarski, R. J., C. L. Ogden, L. M. Grummer-Strawn, K. M. Flegal, S. S. Guo, R. Wei, Z. Mei, L. R. Curtin, A. F. Roche, and C. L. Johnson. 2000. CDC growth charts: United States. *Adv. Data* **134:**1–27.

Ladodo, K. S., and T. E. Borovick. 1992. The use of an isolated soy protein formula for nourishing infants with food allergies, p. 85–89. *In* F. H. Steinke, D. H. Waggle, and M. N. Volgarev (ed.), *New protein foods in human health: nutrition, prevention, and therapy.* CRC Press Inc., Boca Raton, FL.

Lawrence, R. A. 2001. Milk banking: the influence of storage procedures and subsequent processing on immunologic components of human milk. *Adv. Nutr. Res.* **10:**389–404.

Marini, A., M. Agosti, G. Motta, and F. Mosca. 1996. Effects of a dietary and environmental prevention programme on the incidence of allergic symptoms in high atopic risk infants: three years' follow-up. *Acta Paediatr. Suppl.* **414:**1–21.

Nutrition Committee of the Canadian Paediatric Society. 1995. Nutrient needs and feeding of preterm infants. *Can. Med. Assoc. J.* **152:**1765–1785.

Oberlander, T. F., R. G. Barr, S. N. Young, and J. A. Brian. 1992. Short-term effects of feed composition on sleeping and crying in newborns. *Pediatrics* **90:**733–740.

O'Connor, D. L., J. Jacobs, R. Hall, D. Adamkin, N. Auestad, M. Castillo, W. E. Connor, S. L. Connor, K. Fitzgerald, S. Groh-Wargo, E. E. Hartmann, J. Janowsky, A. Lucas, D. Margeson, P. Mena, M. Neuringer, G. Ross, L. Singer, T. Stephenson, J. Szabo, and V. Zemon. 2003. Growth and development of premature infants fed predominantly human milk, predominantly premature infant formula, or a combination of human milk and premature formula. *J. Pediatr. Gastroenterol. Nutr.* **37:**437–446.

O'Connor, D. L., M. Masor, C. Paule, and J. Benson. 1997. Amino acid composition of cow's milk and human requirements, p. 203–213. *In* R. A. S. Welch et al. (ed.), *Milk composition, production and biotechnology*. CAB International, Wallingford, United Kingdom.

O'Connor, D. L., S. Merko, and J. Brennan. 2004. Human milk feeding of very low birth weight infants during initial hospitalization and after discharge. *Nutr. Today* **39**:102–111.

Raiten, D. J., J. M. Talbot, and J. H. Waters. 1998. Assessment of nutrient requirements for infant formulas. *J. Nutr.* **128**:2059S–2293S.

Rassin, D. K. 1994. Essential and non-essential amino acids in neonatal nutrition, p.183–195. *In* N. C. Raiha (ed.), *Protein metabolism during infancy*. Raven Press, New York, NY.

Ryan, A. S., Z. Wenjun, and A. Acosta. 2002. Breastfeeding continues to increase into the new millennium. *Pediatrics* **110**:1103–1109.

Setchell, K. D., L. Zimmer-Nechemias, J. Cai, and J. E. Heubi. 1997. Exposure of infants to phyto-oestrogens from soy-based infant formula. *Lancet* **350**:23–27.

Sharpe, R. M., B. Martin, K. Morris, I. Greig, C. McKinnell, A. S. McNeilly, and M. Walker. 2002. Infant feeding with soy formula milk: effects on the testis and on blood testosterone levels in marmoset monkeys during the period of neonatal testicular activity. *Hum. Reprod.* **17**:1692–1703.

Sheehan, D. M. 1995. The case for expanded phytoestrogen research. *Proc. Soc. Exp. Biol. Med.* **208**:3–5.

Soldin, S. J., B. Brugnara, K. C. Gunter, and E. C. Wong. 2003. *Pediatric reference ranges*, 4th ed. AACC Press, Washington, DC.

Statistics Canada. 2001. National longitudinal survey on children and youth, 1994/95 and 1996/97 data. Statistics Canada, Ottawa, Canada.

Steinberg, L. A., N. C. O'Connell, T. F. Hatch, M. F. Picciano, and L. L. Birch. 1992. Tryptophan intake influences infants' sleep latency. *J. Nutr.* **122**:1781–1791.

Strom, B. L., R. Schinnar, E. E. Ziegler, K. T. Barnhart, M. D. Sammel, G. A. Macones, V. A. Stallings, J. M. Drulis, S. E. Nelson, and S. A. Hanson. 2001. Exposure to soy-based formula in infancy and endocrinological and reproductive outcomes in young adulthood. *JAMA* **286**:807–814.

Telang, S., C. L. Berseth, P. W. Ferguson, J. M. Kinder, M. DeRoin, and B. W. Petschow. 2005. Fortifying fresh human milk with commercial powdered human milk fortifiers does not affect bacterial growth during 6 hours at room temperature. *J. Am. Diet. Assoc.* **105**:1567–1572.

Thorkelsson, T., F. Mimouni, R. Namgung, M. Fernandez-Ulloa, S. Krug-Wispe, and R. C. Tsang. 1994. Similar gastric emptying rates for casein- and whey-predominant formulas in preterm infants. *Pediatr. Res.* **36**:329–333.

Tolia, V., C. H. Lin, and L. R. Kuhns. 1992. Gastric emptying using three different formulas in infants with gastroesophageal reflux. *J. Pediatr. Gastroenterol. Nutr.* **15**:297–301.

von Berg, A., S. Koletzko, A. Grubl, B. Filipiak-Pittroff, H. E. Wichmann, C. P. Bauer, D. Reinhardt, and D. Berdel. 2003. The effect of hydrolyzed cow's milk formula for allergy prevention in the first year of life: the German Infant Nutritional Intervention Study, a randomized double-blind trial. *J. Allergy Clin. Immunol.* **111**:533–540.

Whitington, P. F., and R. Gibson. 1977. Soy protein intolerance: four patients with concomitant cow's milk intolerance. *Pediatrics* **59:**730–732.

World Health Organization. 2001. The optimal duration of exclusive breastfeeding: report of an expert consultation, Geneva, Switzerland, 28 to 30 March 2001. Department of Nutrition for Health and Development, WHO, Geneva, Switzerland. http://www.who.it/child-adolescent-health/New_Publications/NUTRITION/WHO_CAH_01_24.pdf.

World Health Organization. 2007. Safe preparation, storage and handling of powdered infant formula: guidelines. http://www.who.int/foodsafety/publications/micro/pif_guidelines.pdf.

Yogman, M. W., and S. H. Zeisel. 1983. Diet and sleep patterns in newborn infants. *N. Engl. J. Med.* **309:**1147–1149.

Ziegler, E. E., and S. J. Fomon. 1989. Potential renal solute load of infant formulas. *J. Nutr.* **119:**1785–1788.

Enterobacter sakazakii
Edited by Jeffrey M. Farber and Stephen J. Forsythe
© 2008 ASM Press, Washington, D.C.

Powdered Infant Formula in Developing and Other Countries—Issues and Prospects

8

S. Estuningsih and N. Abdullah Sani

INTRODUCTION

The first infant formula was developed by Henri Nestlé in 1860 in response to the high mortality rate among infants born to working-class women in Switzerland who had no time to nurse. It was a combination of cow's milk and cereals and was called Farine Lactee. Infant formula became increasingly popular in developed countries during the 20th century as an alternative to breast-feeding. The medical community supported the use of infant formula, because it believed that artificial feeding could be more easily monitored and the nutrient content of the milk ensured.

The post-World War II baby boom provided a market for the expanding infant formula industry. Between 1946 and 1958, the incidence of breast-feeding halved in the United States, leaving only 25% of infants still being breast-fed at the time of hospital discharge. During the 1960s, when birth rates tapered off, some infant formula companies began marketing campaigns in nonindustrialized countries. Unfortunately, poor sanitation led to an increase in mortality rates among infants fed formula prepared with contaminated drinking water. Organized protests, the most famous of which was the Nestlé boycott of 1977, called for the end of what was felt to be unethical marketing.

In more recent years the use of infant formula, even in developed countries, has come under scrutiny. Infant formula use has been shown to contribute to several infant conditions, including bacterial infections (Kaleida et al., 1993).

S. Estuningsih, Department of Clinical, Reproduction, and Veterinary Pathology, Bogor Agricultural University, Jalan Agatis, IPB Campus, Darmaga, 16681, Bogor, Indonesia. N. Abdullah Sani, Food Science Programme, School of Chemical Sciences and Food Technology, Faculty of Science and Technology, Universiti Kebangsaan Malaysia, 43600 Bangi, Selangor, Malaysia.

Infant formula, like other food products, is sometimes subject to product recalls, usually due to bacterial contamination. Though formula is available without a prescription, generally it is recommended that it be used under medical supervision.

Recent initiatives have begun to encourage a resurgence of breast-feeding. As a result, infant formula companies now are required to preface their product information with statements such as the following from Nestlé: "Breast milk is the best for babies. Before you decide to use an infant formula consult your doctor or clinic for advice." However, infant formula remains a popular infant feeding option. The baby bottle has become a very visible part of Western culture and, increasingly, of other developed and developing nations.

Infant formula is not a sterile product and requires special handling, including the cleanliness of the water used to reconstitute the powder, the storage temperature, and the hold time. In most developing countries, these conditions can be much less than ideal. Nevertheless, in some situations, e.g., in cases where mothers work away from the home, not enough breast milk is produced, or the mother has an infection, infant formula is the best choice.

LABELING—ADVICE ON PREPARATION

The labeling requirements for infant formula are very comprehensive, with elements of information, advice, and warning. The Codex Alimentarius Committee (2005) requires recommendations for the preparation, feeding, and storage of the product as sold, opened, and prepared for consumption. The recommendations for the preparation of infant formula, formula for special medical purposes intended for infants, and follow-up formula used at home are detailed and often accompanied by illustrations.

In Indonesia, guidance is provided by pictures and statements in local languages. Regarding the minimization of bacterial contamination, the suggestions on powdered infant formula (PIF) products in developing countries such as Indonesia are adequate.

MARKETING OF PIF

The fact that more infant formula is sold in Belgium, which has a population of 10 million, than in the whole of sub-Saharan Africa, covering a population of more than 650 million, puts the sale and use of this product in developing countries into context. Even in sub-Saharan Africa, the sale of formula is highly concentrated in the more affluent urban areas, including the approximately 10 million South Africans who have a middle- or upper-class standard of living. The volume of sales of infant formula is low in Africa and is not growing, primarily because few can afford it. The majority of women resort to traditional foods to supplement or replace breast milk.

No detailed data on PIF sales were available for many developing countries. In Indonesia, the distribution of sales and marketing of PIF products and baby foods varies. In large cities, PIF products and baby foods (imported and local products) are easy to find in the supermarket or traditional market and pharmacy. It is still easy to find local PIF products and baby foods in small cities. In the rural regions, the distribution is relatively minimal due to a lack of consumers and to poverty. Most people in rural areas are small-scale farmers.

In the 1970s, many infant formula manufacturing companies distributed free samples to women in developing countries as a way of promoting their products. Many women became dependent solely on these formulas. Mothers would use the free formula for the first few weeks of their infants' life, while letting their own milk supply dry up, and then the free formula would cease. In many cases, women were forced to pay for formula they could not afford, thus jeopardizing the well-being of their entire families. In some cases, women who used this free formula lacked access to clean water or enough fuel to sterilize feeding equipment. These two factors led to numerous cases of infant malnutrition, caused by diluting the formula because of its high cost, and to greater infant mortality from contaminated formula. At hospitals and clinics in Pakistan, some infant formula companies continue to give out free samples and to distribute posters and calendars promoting their formulas (Kwa, 1993).

HIV IN DEVELOPING COUNTRIES AND THE PROMOTION OF PIF

One of the biggest scourges, at least from a health point of view, which has been carried over into the new millennium is HIV-AIDS (human immuno-deficiency virus-acquired immunodeficiency syndrome). In 2004, global figures for HIV were 17.6 million women and 2.2 million children under 15 years of age (Joint United Nations Programme on HIV/AIDS [UNAIDS]; www.unaids.org). In 2005, around 700,000 children under 15 became infected with HIV, mainly through mother-to-child transmission. Nearly 90% of all HIV-infected babies are born in Africa, according to UNAIDS (http://www.unaids.org/en/HIV_data/2006GlobalReport/default.asp).

Mother-to-child transmission occurs when an HIV-positive woman passes the virus to her baby. This can occur during pregnancy, labor and delivery, or breast-feeding. Without treatment, around 15 to 30% of babies born to HIV-positive women will become infected with HIV during pregnancy and delivery. A further 5 to 20% will become infected through breast-feeding. HIV is present in the colostrum, mothers' milk expressed after the first few days of delivery, and in regular breast milk. Not all HIV-positive mothers infect their babies, but vertical transmission is far more likely than sexual transmission or intravenous transmission of HIV.

A safe alternative to breast-feeding in developed countries is bottle feeding. However, in developing countries with high infant mortality rates, the situation is very different. The known protective effects of breast-feeding against gastrointestinal infection are very important. In these countries, the risk of vertical HIV transmission needs to be weighed against the risk of gastrointestinal infection due to the lack of clean water and sanitation, as well as the cost of formula and the potential for stigmatization where breast-feeding is a cultural norm. The United Nations has recently issued a new directive stating that, due to the escalating problem of HIV, all women should be informed of the risk of vertical transmission via breast milk. It recommends that, due to the diversity of situations in developing countries, the decision to breast-feed should be made by the mother.

Recently, a priority of the Thai government has been to decrease vertical transmission while still striving for optimal infant health. In order to deter mother-to-child HIV transmission, the Thai government provides free HIV testing to all pregnant mothers and gives HIV-positive mothers free baby formula powder for 1 year. Preparing formula is safe in most areas of Thailand, since 89% of the urban population and 72% of the rural population have access to clean water.

The solutions being used in Thailand to deter the transmission of HIV from mother to infant have not been implemented in Zambia, one of several African countries where HIV infects up to 25% of the population. Due to its struggling economy, Zambia has been unable to adequately invest in public health programs, i.e., the Zambian government does not have the resources necessary to advance infant health or to prevent vertical transmission of HIV.

Another complicating factor in Zambia is that clean water is not available for most mothers to prepare baby formula. While boiling contaminated water generally makes it safe to use in formula, 60% of Zambian women are illiterate, which makes instructing women on how to properly sterilize water very difficult.

MALNUTRITION

It is estimated that 5 million neonatal deaths occur worldwide, of which 98% are in developing countries, with 92% being in Asia or Africa. One out of every three children under 5 years of age in developing countries is malnourished. This unacceptable state of affairs leads to a great deal of human suffering, both physical and emotional. It is a major drain on developing countries' prospects for development, because malnourished children require more intensive care from their parents and are less physically and intellectually productive as adults. In Malaysia in 1998, 9.6% of infants were born with a low birth weight (<2,500 g). In Bangladesh, the weaning age is between 6 and 23 months. Most babies are underweight, and the underweight babies represent the group that is at risk of consuming contaminated feeds.

The underlying causes of malnutrition vary across regions. In many Asian countries, poverty, the low status of women, poor care during pregnancy, high rates of low birth weight, high population densities, unfavorable child care practices, and poor access to health care are underlying causes. In sub-Saharan Africa, extreme poverty, inadequate caring practices for children, low levels of education, and poor access to health services are among the major factors causing malnutrition. Conflicts and natural disasters in many countries have further exacerbated the situation. The increase in the number of undernourished children in Africa also reflects a rapid rate of population growth. In many countries in Africa, the devastating effects of HIV-AIDS, particularly in the second half of the 1990s, have reversed some of the gains made in this decade's early years.

FACTORS THAT INFLUENCE BREAST-FEEDING, FORMULA FEEDING, AND WEANING

Biological evidence shows that the health benefits of breast-feeding for infants are far greater than the benefits of formula feeding. Breast-feeding during the first 13 weeks of life confers protection against gastrointestinal illness that persists beyond the period of breast-feeding itself (Duffy et al., 1997).

The use of PIF is related to the exclusiveness of breast-feeding. According to UNICEF, the global rate of exclusive breast-feeding increased 18% (from 29% to 46%) between 1989 and 1999 (http://www.unicef.org/newsline/01pr25.htm and http://www.childinfo.org/areas/malnutrition/). However, it is difficult to give a general description of the situation from country to country, especially for developing countries. For example, the reported exclusive breast-feeding rate in Liberia is 73%, while in Kenya it is only 5% (Asian Development Bank; www.adb.org/countries/default.asp). In Jakarta, Indonesia, a survey was conducted on breast-feeding over a 5-year period (1997 to 2002). The results showed a decrease in breast-feeding from the first hour of life from 8.0 to 3.7% while increasing the use of PIF from 10.8 to 32.5% over the 5-year period (Table 1) (Anonymous, 2004). According to the Bangladesh Breast Feeding Foundation, in 2003 the national prevalence of exclusive breast-feeding in

Table 1 Comparison of breast-feeding among infants in Jakarta, Indonesia, over a 5-year period (1997 to 2002)[a]

	% of babies in:	
Feeding regimen	1997	2002
All ages 1 month to 1 year	96.30	95.90
Breast-fed from the first hour	8.00	3.70
Exclusively breast-fed for 4 months	52.0	55.10
Exclusively breast-fed for 6 months	42.40	39.50
Bottle fed infant formula	10.80	32.45

[a]Data are from the Ministry of Health of the Republic of Indonesia (2004).

Bangladesh for 4- to 6-month-old infants was 35.1%, which means that 64.9% of babies are dependent on extraneous feeding, including infant food formula.

There are many factors that influence the use of PIF. These include the economic status of the family, the education level of the mother, health care access, and cultural and social norms. The rapid increase of urbanization in developing countries has led to a new culture among women and an increase in formula feeding. Many young women are employed in industries, and the length of the maternity leave limits the amount of time that they can breast-feed. Cutting (1995) also noted that the promotion of infant formula undermines breast-feeding.

In many developing countries, breast-feeding is the only economically feasible option for feeding infants and is the cultural norm. Poor women breast-feed their babies at least for the first year, and then their babies get substitute food after weaning, although some women still breast-feed until their baby reaches 2 years of age. In Malaysia, a study by Manan (1995) found that the majority of breast-feeding mothers belonged to the groups having incomplete primary schooling or completed primary education only and with a household income below $200 per month.

In Nigeria, a study by questionnaire of 1,845 urban and 349 rural mothers in all regions of the country reported that 99% of all mothers commenced breast-feeding their infants. In urban areas, 77% of infants were given infant formula by 3 months of age; in rural areas, 40% of infants were given infant formula by 3 months of age (Orwell et al., 1984).

In Sarawak, Malaysia, mothers who received prenatal care were more likely to initiate breast-feeding than those who received no prenatal care (84.1% and 75%, respectively) (Kwa, 1993). In Indonesian hospitals, infants are roomed with their mother, and thus hospital staff can encourage mothers to breast-feed their babies.

Sall et al. (1986) surveyed breast-feeding in several cities in Senegal. Three hundred sixty-eight mothers (73.6%) breast-fed exclusively, 117

(23.4%) mixed breast-feeding and bottle feeding, and 15 (3%) bottle-fed exclusively. The duration of breast-feeding ranged from 6 to 24 months and averaged 14 months. In Saint Louis, Kaolack, and Ziguinchor, between 80 and 84% of mothers breast-fed exclusively, but in Peking and Dakar, 39% and 34% of mothers mixed breast and bottle feeding, respectively. Their reasons for doing so included employment of the mother, urbanization, and the impact of advertising. The trend to exclusive bottle feeding is still weak in cities and towns in Senegal, ranging from 1 to 5%. Breast-feeding was prominent in lower social groups, except in Dakar, where wealthier women tended to employ mixed feeding. In Zuinchor, Saint Louis, and Kaolack, breast-feeding predominated at all educational levels. More educated women in Dakar and Peking tended to mix breast-feeding and bottle feeding, whereas women with average educational levels breast-fed. Women who bottle fed did so exclusively for professional or medical reasons.

Employment has little effect on whether women initiate breast-feeding. However, work does appear to have a substantial effect on how long women are able to breast-feed, particularly those with less education (Ryan and Martinez, 1989). Most developing and industrialized countries have maternity leave and other legal provision to support employed mothers. It is a legal requirement for employed mothers to be given a leave of absence after the birth of their baby. In Kenya, the Bahamas, and Bolivia it is between 1 and 11 weeks, in Indonesia it is 12 weeks, while in Sri Lanka it is 84 days for the first two children and 48 days for the third child and onwards. In Australia, Norway, and the United Arab Emirates, the period is more than 26 weeks (http://www.waba.org.my/womenwork/mpc5Sept.pdf). For working mothers in the developing world who can afford it, infant formula can be a vital product, as mothers often have to return to work when their baby is a few months old, and they may be away from their babies from sunrise to sunset. In Indonesia, most babies get PIF when the mother leaves for work, or they get both breast milk and a PIF supplement. Alternatively, the babies may be breast-fed by other women. PIF also is useful for women who, for medical or other reasons, cannot breast-feed. The WHO estimates that 600,000 women die each year from complications related to pregnancy and childbirth. Infant formula often is used as the best alternative to breast milk for babies left without mothers.

In Sri Lanka, the cost of food, electricity, clothing, and education consumes a major portion of a family's income, and it is a necessity rather than a luxury to have two incomes. Therefore, if breast milk is freely available, a mother would not resort to purchasing infant formula unless she had to. In Indonesia, most babies are breast-fed until the age of 3 to 4 months, after which they may continue being fed PIF mixed with breast feeding. This occurs only if the family is able to buy PIF. However, many families are not financially stable and cannot afford to replace breast-feeding with PIF.

When formula promotion was stopped in Indonesia, infant mortality decreased from 51.6 to 33.4/1,000 live births, and cases of diarrhea decreased from 40.2 to 5/1,000 live births. This result was obtained only with upper- and middle-economic class families. It was not observed among poor families, where babies sometimes are born at home by trained birth helpers or nurses and mothers make their own decisions regarding whether or not to breast-feed their babies.

Although breast-feeding is widely practiced, its duration has dwindled, and the early introduction of solid foods is widespread. Even though infants up to 4 to 6 months should receive only breast milk to remain as healthy as possible, infants less than 4 to 6 months often receive other milk or gruels before the infants are ready to be weaned. In areas where PIF can be obtained, once the mother fails to breast-feed her baby, she will feed her baby PIF. However, in rural areas, these babies will not be fed PIF, either because it is not available, there is lack of clean water, or for financial reasons. The vast majority of mothers in developing countries do not have the means to buy infant formula and therefore feed their baby inferior traditional substitutes for breast milk, including water sweetened with sugar or honey, sweet tea, rice milk, cow milk, or cassava flour and water. In Malaysia, a substantial number of breast-fed babies were given weaning foods in the form of a porridge mixture (rice with egg, vegetables, meat, fish, and cereals) between the ages of 0 and 3 months. A survey by Manan (1995) found that 2.7% of Malaysian children were fed sweetened condensed milk, with 62.5% of them being on this for the first 3 months. These substitutes can be dangerous, because they lack the required nutritional content and balance.

Increasing use of PIF in developing countries does not parallel an increase of mothers' knowledge of food-borne diseases and how to prepare PIF. This is one of the reasons that infants develop infections. Therefore, education of mothers is very important in developing countries.

FACTORS AFFECTING THE SAFETY OF PIF

Water Quality
Water quality is one of the social concerns in developing countries, and some countries still have problems providing clean water. Other countries may have limited areas with a clean water supply, while other parts still lack clean water. Although *Enterobacter sakazakii* appears not to be found in drinking water of poor quality, safe drinking water always should be used to reconstitute PIF. To

Table 2 Access to clean water in urban and rural areas in some developing countries[a]

Country	% Access to:	
	Clean water	Adequate sanitation
Cambodia	30	17
China	75	40
Indonesia	78	55
Nepal	88	28
Pakistan	90	62
Sri Lanka	77	94
Thailand	84	96
Vietnam	77	47
Bangladesh	97	48

[a]Data source: Plan UK (http://plan-uk.org).

reconstitute PIF, boiled water, as suggested on every label of PIF container, should be used. However, one problem is that some mothers mix boiled water with normal water, which still has the possibility of microbial contamination. A comparison of water and sanitation among developing countries in Asia is described in Table 2. In Indonesia, access to clean water is still a serious problem. Only 78% of the population in urban areas and only 54% of the population in rural areas had access to safe water during the period from 1990 to 1996, while in the same period in Bangladesh the raters were 77% and 30%, respectively (Asian Development Bank; www.adb.org/countries/default.asp).

In Indonesia, middle-class and upper-class families have access to clean water. The water sources can be from clean water installations, bored wells, hand pumps, or natural sources. Every family is equipped with a kitchen containing a stove or burner. In urban areas, many women are employed and are able to care for their babies for only 3 months. The rest of the time they have to replace breast-feeding with PIF or other substitutes.

Preparation and Storage of PIF

Every family has their own way of preparing the family food, including PIF. Some mothers are not aware of how to prepare PIF, even though preparation instructions always are put on the packaging label. There are several ways that mothers prepare PIF.

1. PIF is poured into a clean bottle and some amount of hot water is added (boiled or dropped from water dispenser), and then it is shaken and cold or room-temperature water is added.

2. PIF is poured into a clean bottle, some cold water (boiled or from a water dispenser) is added, the mixture is shaken, and hot water mixed with room-temperature water is added, and then it is shaken again.

3. Mothers put some hot water in a bottle, let it cool, add PIF as instructed, and then shake it well.

The reconstituted milk normally is given immediately to the baby. In cases where the feed is not finished, the mother will keep this milk until the next feeding time, which may be several hours later. Some mothers or baby sitters understand that used bottles have to be cleaned or washed and then boiled to sterilize them before the next feeding. To feed while traveling, some mothers carry hot water in a thermos flask and reconstitute the PIF properly. However, many others just put the PIF in a bottle, add some water, shake it, and feed the baby.

Reconstituted PIF normally is stored at room temperature (22 to 32°C) for several hours. This occurs in families without a refrigerator and, because of a limited knowledge of hygiene, even in families that have a refrigerator. Many young mothers have little knowledge about PIF and have a poor understanding of how to prepare, handle, and store it. Practical knowledge or information comes from the physician or other child-bearing women. Consultation with a physician at a health center seldom occurs. Many women in rural areas have low levels of literacy, and this is the main reason why, even though the information is written on the package of PIF, their preparation knowledge is still limited.

The duration of the holding time after rehydration normally varies considerably. For babies who like to drink PIF and are able to drink, there is normally no unused feed. However, some babies drink very slowly and in small amounts, so there is often some feed remaining in the bottle. This leftover milk may be kept for several hours at room temperature (22 to 32°C). Normally, the holding time of reconstituted PIF is between 2 and 4 h. Sometimes, it will be disposed of after physical changes are seen, such as when it starts to smell or the PIF starts to clump. Among educated women, the situation can be better. Handling information is relatively available, including at supermarket sales where the mother purchases the PIF. For women in rural areas, this situation does not exist.

Storage of the PIF after opening the container is very important. PIF that is packed in tins is safer. Product can be stored in tins, with some products including a plastic closure. In contrast, PIF that is vacuum-packed in aluminium foil and cardboard boxes may have additional risks. Many mothers do not realize that the method of storage of PIF can lead to microbial contamination. After taking the required amount from the aluminium bag, it may be left open. The package often stands in the kitchen, in a drawer, or the cupboard. Some mothers who have good hygienic knowledge will transfer the package, after opening it, to a special plastic box or to glassware. The storage system can lead to an increased risk in families with inadequate kitchen facilities. In the standard family, the kitchen is equipped with a refrigerator, but many families have fewer amenities, i.e., kitchens are not equipped with refrigerators or storage facilities, and poor sanitation practices exist.

Hospital Preparation Conditions

Due to high birth rates, many obstetric hospitals are being built in cities and urban areas. In the cities and some other urban areas, the hospital has an area for a mother and her newborn baby, with special equipment and facilities for the bathing of babies as well washing of the bottles. The nurses are educated to know procedures related to baby care, including milk preparation. However, bacterial contamination still is possible due to a lack of hygienic conditions, including potable water.

In hospitals, practices vary according to local arrangements and the availability of trained personnel and adequate facilities. Centralized preparation of ready-to-feed formula and on-ward preparation are possible, and both have advantages and disadvantages. For both, the availability of safe (sterile) water and aseptic conditions for preparation are required. The transport of ready-to-feed preparations to the wards under sustained refrigeration and refrigeration on the ward up to the feeding time are important factors to control.

The situation in rural areas is very different. The mothers go to the obstetric nurse, who has a licence for doing practical work regarding helping the mother give birth. They normally perform home births with simple facilities: a normal bathroom and kitchen, and a porcelain bench for the use of the baby. This birth process is only for natural delivery of the baby. When the mother has a medical complaint, they will send the mother to the nearest hospital, which has more advanced facilities, e.g., a vacuum extractor or a surgery room for cesarean section.

SURVEY OF *E. SAKAZAKII* IN DEVELOPING COUNTRIES

There is a dearth of information on the contamination of PIF sold in developing countries, and there also has been no surveillance of the disease burden resulting from consumption of contaminated PIF in developing countries. The occurrence of *E. sakazakii* infection in neonatal babies in developing countries is not known.

The potential risks of contamination cannot be ruled out, given that studies from developed countries have shown that some batches of PIF are contaminated. Since many developing countries import PIF, the incidence and levels of *E. sakazakii* are likely to be the same as those in products in the exporting countries and those reported in published surveys. The levels would not increase during transport and distribution.

E. sakazakii contamination in PIF purchased at retail stores in an Asian country showed that 3 out of 100 dehydrated cans of PIF contained *E. sakazakii*. According to most-probable-number methods, the concentrations of *E. sakazakii* were calculated to be 0.0092 per g, 0.46 per g, and 0.0036 per g for the three contaminated cans (FAO-WHO, 2006b). A survey of the occurrence of *E. sakazakii* in milk-based infant formula in Thailand by the Center for Food Safety and Applied Nutrition in 2002 showed that 5.97% of the samples were positive. A survey also was done at the import checkpoint and showed a 1.87% positive rate (2 samples from 107 total samples). In retail stores, 9 out of 86 samples (10.47%) were positive, while 12 out of 201 samples (5.97%) from the manufacturer were positive (Anonymous, 2006). In Fuyang City, China (2004), a survey was conducted on 87 PIF samples for neonates, and 12.64% of the samples analyzed were found to be positive. Furthermore, there was a survey of 169 samples of PIF that were collected from 11 provinces in China. The products were made in 63 companies from 15 provinces as well as 6 countries outside China. The rate of positive samples was 2.96%.

A total of 74 infant foods purchased in Indonesia in 2003 (*n* = 74) were analyzed for *Enterobacteriaceae*, with a special emphasis on *Salmonella* and *Shigella*. While all samples were found to be negative for *Salmonella* and *Shigella*, other *Enterobacteriaceae* that formed yellow-pigmented colonies on tryptic soy agar were isolated with a high frequency. Further characterization showed that 10 samples (13.5%) were contaminated with *E. sakazakii* (Estuningsih et al., 2006b). In 2004, a survey on the occurrence of contamination of PIF by *E. sakazakii* in Jakarta, Indonesia, showed that 5 out of 40 samples (12.5%) were positive (Estuningsih et al., 2006a).

The presence of *E. sakazakii* and other *Enterobacteriaceae* was determined in 72 PIF from six brands in Malaysia. Presumptive *E. sakazakii* colonies were confirmed by using rapid biochemical tests and by real-time PCR. *E. sakazakii* was isolated from 9/72 PIF from three brands, with an incidence of 12.5%. Presumptive *E. sakazakii* colonies were confirmed by real-time PCR. The most frequently isolated *Enterobacteriaceae* were *Enterobacter cloacae* (from 38 samples), *Serratia liquefaciens* (from 9 samples), *E. sakazakii* (from 9 samples), and *Salmonella* group 1 (from 5 samples). Growth kinetics of three isolated *E. sakazakii* strains from PIF and the type strain ATCC 51329[T] at various temperatures was evaluated in reconstituted PIF and tryptic soy broth. All strains had a shorter generation time in infant formula than in tryptic soy broth. The organism grew optimally at 37 to 45°C, and no growth was observed at 4 or 50°C. The 12.5% incidence of *E. sakazakii* in PIF as well as a short generation time of 29.92 min at room temperature (25°C) is a cause for concern (Abdullah Sani et al., 2007).

SUMMARY

In many developing countries, the proportion of special subpopulations consisting of low-birth-weight infants and infants of HIV-infected mothers is higher than it is in developed countries. Therefore, the use of PIF in these circumstances may be increasing. The basis of the higher demand for PIF is the recommendation for infants of HIV-positive mothers that, where replacement feeding is acceptable, feasible, affordable, sustainable, and safe, all breast-feeding should be avoided (WHO, 2001).

Human milk fortifiers are required to compensate for the nutritional needs of very-low-birth-weight infants. In circumstances where the mother cannot breast-feed or chooses not to breast-feed for any reason, special PIF may be required for the feeding of low-birth-weight infants. Therefore, well-controlled studies need to be conducted to assess the extent of risk associated with contaminated PIF for infants in developing countries.

Many consumers, including those directly involved in caring for infants, are not aware that PIF is not a sterile product and may be contaminated with pathogens that can cause serious illness, and they lack information on how handling, storage, and preparation practices can influence the risk. Effective risk communication efforts for both the public and health professionals are needed. Information and education about basic hygiene practices in connection with food handling, storage, and preparation at home also need to be emphasized.

REFERENCES

Abdullah Sani, N., M. Ghassem, and A. S. Babji. 2007. *Enterobacter sakazakii* and growth characteristics in infant formula milk, FS-7:98. 10th ASEAN Food Conference, 21 to 23 August 2007, Kuala Lumpur, Malaysia.

Anonymous. 2004. Directorate of Public Nutrition, Ministry of Health of Republic of Indonesia.

Anonymous. 2006. Survey on exclusive breast-feeding in Jakarta. Room document from call for data for FAO meeting.

Codex Alimentarius Commission. 2005. Agenda item 4, CX/FH/05/37/4. *Rep. 37th Sess. Codex Comm. Food Hyg.* Buenos Aires, Argentina.

Cutting, W. 1995. Breast feeding is best feeding. *Dialogue Diarrhoea* **59:**1. http://www.Rehydrate.org/dd/dd59.htm.

Duffy, L. C., H. Faden, R. Wasielewski, J. Wolf, D. Krystofik, and T. Williamsville. 1997. Exclusive breastfeeding protects against bacterial colonization and day care exposure to otitis media. *Pediatrics* **100:** E7.

Estuningsih, S., C. Kress, A. A. Hassan, Ö. Akineden, E. Schneider, and E. Usleber. 2006b. *Enterobacter sakazakii* and other *Enterobacteriaceae* in infant formula manufactured in Indonesia and Malaysia. *J. Food Prot.* **69:**3013–3017.

Food and Agriculture Organization-World Health Organization (FAO-WHO). 2004. *Enterobacter sakazakii* and other microorganisms in powdered infant formula: meeting report. *Microbiological risk assessment series 6.* World Health Organization-Food and Agriculture Organization of the United Nations, Geneva and Rome. WHO Press, Geneva, Switzerland. http://www.who.int/foodsafety/publications/micro/mra6/en/index.html.

Food and Agriculture Organization-World Health Organization (FAO-WHO). 2006a. *Enterobacter sakazakii* and *Salmonella* in powdered infant formula: meeting report. *Microbiological risk assessment series 10.* World Health Organization-Food and Agriculture Organization of the United Nations, Geneva and Rome. WHO Press, Geneva, Switzerland. http://www.who.int/foodsafety/publications/micro/mra10/en/index.html.

Food and Agriculture Organization-World Health Organization (FAO-WHO). 2006b. Room document from call for data. *FAO-WHO 2nd Risk Assess. Workshop.* 16 to 20 January 2006, Rome, Italy. http://www.fao.org/ag/agn.

Kaleida, P. H., D. G. Nativio, H. P. Chao, and S. N. Cowden. 1993. Prevalence of bacterial respiratory pathogens in the nasopharynx in breast-fed versus formula-fed infants. *J. Clin. Microbiol.* **31:**2674–2678.

Kwa, S. K. 1993. Breastfeeding and the use of maternal health services in Sarawak. *Malays. J. Reprod. Health* **11:**8–19.

Manan, W. A. 1995. Breast-feeding and infant feeding practices in selected rural and semi-urban communities in Malaysia. *Malays. J. Nutr.* **1:**51–61.

Orwell, S., D. Clayton, and A. E. Dugdale. 1984. Infant feeding in Nigeria. *Ecol. Food Nutr.* **15:**129–141.

Ryan, A. S., and G. A. C. Martinez. 1989. Breastfeeding and working mother. A profile. *Pediatrics* **83:**524–531.

Sall, M. G., N. Kualuvi, H. D. Sow, A. Ngom, A. Sanoko, and B. Wade. 1986. Result of a survey on breast feeding in several Senegal cities. *Afr. Med.* **25:**477–478.

World Health Organization (WHO). 2001. New data on the prevention of mother-to-child transmission of HIV and their policy implications: conclusions and recommendations. WHO Technical Consultation on behalf of the UNFPA/UNICEF/WHO/UNAIDS Inter-Agency Task Team on Mother-to-Child Transmission of HIV. Report no. WHO/RHR/01.28. WHO, Geneva, Switzerland.

Enterobacter sakazakii
Edited by Jeffrey M. Farber and Stephen J. Forsythe
© 2008 ASM Press, Washington, D.C.

Regulatory Aspects

9

*Jeffrey M. Farber, Franco Pagotto, and
Jean-Louis Cordier*

INTRODUCTION

Microbiological sampling plans and limits for products intended for infants
and young children were discussed for the first time, to our knowledge, by
the International Commission on Microbiological Specifications for Foods
(ICMSF) (1974); they also discussed criteria for a number of diverse dried
foods. These products were included in the category of special dietary foods
to be eaten by a high-risk category of consumers, i.e., infants and elderly
people (Table 1).

These plans and limits, covering a wide variety of products designed for
very different consumers, were reviewed during the following years, and new
recommendations were published by the ICMSF (1986). These recommendations were limited to foods for infants and children, thus covering several
types of infant formulae, as described in chapter 6 of this book, as well as infant cereals and weaning foods fed after the age of 4 to 6 months, depending on the recommendations in force at the time (Table 2). The
recommendations of 1986 were much more focused than the initial ones and
included only the necessary and relevant parameters, i.e., aerobic mesophilic
counts and coliforms as process hygiene indicators and *Salmonella* as the epidemiologically relevant pathogen. For *Salmonella* case 15, defining the strin-
gency of the sampling plan (ICMSF, 2002), it was proposed that the severity
of the hazard, as well as the potential growth after reconstitution of the dry
products, needed to be taken into consideration. Other microorganisms,
such as *Staphylococcus aureus*, *Clostridium perfringens*, and *Bacillus cereus*,

JEFFREY M. FARBER AND FRANCO PAGOTTO, Bureau of Microbial Hazards, Health Products and Food
Branch, Health Canada, Ottawa, Ontario, Canada. JEAN-LOUIS CORDIER, Nestlé Nutrition, Avenue
Reller 22, CH-1800 Vevey, Switzerland.

Table 1 Microbiological sampling plans and recommended limits for special dietary foods (ICMSF, 1974)

Microorganism(s)	n	c	Microbiological limit (MPN/g)[b]	
			m	M
Aerobic mesophilic counts	5	1	10^4	10^6
Escherichia coli	5	2	<3	10
Staphylococcus aureus	10	1	10	100
Bacillus cereus	10	1	10^2	10^4
Clostridium perfringens	10	1	10^2	10^3
Salmonella (in 25-g samples)	60	0	0	NA[a]

[a]NA, not applicable.
[b]*n* is the number of samples to be analyzed per lot, *c* is the number of samples allowed between "*m*" and "*M*" values, *m* is the microbiological limit that separates good from marginally acceptable quality, and *M* is the microbiological limit that separates marginally acceptable from unacceptable quality.

Table 2 Microbiological sampling plans and recommended limits for food for infants and children (ICMSF, 1986)

Microorganism(s)	n	c	Microbiological limit (MPN/g)[b]	
			m	M
Aerobic mesophilic counts	5	1	10^4	10^5
Coliforms	5	1	10	10^2
Salmonella (in 25-g samples)	60	0	0	NA[a]

[a]NA, not applicable.
[b]*n* is the number of samples to be analyzed per lot, *c* is the number of samples allowed between "*m*" and "*M*" values, *m* is the microbiological limit that separates good from marginally acceptable quality, and *M* is the microbiological limit that separates marginally acceptable from unacceptable quality.

were discussed as well, but they were considered to represent only a moderate direct health hazard (ICMSF, 1986). Additional epidemiological, scientific, and technical principles for the establishment of microbiological criteria, which were published much later (CAC, 1997), reinforced the fact that there was no need to include them in any recommendations. However, two-class plans ($n = 5$, $c = 0$, $m = 100$) have nevertheless been discussed as a potential option in case there was an urgent need to define limits for these bacteria. The rationale for such two-class plans was based on the fact that not all of these moderate pathogens represent a direct risk of infection but would require substantial growth to high levels to cause illness.

The initial 1974 sampling plans and recommended limits formed the basis for discussions by the Codex Alimentarius Commission and helped to develop the specifications included in the recommended Code of Hygiene for infant foods (CAC, 1979), which is summarized in Table 3.

Table 3 Microbiological specifications recommended by the Codex Alimentarius Commission (CAC, 1979)

Microorganism(s)	n	c	Microbiological limit (MPN/g)[b]	
			m	M
Aerobic mesophilic counts	5	1	10^3	10^4
Coliforms	5	1	<3	20
Salmonella (in 25-g samples)	60	0	0	NA[a]

[a]NA, not applicable.
[b]n is the number of samples to be analyzed per lot, c is the number of samples allowed between "m" and "M" values, m is the microbiological limit that separates good from marginally acceptable quality, and M is the microbiological limit that separates marginally acceptable from unacceptable quality.

Recommendations such as the ones published by the ICMSF (1974, 1986) and the Codex Alimentarius Commission (1979) have been widely adopted in national regulations, either as is or with slight modifications, such as slightly more stringent limits for coliforms or *Enterobacteriaceae*, e.g., an absence in 1-g samples, as is the criterion in Belgium (Van Acker et al., 2001).

Although sporadic isolated outbreaks of *Enterobacter sakazakii* already had been reported, no specific limits were implemented by public health authorities for this emerging opportunistic pathogen. It was not until 2001 and 2002 that discussions regarding *E. sakazakii* and appropriate control measures, including microbiological criteria, were initiated in various countries.

CODEX ALIMENTARIUS COMMISSION

Specifications recommended early on by the Codex Alimentarius Commission (1979) were considered to be valid and appropriate until very recently. It was not until 2003, when the United States and Canada drafted a risk profile on *E. sakazakii*, which was tabled during the 35th Session of the Codex Alimentarius Food Hygiene Commission (CCFH, 2003), that the question relating to the microbiological safety of powdered infant formulae was raised at the international level.

The risk profile summarized the knowledge of *E. sakazakii* available at the time and increased the evidence of links to the consumption of powdered infant formulae. The document highlighted the issues as well as existing gaps; in particular, the lack of knowledge regarding the susceptible infant populations, differences in virulence, thermal tolerance, and growth kinetics, and the dose response of this emerging hazard. The group urged the Codex Committee on Food Hygiene to undertake risk management activities and, in particular, to initiate the revision of the Code for Foods for Infants and Children.

This initiative resulted in the organization, jointly by the Food and Agriculture Organization of the United Nations (FAO) and the World

Health Organization (WHO), of an expert meeting to assess the situation with respect to *E. sakazakii* and other microorganisms in powdered infant formulae. The aim was to provide input to the revision of the Recommended International Code of Hygienic Practice for Foods for Infants and Children (CAC, 1979), chaired by Canada. The outcome and conclusions of this expert meeting were published as an FAO-WHO risk assessment report (2004).

During this meeting, a thorough assessment was performed on all potential microbiological hazards, which in the end were categorized into three groups. *Salmonella enterica* and *E. sakazakii* were the only microorganisms for which a well-established cause of illness in infants and a link with powdered infant formula was identified, thus they were placed in Category A. Although sporadically found in powdered infant formulae, other *Enterobacteriaceae* were included in Category B, because there was no clear evidence of association between these organisms and illness. Other organisms, including *Bacillus cereus*, *Clostridium difficile*, and *Clostridium botulinum*, were assigned to group C, since they were not identified in powdered infant formulae, or, if they were identified, no causal association between illness and product was demonstrated.

The key findings were that intrinsic contamination of powdered infant formula with *E. sakazakii* or *Salmonella* can cause infection and illness in infants. Based on the information from reported cases, it was deduced that infants up to 1 year of age were at risk, but that infants at the greatest risk were neonates (≤28 days), in particular preterm infants, low-birth-weight infants, and immunocompromised infants. It was also recognized that the risk of acquiring illness increased rapidly if growth of *E. sakazakii* was allowed, such as in the case of poor hygienic practices during preparation, handling, or storage.

Although a reduction in occurrence and levels of bacteria was deemed possible through reinforced hygienic measures during manufacture, it also was acknowledged that it was not possible to manufacture sterile powders with the current technologies. Therefore, only by using a combination of intervention measures, such as the use of microbiological criteria and recommendations for the hygienic preparation of infant feeds, would one achieve the greatest impact.

Based on the outcome of this report, the revision of the Recommended International Code of Hygienic Practice for Foods for Infants and Children (CAC, 1979) has been undertaken under the chairmanship of Canada. The ICMSF has been requested to contribute by proposing risk-based microbiological criteria as one of the control measures, and the proposal currently annexed in the revised draft code is presented in Table 4.

Over the last 2 to 3 years, three drafts of the revised code have been prepared by a specific working group and discussed by the Codex Committee for Food Hygiene, and these discussions are continuing. The main stumbling blocks identified during the preparation of the initial drafts certainly is the scope of the new code, i.e., the types of products to be included in relation to

Table 4 Draft microbiological criteria for powdered infant formula, follow-up formula up to 12 months, formula for special medical purposes, and human milk fortifiers as proposed for the revised Recommended International Code of Hygiene Practice for Foods for Infants and Children (CCFH, 2007)[a]

Microorganism(s)	n	c	m	M
Criteria for pathogenic organisms				
E. sakazakii (in 10-g samples)	30	0	0	NA[b]
Salmonella (in 25-g samples)	60	0	0	NA
Criteria for process hygiene				
Aerobic mesophilic counts	5	2	500/g	5,000/g
Enterobacteriaceae (in 10-g samples)	10	2	0	NA

[a]It should be noted that further scoping and discussion of these criteria will continue during subsequent meetings of the working group drafting the revised code, as well as during the annual CCFH meetings.
[b]NA, not applicable.

the existing definitions of infants (children up to 12 months) and the definitions of infant formulae, which may vary depending on the country. In order to overcome issues related to existing definitions but nevertheless to take into consideration epidemiological data and feeding practices as well as uncertainties in risk management, two different sets of criteria have been proposed, basically splitting follow-up formula into two groups, i.e., for infants up to 12 months and for young children (>12 months). Criteria for the first group are presented in Table 4. In the case of powdered follow-up formula and formula for special medical purposes for young children, no criterion is included for *E. sakazakii*, whereas the other three parameters are identical to the ones in Table 4. This proposal will be discussed at the upcoming meeting of the Codex Committee on Food Hygiene in October 2007.

In 2006, a second expert meeting was organized in order to review new epidemiological data and information on *E. sakazakii*, as well as to discuss in more detail whether the current control measures for *Salmonella* still were appropriate. This latter part of the discussion was triggered by the occurrence of two outbreaks in France (Anonymous, 2005a; Brouard et al., 2007). This second FAO-WHO report basically confirmed the findings and conclusions of the first report, as not much new information was available. However, emphasis was placed on the discussions regarding control measures, i.e., the microbiological criteria as well as the preparation, handling, and storage of infant feeds. The relevance of the proposed microbiological criteria was confirmed, and the risk model on the different scenarios for the preparation of infant feeds was developed in much greater detail (FAO-WHO, 2006), confirming that poor practices can lead to a dramatically increased risk of infection. As the model is an integral part of the mitigation strategy, it will be available on the WHO website in the future.

UNITED STATES

An outbreak of *E. sakazakii* in Tennessee following the consumption of contaminated powdered infant formula (Anonymous, 2005b) triggered several actions by the FDA to address this emerging issue. For example, health care professionals were alerted to the risk related to the use of powdered infant formulae (FDA, 2002).

During subsequent years, surveys of commercialized infant formulae have been performed and products have been tested for *E. sakazakii*, initially in four samples amounting to 333 g but later only once in 333 g, using a most-probable-number (MPN) technique (2002b). The only reference to what could be considered a sampling and compositing plan was discussed by Zink (2003).

In 2006, the revised GMPs for infant formula were posted for consultation (Federal Register, 2006). A noticeable change was the split of the criteria into food safety parameters and hygiene indicators, as well as the introduction of a criterion for *E. sakazakii* (for a 10-g sample, $n = 30$, $c = 0$, $m = 0$), identical to the one enforced by the European Union in January 2006 and the one proposed in the draft revised code of the Codex Alimentarius. In addition, the introduction of *Enterobacteriaceae* (in 10-g samples) has been envisaged as a parameter for process hygiene. The release of the final GMP document is expected in 2007 or 2008.

EUROPEAN UNION

In Europe in 2003, the European Food Safety Authority (EFSA) was mandated to review and assess the situation with respect to the microbiological safety of infant formulae and follow-on formulae. The results, conclusions, and recommendations of this assessment, performed by the Scientific Panel on Biological Hazards, was published in 2004 (EFSA, 2004).

The panel concluded that *Salmonella* and *E. sakazakii* were indeed the microorganisms of greatest concern in formulae for special medical purposes, for infant formulae, and for follow-on formulae, although there was no history of illness for the latter.

The opinion also acknowledged the fact that *E. sakazakii* is widespread, suggesting that the consumption of small numbers of the organism would not lead to illness of healthy infants and children, while infants of less than 4 to 6 weeks of age and, in particular, of preterm babies, underweight babies, or babies that were immunocompromised or were from immunocompromised mothers, were at particular risk.

In terms of management options, the Scientific Panel on Biological Hazards recommended the introduction of a performance objective aimed at achieving very low levels of both relevant hazards. Although an absence of the pathogen in 1, 10, or 100 kg of product has been suggested, this recommendation has not been discussed in great detail. However, it is clear that adherence to good manufacturing practices and good hygiene practices was considered key in achieving this objective, considering that testing for *Enterobacteriaceae* in environmental and product samples (absence in 10-g samples) would be an appropriate tool to verify adherence to hygienic practices and compliance with the performance objective.

As in the case of the FAO-WHO reports (2004, 2006), management of the safety of infant formulae was focused not only on end product testing against microbiological criteria. The EFSA opinion clearly recognized the role of poor handling practices during preparation of the formula as an important element contributing to outbreaks. It has therefore recommended the development of guidelines for the preparation, handling, storage, and use of infant formulae in homes and health care facilities.

The European Commission has, based on this opinion, established microbiological criteria for formulae for special medical purposes and for infant formulae (for infants up to 6 months of age) that include two food safety parameters (*Salmonella* and *E. sakazakii*), with detection of other *Enterobacteriaceae* as a process hygiene parameter (European Commission, 2005; Table 5). It should be noted that the reference method for *E. sakazakii* corresponds to the recently issued International Organization for Standards (ISO) Technical Standard 22964 (2006).

Recently, the EFSA BIOHAZ Panel has undertaken the review of the relationship between *E. sakazakii, Salmonella*, and other *Enterobacteriaceae* in infant and follow-on formulae. The panel in charge of the assessment concluded that there is no correlation between *Salmonella* and other *Enterobacteriaceae*, but that a relationship between *E. sakazakii* and other *Enterobacteriaceae* was apparent in some of the available data. It recommended further investigations using standard protocols and analytical

Table 5 Current European Union microbiological criteria for formulae for special medical purposes and for formulae for infants up to 6 months of age

Microorganism(s)	*n*	*c*	*m*	*M*

Microorganism(s)	*n*	*c*	*m*	*M*
Enterobacteriaceae (in 10-g samples)	10	0	0	NA[a]
Enterobacter sakazakii (in 10-g samples)	30	0	0	NA
Salmonella (in 25-g samples)	30	0	0	NA

(Microbiological limit spans columns *m* and *M*)

[a]NA, not applicable.

methods in order to enable a correct comparison and to confirm whether a relationship exists (EFSA, 2007).

In July 2007, the Standing Committee on Food Chain and Animal Health (SCoFCAH) approved amendments to the EC regulation 2073/2005 by deleting the link (in terms of analyses) between the hygiene parameter *Enterobacteriaceae* and the safety parameters *Salmonella* and *E. sakazakii* for powdered infant formulae; i.e., now both parameters are being considered at the same level. In addition, *Salmonella* ($n = 30$, $c = 0$, $m = 0$ [in 25 g]) have been introduced for follow-on formulae as well as a stringent hygiene criterion, i.e., *Enterobacteriaceae* with $n = 5$, $c = 0$, $m = 0$ (in 10 g). The technical vote will be confirmed later in 2007, and amendments are thus likely to be implemented by the end of 2007 or the beginning of 2008.

OTHER COUNTRIES

Detailed information on the regulatory situation in other countries is scarce, and the relevant references are often difficult to collect and also difficult to cross-check with respect to validity, updates, age groups covered, etc. (Table 6). The terminology used in the column "parameter" has been unified as far as possible and may thus not always correspond to the terminology used in the original language of the specification. However, it is known that some countries already have modified their existing legislation, e.g., Argentina (Anonymous, 2006). Other countries, e.g., Switzerland, are considering a change in order to align their regulations more closely with the European Union legislation. It is also known that several countries, e.g., Thailand, China, the United Arab Emirates, and some countries in Latin America, are applying the same stringent requirements as those that are applied in the United States or the European Union.

PREPARATION AND HANDLING OF FEEDS

Although not a regulatory issue, guidelines for the preparation of infant feeds play an important role in ensuring their safety. The importance of the adherence to good hygiene practices during preparation has been stressed in all assessments performed, which are discussed in the previous sections. In every case, it has been considered very important to combine different management options, such as microbiological criteria and guidelines for the preparation of powdered infant formulae, to address this issue and to ensure the safety of the consumed feeds.

While the establishment of microbiological criteria has progressed rapidly and has led to the creation and even the implementation of very

Table 6 Microbiological criteria for powdered infant formula from various countries[a]

Country, product type, and infant group by age	Parameter (or analytical method)	n	c	m	M
Argentina					
Products for infants up to 6 months	Aerobic mesophilic count at 37°C	5	2	10^3 CFU/g	10^4 CFU/g
	Enterobacteriaceae	10	0	0/10 g	NA
	Salmonella	30	0	0/25 g	NA
	Enterobacter sakazakii (according to ISO 22964)	30	0	0/10 g	NA
Products for infants (>6 months) and young children	Aerobic mesophilic count at 37°C				3×10^4 CFU/g
	Coliforms at 37°C				<3 MPN/g
	Escherichia coli				0/5 g
	Salmonella				0/100 g
	Coagulase-positive staphylococci				0/1 g
	Yeasts and molds				10^2/g (dairy products), 10^3/g (cereal based)
Brazil					
Premature infants	Coliforms				0/g at 45°C and 10 CFU/g at 35°C
	Staphylococcus				0/g
	Bacillus cereus				50 CFU/g
	Salmonella				0/g
0–12 months	Coliforms				0/g at 45°C and 10 CFU/g at 35°C
	Staphylococcus				0/g
	Bacillus cereus				100 CFU/g
	Salmonella				0/g

(*continued*)

Table 6 Microbiological criteria for powdered infant formula from various countries[a] (*continued*)

Country, product type, and infant group by age	Parameter (or analytical method)	n	c	Microbiological limit m	M
Brunei					
Up to 12 months	Aerobic mesophilic count				10^5 CFU/g at 37°C for 48 h
	Coliforms				50 CFU/g
Canada[b]					
Up to 12 months	ACC (according to MFHPB-18)	5	2	10^3 CFU/g	10^4 CFU/g
	Escherichia coli (according to MFHPB-19)	10	1	<1.8 CFU/g	10^1 CFU/g
	Salmonella (according to MFHPB-20)	20	0	0/g	NA
	Staphylococcus aureus (according to MFHPB-21)	10	1	10^1 CFU/g	10^2 CFU/g
	Bacillus cereus (according to MFLH-42)	10	1	10^2 CFU/g	10^4 CFU/g
	Clostridium perfringens (according to MFHPB-23)	10	1	10^2 CFU/g	10^3 CFU/g
Central America[c] (Costa Rica, Guatemala, Panama, El Salvador, Nicaragua)					
Dried and instant products 0–12 months	Aerobic mesophilic counts	5	2	10^3 CFU/g	10^4 CFU/g
	Coliforms	5	1	<3 CFU/g	20 CFU/g
	Salmonella	60	0	0/25 g	
Chile					
0–12 months	Aerobic mesophilic counts	5	1	10^3 CFU/g	10^4 CFU/g
	Coliforms	5	1	<3 MPN/g	20 MPN/g
	Escherichia coli	5	0	<3 CFU/g	
	Bacillus cereus	5	0	<10 CFU/g	

	n	c	m	M
Salmonella	10	0	0/25 g	
Staphylococcus aureus	5	0	<3 CFU/g	
China				
Up to 12 months				
Molds and yeasts				50 CFU/g
Aerobic mesophilic counts (37°C)				30,000 CFU/g
Coliforms				40 CFU/100 g
Pathogenic bacteria				Absent
Colombia				
0–12 months				
Aerobic mesophilic counts	3	1	5,000 CFU/g	10,000 CFU/g
Coliforms	3	1	<3 MPN/g	11 MPN/g
Fecal coliforms	3	0	<3 MPN/g	
Staphylococcus			<100 CFU/g	
Bacillus cereus	3	1	100 CFU/g	200 CFU/g
Salmonella	3	0	0/100 g	
Molds and yeasts	3		100/g	300 CFU/g
Staphylococcus-positive coagulase	3	0	<100 CFU/g	
Sulfite-reducing spores	3	1	10 CFU/g	100 CFU/g
	3	1	5,000 CFU/g	10,000 CFU/g
Dominican Republic[c]				
Dried and instant products (0–12 months)				
Aerobic mesophilic counts	5	2	10^3 CFU/g	10^4 CFU/g
Coliforms	5	1	<3 CFU/g	20 CFU/g
Salmonella	60	0	0/25 g	
Dried products requiring heating prior to consumption (0–12 months)				
Aerobic mesophilic counts	5	3	10^4 CFU/g	10^5 CFU/g
Coliforms	5	2	10 CFU/g	100 CFU/g
Salmonella	5	0	0 CFU/25 g	

(continued)

Table 6 Microbiological criteria for powdered infant formula from various countries[a] (*continued*)

Country, product type, and infant group by age	Parameter (or analytical method)	n	c	Microbiological limit		
				m	M	
Ecuador						
0–12 months	Aerobic mesophilic counts	5	2	10^3 CFU/g	10^4 CFU/g	
	Coliforms	5	1	<3 MPN/g	20 MPN/g	
	Molds and yeasts	5	2	30 CFU/g	100 CFU/g	
	Salmonella	5[d]	0	0 CFU/g		
European Union						
0–6 months	*Enterobacteriaceae* (according to ISO 21528-1)[c]	10	0	0/10 g	NA	
	Enterobacter sakazakii (according to ISO/DTS 22964)	30	0	0/10 g	NA	
	Salmonella (according to EN/ISO 6579)[y]	30	0	0/25 g	NA	
	Listeria monocytogenes (according to EN/ISO 11290-1)	10	0	0/25 g	NA	
Honduras						
0–12 months	Aerobic mesophilic counts				50,000 CFU/g	
	Coliforms				90 CFU/g	
	Escherichia coli				0/g	
Malaysia						
Up to 12 months	Aerobic mesophilic counts				10^4/g at 37°C for 48 h	
	Coliforms				10/g at 37°C for 48 h	
	Escherichia coli				Not specified; must comply with GMP	

Mexico			
0–12 months	Aerobic mesophilic counts		2,500 CFU/g
	Coliforms		20 MPN/g
	Escherichia coli		Negative in 1 g
	Salmonella		Absent in 25 g
	Staphylococcus		Negative in 1 g
Peru			
0–12 months	Aerobic mesophilic counts	5	2	10^3 CFU/g
	Enterobacteria	5	1	10 CFU/g
	Bacillus cereus	10	1	10^2 CFU/g
	Molds	5	1	10^2 CFU/g
	Staphylococcus aureus	5	1	10 CFU/g
	Salmonella	60	0	0/25 g
Singapore			
Up to 12 months	Aerobic mesophilic counts		10^5 CFU per g at 37°C for 48 h
	Coliforms		50 CFU/g
	Enterobacter sakazakii		Absent in 1 g
South Korea			
Up to 6 months	Aerobic mesophilic counts		<20,000 CFU/g
	Enterobacter sakazakii		Negative
	Coliforms		Negative

(continued)

Table 6 Microbiological criteria for powdered infant formula from various countries[a] *(continued)*

Country, product type, and infant group by age	Parameter (or analytical method)	n	c	Microbiological limit m	M
Thailand					
0–12 months	Aerobic mesophilic counts				10^4 CFU/1 g
	Escherichia coli				Absent in 0.1 g
	Bacillus cereus				10^2 CFU/1 g
	E. sakazakii				Absent in 333 g
	Pathogenic microorganisms				Absent
United States[g]					
0–12 months	Enterobacter sakazakii	30	0	0/10 g	NA
	Salmonella	60	0	0/25 g	NA
Uruguay					
	Enterobacter sakazakii (according to the FDA method, 2002b)				Absent in 333 g
Venezuela					
0–12 months	Aerobic mesophilic counts	5	1	5×10^3 CFU/g	1×10^4 CFU/g
	Coliforms	5	2	<3 MPN/g	7 MPN/g
	Staphylococcus aureus	5	1	10 CFU/g	10^2 CFU/g
	Bacillus cereus	5	2	10^2 CFU/g	10^3 CFU/g
	Salmonella (routine control)	10	0	0/25 g	
	Salmonella (special)	30	0	0/25 g	
	Molds	5	2	10^2 CFU/g	10^3 CFU/g
	Listeria monocytogenes	5	0	0/25 g	
	Clostridium perfringens	5	2	10^2 CFU/g	10^3 CFU/g

Vietnam

Up to 12 months	
Aerobic mesophilic counts	10^4 CFU/g
Coliforms	10 CFU/g
Escherichia coli	Absent in 1 g
Staphylococcus aureus	Absent in 1 g
Bacillus cereus	10^2 CFU/g
Yeasts and molds	Absent in 1 g
Salmonella	Absent in 25 g

aSeveral countries follow criteria that do give specific n, c, m, or M values, other do not. For Hong Kong, no specific microbiological standard is specified. However, for contaminants, to the extent possible in GMP, the product should be free from contaminants or objectionable matter. For Indonesia, no microbiological standard is specified. For the Phillipines and South Africa, for infants 0 to 12 months old, the following standards apply: When tested by appropriate methods of sampling, and examination, the product (i) shall be free from pathogenic microorganisms; (ii) shall not contain any substances originating from microorganisms in amounts that may represent a hazard to health; and (iii) shall not contain any other poisonous or deleterious substances in amounts that may represent a hazard to health. The product shall be prepared, packed, and held under sanitary conditions and should comply with the recommended International Code of Hygienic Practice for Foods for Infants and Children (CAC, 1979).

bCanada's criteria are under review.

cFollows the Codex criteria.

dCorresponding to 20 samples of 25 g each, grouped in five sets of 4 samples each.

eHygienic indicator.

fSafety indicator.

gThe criteria for the USA are only a proposal at this stage.

similar stringent criteria throughout the world, the establishment of guide-lines for the safe preparation, handling, and storage of infant feeds has lagged behind. Contrary to the microbiological criteria, the guidelines proposed or published vary widely in their content and approach. For example, the guidelines developed in France allow for the use of either unboiled tap water or mineral water, and they recommend the reconstitution of the powdered infant formula at a low temperature in order to avoid microbial growth (AFSSA, 2005), while in other cases reconstitution at 70°C is recommended, e.g., by the United Kingdom authorities and by FAO-WHO.

It is certainly of the utmost importance that such guidelines and guidance documents are published. They should, however, provide sufficient flexibility, taking into account national or regional differences and habits as well as available resources. It certainly will not be of benefit to impose certain practices if they are not followed due to a lack of infrastructure, facilities, or trained personnel. In fact, as was shown in the first report by the FAO-WHO (2004), a lack of adherence to recommendations, such as the reconstitution at 70°C, by a small percentage of users would lead to a dramatic drop in the efficacy of the overall risk reduction. While an important step, reconstitution is not the only step that may jeopardize the safety of the feeds; poor cleaning practices, poor layout of premises, which increase the risk of cross-contamination, and poorly informed and trained personnel are all very important factors. Prevention should be considered more globally as, for example, in the guidelines issued by the American Dietetic Association, which considered all elements of health care facilities during the preparation of formula and breast milk (ADA, 2004).

CONCLUSIONS

It is evident that, as far as regulatory approaches to control *E. sakazakii* are concerned, a multipronged approach is the best one. This will require continued cooperation and collaboration between hospitals, industry, and governments.

Current industry efforts to control this organism have focused on improving good hygienic practices, along with environmental monitoring and end-product testing for the organism. Industry has made good progress in reducing the incidence of *E. sakazakii* both in the environment and in the end product. This will need to continue in the future.

With respect to *E. sakazakii* and microbiological criteria in general, efforts will need to be made at the international level to standardize the methods for isolating the organism from powdered infant formula, including the amount of product that is tested. There appears to be a growing consensus at the international level that companies can produce product wherein *E. sakazakii* is absent in 30 random 10-g samples taken from a lot.

There are countries that have microbiological criteria for powdered infant formula for organisms other than *Salmonella* and *E. sakazakii* (Table 6). Although, in most cases, these can be thought of as indicator microorganisms, it does become somewhat difficult to identify the specific level of these organisms, above which the product would be unsafe. This does not mean, however, that countries without specific criteria for these organisms could not take action against a product that contained any of these potential pathogens. It would be better for countries to focus on the most likely microbial hazards occurring in powdered infant formula, i.e., *Salmonella* and *E. sakazakii*.

The last component of the three-pronged approach includes the use of labeling and education. Simpler and more directed information is needed on the cans of formula. In addition, better educational material directed at hospitals and caregivers is needed.

ACKNOWLEDGMENT

We acknowledge the great help of Dan March in helping to compile much of the information in Table 6.

REFERENCES

Agence Française de Sécurité Sanitaire des Aliments (AFSSA). 2005. Recommendations d'hygiène pour la préparation et la conservation des biberons, Juillet 2005. http://www.sante .gouv.fr/htm/actu/biberon/rapport_afssa.pdf.

American Dietetic Association (ADA). 2004. Infant feedings: guidelines for preparation of formula and breastmilk in health care facilities. American Dietetic Association, Chicago, IL.

Anonymous. 2005a. Epidémie de salmonellose à *Salmonella enterica* sérotype Agona chez des nourrissons, France, Janvier–Avril 2005. Point final de l'investigation au 10 Juin 2005. www .invs.sante.fr/presse/2005/le_point_sur/salmonella_agona_150605/index.html.

Anonymous. 2005b. *Enterobacter sakazakii* infections associated with the use of powdered infant formula—Tennessee, 2001. *Morb. Mortal. Wkly. Rep.* **51**:297–300.

Anonymous. 2006. Expediente no 1-47-2110-3888-06-1: alimentos para lactantes e infantes. http://www.alimentosargentinos.gov.ar/programa_calidad/Marco_Regulatorio/CONAL/ Resoluciones_Conjuntas/1_47_2110_3888_06_1_sakazakii.pdf.

Brouard, C., E. Espié, F.-X. Weill, A. Kérouanton, A. Brisabois, A.-M. Forgue, V. Vaillant, and H. de Valk. 2007. Two consecutive large outbreaks of *Salmonella enterica* serotype Agona infections of infants linked to the consumption of powdered infant formula. *Pediatr. Infect. Dis. J.* **26**:148–152.

Codex Alimentarius Commission (CAC). 1979. Recommended international code of hygienic practice for foods for infants and children. CAC/RCP 21-1979. Food and Agriculture Organization, Rome, Italy.

Codex Committee on Food Hygiene (CCFH). 2003. Risk profile of *Enterobacter sakazakii* in powdered infant formula—prepared by the United States of America and Canada. ftp://ftp.fao.org/codex/ccfh35/fh03_13e.pdf.

Codex Committee on Food Hygiene (CCFH). 2007. CX/FH07/39/4. Codex Committee on Food Hygiene, 39th session, New Delhi, India. ftp://ftp.fao.org/codex/ccfh39/fh39_04e.pdf.

European Commission (EC). 2005. Commission regulation (EC) no. 2073/2005 of 15 November 2005 on microbiological criteria for foodstuffs. *Official J. Eur. Union* L338/1-26.

European Food Safety Authority (EFSA). 2004. Opinion of the Scientific Panel on Biological Hazards on the request from the Commission related to the microbiological risks in infant formulae and follow-on formulae. *EFSA J.* 113:1–35.

European Food Safety Authority (EFSA). 2007. Scientific opinion of BIOHAZ Panel on the request from the Commission for review of the opinion on microbiological risks in infant formulae and follow-on formulae with regard to *Enterobacteriaceae* as indicators. *EFSA J.* 444:1–14.

Federal Register. 2006. Current good manufacturing practice, quality control procedures, quality factors, notification requirements and records and reports for the production of infant formula; reopening of the comment period. Department of Health and Human Services, USFDA. *Fed. Regist.* 71:43392–43398.

Food and Agriculture Organization-World Health Organization (FAO-WHO). 2004. *Enterobacter sakazakii* and other microorganisms in powdered infant formula: meeting report. *Microbiological risk assessment series 6.* World Health Organization-Food and Agriculture Organization of the United Nations, Geneva and Rome. WHO Press, Geneva, Switzerland. http://www.who.int/foodsafety/publications/micro/mra6/en/index.html.

Food and Agriculture Organization-World Health Organization (FAO-WHO). 2006. *Enterobacter sakazakii* and *Salmonella* in powdered infant formula: meeting report. *Microbiological risk assessment series 10.* World Health Organization-Food and Agriculture Organization of the United Nations, Geneva and Rome. WHO Press, Geneva, Switzerland. http://www.who.int/foodsafety/publications/micro/mra10/en/index.html.

Food and Drug Administration (FDA). 2002a. FDA warns about possible *Enterobacter sakazakii* infections in hospitalized newborns fed powdered infant formulas. April 2002. http://www.cfsan.fda.gov/%7Elrd/tpinf.html.

Food and Drug Administration (FDA). 2002b. Isolation and enumeration of *Enterobacter sakazakii* from dehydrated powdered infant formula. July 2002, revised August 2002. http://www.cfsan.fda.gov/~comm/mmesakaz.html.

International Commission on Microbiological Specifications for Foods (ICMSF). 1974. *Microorganisms in foods*, vol. 2. *Sampling for microbiological analysis: principles and specific applications.* University of Toronto Press, Toronto, Canada.

International Commission on Microbiological Specifications for Foods (ICMSF). 1986. *Microorganisms in foods*, vol. 2. *Sampling for microbiological analysis: principles and specific applications,* 2nd ed. University of Toronto Press, Toronto, Canada.

International Commission on Microbiological Specifications for Foods (ICMSF). 2002. *Microorganisms in foods*, vol. 7. *Microbiological testing in food safety management.* Kluwer Academic/Plenum Publishers, New York, NY.

International Organization for Standardization (ISO). 2006. Milk and milk products—detection of *Enterobacter sakazakii*. Technical specification ISO/TS 22964 // IDF/RM 210. International Organization for Standardization, Geneva, Switzerland.

Van Acker, J., F. de Smet, G. Muyldermans, A. Naessens, and S. Lauwers. 2001. Outbreak of necrotizing enterocolitis associated with *Enterobacter sakazakii* in powdered milk formula. *J. Clin. Microbiol.* **39**:293–297.

Zink, D. 2003. FDA field survey of powdered infant formula finished products and selected ingredients for possible *Enterobacter sakazakii* contamination. Contaminant and Natural Toxicants Subcommittee of the Food Advisory Committee Center for Food Safety and Applied Nutrition (CFSAN) Food and Drug Administration (FDA), U.S. Department of Agriculture, APHIS, Riverdale, MD. http://www.fda.gov/ohrms/dockets/ac/03/minutes/3939m1_summary%20Minutes.htm and http://www.fda.gov/ohrms/dockets/ac/03/slides/3939s1.htm.

Enterobacter sakazakii
Edited by Jeffrey M. Farber and Stephen J. Forsythe
© 2008 ASM Press, Washington, D.C.

Enterobacter sakazakii—Personal Perspectives and Reminiscences from a 32-Year History†

10

J. J. Farmer III

I am very pleased that the editors asked me to contribute a personal narrative to this first book devoted to *Enterobacter sakazakii*. I remember reading the first book on *Vibrio parahaemolyticus* in which Miwatani and Takeda (1976) provided invaluable personal information about the organism that was not published and not available from other sources. They have been my model, and I hope that my reminiscences and musings will be interesting or helpful. My history and interest with *E. sakazakii* extend from the mid-1970s (Farmer et al., 1980a) to the present (Iversen et al., 2006), so this historical perspective may be unique. I especially want to thank the editors of this volume, Steve Forsythe and Jeff Farber, each of the contributors, and finally the American Society for Microbiology for making its publication a reality. If, indeed, "information is the currency of democracy," then all of you have made significant deposits into the *E. sakazakii* account. I was invited by the editors to write a personal narrative, thus the use of "I"; however, I would be remiss if I did not acknowledge the contribution of the other authors of the original and subsequent papers, including all the original participants of the *Enterobacteriaceae* Study Group that isolated and identified strains of *E. sakazakii*.

In the 1980 paper (Farmer et al., 1980a), I wrote a concluding paragraph:

Some microbiologists, particularly clinical microbiologists, have questioned the wisdom of describing new species. They fail to realize that the first step in

JOHN J. (JIM) FARMER III, Scientist Director, U.S. Public Health Service (Retired), 1781 Silver Hill Road, Stone Mountain, GA 30087.

†Dedications: This chapter is dedicated to the memory of Riichi Sakazaki (21 August 1920 to 11 January 2002), for whom this organism is named, and to Harry Muytjens for his pioneering work in uncovering its ecology and epidemiology in cases of neonatal meningitis.

255

understanding any entity—whether it is an organism, observation, event, or phenomenon—is to describe and name that entity. There is nothing wrong with the vernacular name "yellow-pigmented *Enterobacter cloacae*"; however, we are convinced that an organism is seldom fully understood or fully appreciated until it is given a scientific name. Up to the present, very little has been written about "yellow-pigmented *Enterobacter cloacae*." We hope this situation will change now that the organism has been given a scientific name. We know, from letters and telephone conversations, that information is forthcoming which will add considerably to our understanding of this new species.

This was in response to clinical microbiologists and physicians who complained about the publication of new species and complained about changes in classification made necessary by the accumulation of new knowledge and data. Little did I realize that the statement "information would be forthcoming" would be such an understatement, and that *E. sakazakii* would ultimately become the subject of an entire book. After being away from active work on *E. sakazakii* for some years, I did a Google search and used the term "*Enterobacter sakazakii*," expecting several hundred hits. There were over a hundred times that many! A recent Google search gave 84,900 hits, and a PubMed search yielded 140 citations. When I first decided to name it as a new species, I did not envision that it would generate this amount of interest and resulting accumulation of information. In this narrative, I will share some of the facts and events that began the process that has culminated in 84,900 Google hits. This is part of a longer narrative describing my 45 years with the *Enterobacteriaceae* that will be titled *Voyage of the Beagle—3* (in preparation).

DISCOVERY

Who discovered *E. sakazakii*? The answer depends on the definition of "discovered," which was why I described the early work on "yellow-pigmented *E. cloacae*" in considerable detail (Farmer et al., 1980a). The first definitive description of neonatal meningitis due to "yellow-pigmented *Enterobacter cloacae*" was in 1961 by Urmenyi and White-Franklin (1961), and it was followed 4 years later by a similar description by Jøker et al. (1965). Earlier accounts probably were published in the literature, and readers are hereby challenged to make a convincing and scientific argument for an earlier report. The prize for finding an earlier report will be a copy of this book that I will present. Those interested might search under the following names: "coliform," "yellow coliform," "pigmented cloacae A," "*Serratia* species," "*Enterobacter* species," "*Erwinia* species," "*Chromobacterium typhiflavum*," "*Chromobacterium* species," "unidentified *Enterobacteriaceae*," and other names. The credit for realizing that "yellow-pigmented *E. cloacae*" should not continue to be classified in the species *E. cloacae* goes to Don Brenner and coworkers who, in the 1970s, used DNA-DNA hybridization to show that the two organisms' relatedness was below the level of species (Fig. 1; also see the citations in Farmer et al. [1980a]). Interestingly, the type strain of *E. cloacae* was not used in these early comparisons, and it was not used in our 1980 study that named it (Farmer et al., 1980a). This limitation has, to my knowledge, not been pointed out. Based on my studies with the genus *Serratia*, I reasoned that if *E. cloacae* and "yellow-pigmented *E. cloacae*" truly were different species, additional phenotypic differences should be present, in addition to the only accepted difference of yellow pigment production. With much additional work and the use of some phenotypic tests not previously used in CDC reference laboratories, six additional differences eventually were found (Farmer et al., 1980a). This solidified our argument that "yellow-pigmented *E. cloacae*" should be reclassified as a separate species. Thus, our 1980 paper was the first paper to use the scientific name *Enterobacter sakazakii*, although the name itself was used in a paper presented at the American Society for Microbiology General Meeting in May 1977. However, the scientific name did not appear in the published abstract. Thus, who discovered *E. sakazakii* (by any of its names) is a matter of opinion and definition, but my vote goes to Urmenyi and White-Franklin (1961), because their description of the organism is detailed and the clinical picture of neonatal meningitis is so typical.

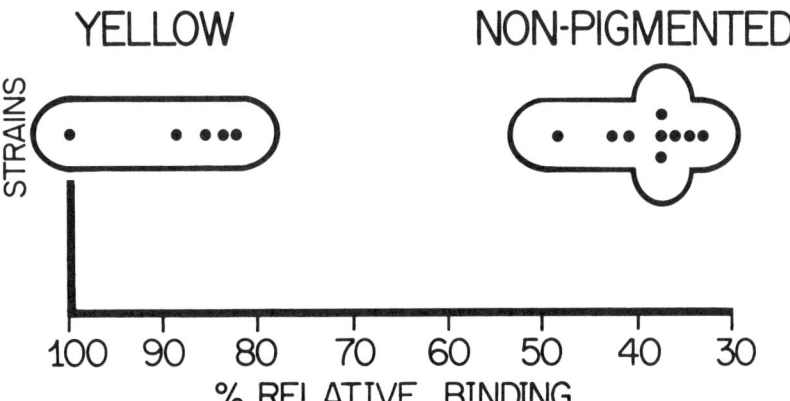

Figure 1 DNA-DNA hybridization data that showed that *E. sakazakii* is a distinct species and not a biogroup of *E. cloacae*. The type strain of *E. sakazakii* (the first strain, indicated by a dot, on the far left) was 83 to 89% related to four other *E. sakazakii* strains but was only 31 to 49% related to nine strains of *E. cloacae*. (CDC Archive, figure no. B77-1390.)

PRONUNCIATION

How is *E. sakazakii* referred to and pronounced? The pronunciation of *"Enterobacter"* is never an issue. However, the pronunciation of "sakazakii" is a story that evolved over time and caused heated dissension. I decided to name the organism to honor Riichi Sakazaki (Fig. 2) for his many and outstanding contributions to our knowledge of the organisms now classified in the families *Enterobacteriaceae*, *Vibrionaceae*, and *Aeromonadaceae* (Farmer et al., 1980a). His family name is pronounced "sah kah zak′ key," but the rules for forming Latin species names required an additional "i," resulting in *sakazakii*, or "sah kah zak′ key eye." The staff of the Enteric Reference Laboratories at the CDC seemed to find that last syllable to be one too many, so there was an unplanned evolution in the late 1970s to the pronunciation "sah kah zak′ key." This pronunciation went against the rules but was the one given in the manuscript that became the 1980 paper (Farmer et al., 1980a). This proposed aberrant pronunciation annoyed the journal editor, so he inserted the statement: "(Editor's Note: This proposal contravenes the rules of pronunciation and thus is not endorsed by the Editor.)" Over the years my pronunciation has evolved, and I am now back to the "correct" pronunciation of "sah kah zak′ key eye." More recently, I tend to like and use the term "Esak," which seems to have come into general usage. If *E. sakazakii* is classified in a new genus in the future, it will be interesting to see what usage

Figure 2 Photograph of Riichi Sakazaki (left) and Jim Farmer taken in October 1980.

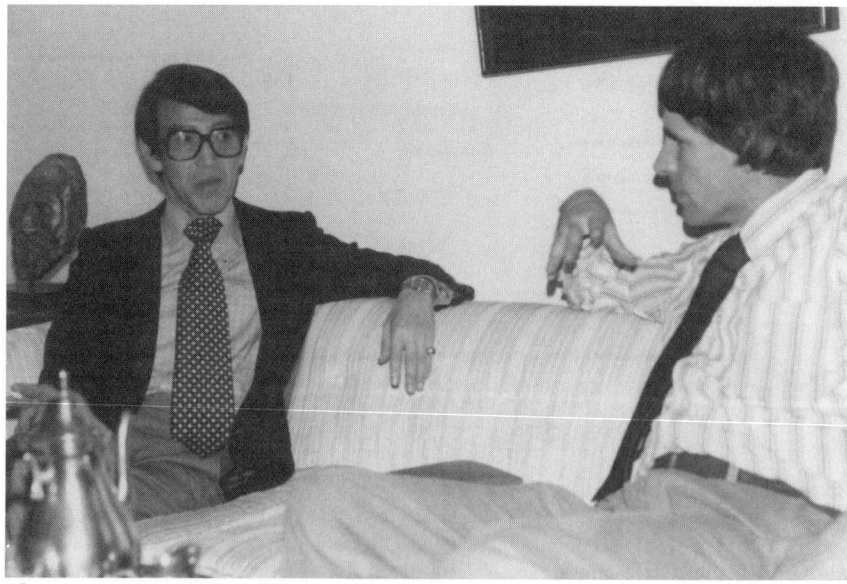

develops. The first letter of the proposed new genus name could yield yet another nickname ("Bsak," "Gsak," etc.). Usage tends to trump all rules formed by taxonomists with the best of intentions. Riichi Sakazaki never discussed with me the fact that I had named this species in his honor, but I understand from his colleagues that he was very pleased. I included his photograph in the manuscript of the 1980 paper, but it was deleted because of an editorial policy. When reprints were printed at the CDC, his picture was put back in. A photograph of Riichi Sakazaki is included as Fig. 2, and I dedicate this personal article to his memory. He died in 2002, and an obituary and another photograph can be found at http://ijs.sgmjournals.org/cgi/reprint/52/5/1435.pdf.

CLASSIFICATION

The 1980 paper (Farmer et al., 1980a) speculated that *E. sakazakii* represented more than one "species." Clearly, there was phenotypic diversity in this species, which led to the designation of 15 biogroups. Biogroup 15 was the most distinct and had four distinguishing characteristics. It also was the only biogroup that fermented α-methyl-D-glucoside. At the time, I suspected that *E. sakazakii* really was a new genus with at least two species. However, there was always a wait in Don Brenner's laboratory to do another round of DNA-DNA hybridization experiments, so the original paper (Farmer et al., 1980a) had to go with the hybridization data for only five *E. sakazakii* strains (Fig. 1). This was compensated for by extensive phenotypic data on 57 strains (actually, more than 100 by the time the paper was published). Recent studies confirm that the observed phenotypic diversity actually reflects evolutionary divergence (Iversen et al., 2006). If a new genus is proposed, I would want to be sure that all named species/subspecies also have precise operational definitions based on phenotype or another readily available method. Those that do not meet this criterion could be kept unnamed (genomospecies), pending the discovery of simple differential tests. This would be similar to the current situation with the *Enterobacter cloacae* complex, in which most of the species in the complex have not been given a scientific name. The term "*E. sakazakii* complex" would be available for laboratories that do not have the resources to do complete differentiation. This term also might prove useful for commercial identification systems, which probably would not include the tests needed for complete differentiation within the complex. When asked by a cleric if his study of biology had taught him anything about the mind of God, Thomas Henry Huxley is famously reported to have said that "it had taught him that God is inordinately fond of beetles." It might be amended with "He is also inordinately fond of

Enterobacteriaceae, as evidenced by the '*Enterobacter cloacae* complex' and '*Enterobacter sakazakii* complex.'"

IDENTIFICATION

Molecular methods are widely used today for identification, and it seems out of fashion not to use them. I would point out that *E. sakazakii* has several distinguishing characteristics that make it easy to recognize and identify. The first is the characteristic colonies that it forms on trypticase soy agar after 2 days of incubation at 25°C (Fig. 3). In my experience, most strains, when freshly isolated from clinical specimens, produce a very unusual colony type: "large, dry or mucoid, crenated (notched or scalloped), and rubbery when touched with a loop (very little growth was removed and the colony snapped back when touched)" (Farmer et al., 1980a) (Fig. 3A). A smaller and smooth colony type, typical of other *Enterobacteriaceae,* often is present (Fig. 3B). Studies on pathogenicity and virulence should always include both colony types, and it is my hypothesis that colony type A is the one that is selected for in human infections. This unusual colony morphology, along with the delayed reaction with the DNase test (100% positive after 7 days but only 2% positive at 1 day), is almost definitive for its differentiation from other organisms in the family *Enterobacteriaceae.* Other useful screening tests and tests for definitive identification were given in the original description (Farmer et al., 1980a). Its cellular morphology and flagellation are not unusual (Fig. 4).

Sixteen digital and high-resolution color figures that relate to the laboratory aspects of *E. sakazakii* can be downloaded without charge (and can be

Figure 3 Colonies of *E. sakazakii* strain 3594-76 (strain 33 in Farmer et al., 1980a) on trypticase soy agar (48 h, 25°C) showing unique size, shape, and consistency, large-stiff-rubbery colony type A, and smaller, smooth colony type B (CDC Archive, photograph no. NP 78-90).

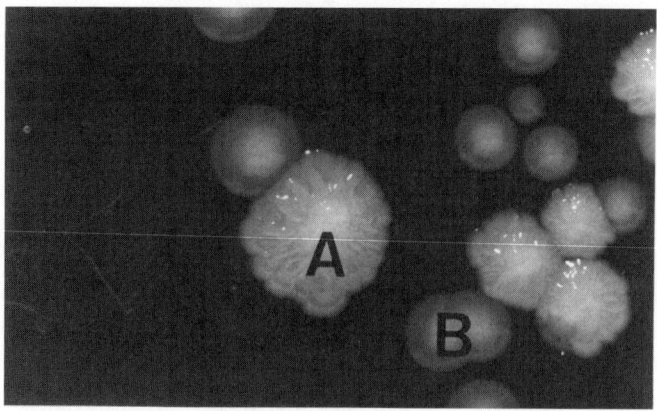

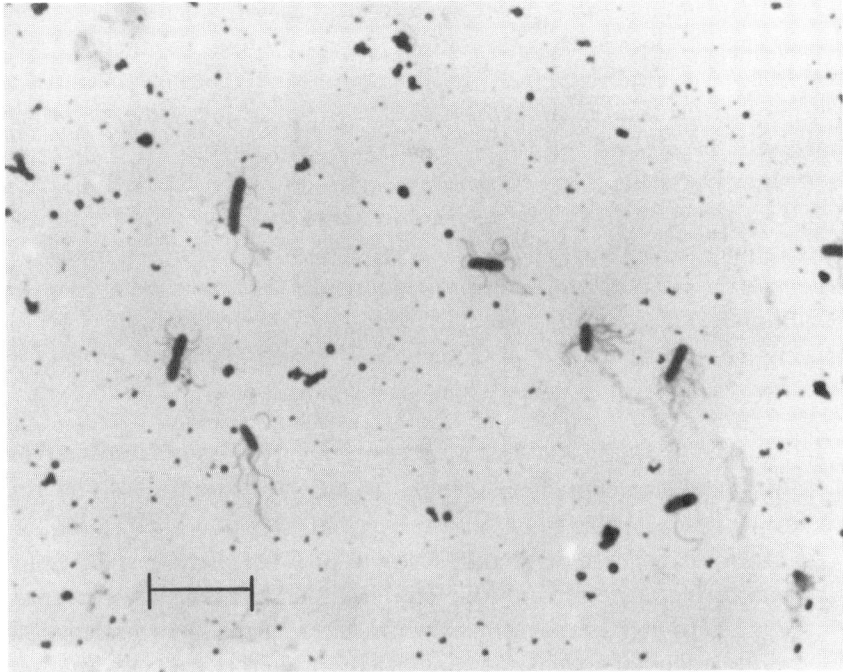

Figure 4 Flagella stain (simplified Leifson) of the type strain of *E. sakazakii* showing that it grows as small rod-shaped cells with numerous peritrichous flagella (CDC Archive, photograph no. B78-2424).

used without copyright restrictions) from CDC's Public Health Image Library, http://phil.cdc.gov/phil/home.asp (under "search" type sakazakii). I deposited most of these images in the late 1970s. For example, a color photograph of the delayed DNase reaction (see above) of *E. sakazakii* is cataloged as ID#:3036. Most of the figures from Farmer et al. (1980a) can be downloaded from the archive. Figures 1 through 4 of this chapter apparently are not yet available. Identifying numbers of three of these figures are listed in the event that they are still available at CDC in some format. These are the designations that were originally assigned for cataloging in the CDC Still Picture Archive.

HUMAN INFECTIONS AND THE EPIDEMIOLOGY AND ECOLOGY OF HOSPITAL INFECTIONS

Our 1980 paper (Farmer et al., 1980a) and the publications it cited gave several early insights on human infections and the epidemiology and ecology of hospital infections. The causal role of *E. sakazakii* in neonatal meningitis was well documented by 1980, but it was not clear how frequently these infections occur. Several papers soon followed to clarify this, and many of these showed an association with the oral ingestion by neonates of powdered infant formula that had been reconstituted with water. Just after the original paper on *E. sakazakii*, I was invited by H. L. Muytjens to assist in the analysis of eight Dutch cases of neonatal meningitis (Muytjens et al., 1983). One puzzling finding in this study has been with me ever since, and it continues to be a subject for thought. The strains isolated from prepared formula were different (via plasmid profile analysis) from the strains isolated from the spinal fluid of infants with meningitis who had ingested formula. This discrepancy of multiplicity of strains has been observed in other investigations. Over the years, several possible explanations have evolved.

1. There is no causal relationship between ingestion of reconstituted formula and the subsequent meningitis; i.e., association does not mean causation.

2. The formula contained multiple strains of *E. sakazakii*, but the laboratory method detected only one. In this explanation, the intestine and/or immune system of the neonate would select for the more virulent strain at the expense of nonpathogenic or less virulent strains (see the previous discussion on the unusual colony type that might have an advantage for this type of selection). Data from *Escherichia coli* O157:H7 epidemiological investigations might be a good model in this regard and illustrate some of the complexities of sampling and isolation.

3. The formula contained a "viable but nonculturable" strain of *E. sakazakii* that caused the meningitis, but it also contained a culturable strain of *E. sakazakii* that was the one isolated in the laboratory. (For additional discussion of this possibility, see http://www.fda.gov/ohrms/dockets/AC/03/transcripts/3939t2.doc.) Laboratory methods that depend on growth and selection can detect only strains that are culturable. In the viable but nonculturable explanation, the intestine of the neonate would revive and then selected for the more virulent strain that was viable but nonculturable, which would then cause the meningitis.

4. The strains with different properties (molecular and/or phenotypic) are really the same clone, but the clone is evolving as it multiplies in the gut and in the rest of the infant's body. I have been involved in several outbreak investigations of other enteric pathogens that had evolving clones (Farmer, 1982) and have even seen this happen on a single Petri dish of a nutrient medium.

Our original paper (Farmer et al., 1980a) also discussed an outbreak of 29 cases of respiratory tract colonization at one hospital. Interestingly, an isolate from a different hospital was from a physician's stethoscope (Farmer et al., 1980a), which immediately suggested a mechanism by which *E. sakazakii* could move from patient to patient, i.e., cross-infection.

E. sakazakii now has been isolated from many different body sites and different human infections, but neonatal meningitis continues to be the main infection of concern. The most important finding since the 1980 study is that powdered infant formula serves as an important reservoir for the strains that can cause neonatal meningitis. The pioneering work of Harry Muytjens deserves special recognition in this regard.

ANNOUNCEMENT: FORMATION OF A NONPROFIT FOUNDATION TO FURTHER WORK AND KNOWLEDGE ON *E. SAKAZAKII*

The need for a foundation to further research of *E. sakazakii* came from several recent discussions and events. The first was information from parents who have suffered the devastating effects of having a child die or become severely retarded for life following *E. sakazakii* infection. There is a clear need to assist them with information and resources, particularly in helping them understand all the possible sources of the infecting strain. There also is the concern that *E. sakazakii* infections are not being accurately identified and reported, and that sources and reservoirs of the infecting strains are not being investigated and reported. Because of privacy and legal issues, hospitals and health authorities have been reluctant to provide information about patients infected with *E. sakazakii*.

Input on the foundation's mission will be solicited from the public, industry, physicians, microbiologist, other scientists, government and regulatory agencies, and from anyone else who wishes to give input. Some items and issues already mentioned include a mission statement, an Internet site, experts to give advice, a database of literature, an archive of photographs and other historical items, a culture collection to further research, research funding, and an international registry of clinical cases and a similar registry of isolates from food, commercial products, the environment, and other sources. The mission statement of the foundation would indicate that efforts will be directed toward learning more about *E. sakazakii* and its role in causing human infections. Those interested in helping with this organizing effort or learning more about it can contact me by letter.

REFERENCES

Centers for Disease Control and Prevention. Public Health Image Library, http://phil.cdc.gov/phil/home.asp

Farmer, J. J., III. 1982. The definition of a bacterial clone for epidemiological analysis: problem associated with different typing methods. National Institutes of Health International

Workshop: The Clone Concept in the Epidemiology, Taxonomy and Evolution of the *Enterobacteriaceae*. Fogarty International Center, National Institutes of Health, Bethesda, MD.

Farmer, J. J., III, M. A. Asbury, F. W. Hickman, D. J. Brenner, and The *Enterobacteriaceae* Study Group. 1980a. *Enterobacter sakazakii:* a new species of "*Enterobacteriaceae*" isolated from clinical specimens. *Int. J. Syst. Bacteriol.* **30:**569–584.

Farmer, J. J., III, D. J. Brenner, and W. H. Ewing. 1980b. *Enterobacteriaceae:* judicial action has been proposed which would make it a rejected name and prevent its use in the literature. *ASM News* **46:**275–279.

Iversen, C., M. Waddington, J. J. Farmer III, and S. J. Forsythe. 2006. The biochemical differentiation of *Enterobacter sakazakii* genotypes. *BMC Microbiol.* **6:**94.

Jøker, R. N., T. Nørholm, and K. E. Siboni. 1965. A case of neonatal meningitis caused by a yellow *Enterobacter*. *Dan. Med. Bull.* **12:**128–130.

Miwatani, T., and Y. Takeda. 1976. *Vibrio parahaemolyticus*—a causative bacterium of food poisoning. Saikon Publishing Co., Tokyo, Japan.

Muytjens, H. L., H. C. Zanen, H. J. Sonderkamp, L. A. Kollee, I. K. Wachsmuth, and J. J. Farmer III. 1983. Analysis of eight cases of neonatal meningitis and sepsis due to *Enterobacter sakazakii*. *J. Clin. Microbiol.* **18:**115–120.

Urmenyi, A. M. C., and A. White-Franklin. 1961. Neonatal death from pigmented coliform infection. *Lancet* **i:**313–315.

Index